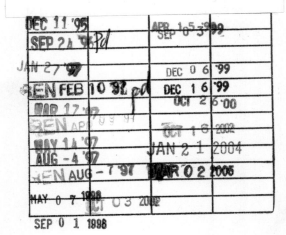

ESSENTIALS OF TRIGONOMETRY

ESSENTIALS OF TRIGONOMETRY

Fifth Edition

Irving Drooyan
Charles C. Carico
Emeriti, Los Angeles Pierce College

Macmillan Publishing Company
NEW YORK

Maxwell Macmillan Canada
TORONTO

Maxwell Macmillan International
NEW YORK OXFORD SINGAPORE SYDNEY

Copyright © 1992 by Macmillan Publishing Company, a division of Macmillan, Inc.

Printed in the United States of America

Earlier editions copyright © 1971, 1977, 1981, and 1986 by Macmillan Publishing Company. A portion of this material has been adapted from *Trigonometry: An Analytic Approach* copyright © 1967, 1973, 1979, 1983, 1987, 1991 by Macmillan Publishing Company.

Macmillan Publishing Company
866 Third Avenue, New York, New York 10022

Macmillan Publishing Company is part of the Maxwell Communication Group of Companies.

Maxwell Macmillan Canada, Inc.
1200 Eglinton Avenue, E.
Suite 200
Don Mills, Ontario M3C 3N1

Library of Congress Cataloging-in-Publication Data

Drooyan, Irving.
  Essentials of trigonometry/Irving Drooyan, Charles C. Carico—
  5th ed.
    p. cm.
  Includes index.
  ISBN 0-02-330575-4
  1. Trigonometry, Plane. I. Carico, Charles C. II. Title.
  QA533.D74 1991
  516.24′2—dc20

90–1466
CIP

Printing: 1 2 3 4 5 6 7 8 9 Year: 2 3 4 5 6 7 8 9 0 1

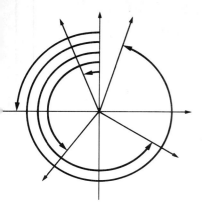

PREFACE

This fifth edition of *Essentials of Trigonometry* reflects twenty years of classroom experience with the earlier editions. Like its predecessors, this edition is designed for use in a one-semester or one-quarter course in trigonometry. The organization of the material permits flexibility in courses that differ in the time available.

The major change in this edition is the "right-triangle approach" in Chapter 1 that has replaced the general angle approach in earlier editions. We have also given increased attention to the circular functions and their graphs and to applications of periodic phenomena. The treatments of inverse functions and parametric equations have been rewritten and included in Chapter 5. The changes in the sequence of topics to improve the presentation are reflected in the following Contents.

Although it is assumed that students undertaking the study of trigonometry have had prerequisite work in algebra and geometry, Appendix A includes a review of topics in these subjects that are needed as a background for the course. Students are referred to a section of this review when the material can be helpful in understanding a particular trigonometric concept.

The following pedagogical features of the previous editions have been retained.

- Approximations to function values are primarily obtained by use of a scientific calculator. However, a discussion on the use of tables is available in Appendix C. The tables include measures of angles in radians and in tenths of degrees as well as degrees and minutes.

- Emphasis is given to obtaining exact function values of 30° and 45° and integral multiples thereof.

- All exercise sets include basic exercises designated A. Examples, which supplement the examples in the text discussions, are included with these exercises.
- Many exercise sets include more challenging exercises designated B, which provide the instructor with flexibility in making assignments, depending on the time available and the objectives of the course.
- Subject matter is continually reviewed through the use of chapter summaries and review exercises.
- A detailed summary of the important topics in the text that appears inside the back cover can serve as a convenient reference for students completing assignments or studying for examinations.

An Instructor's Manual accompanies the text and includes

- Solutions to even-numbered exercises.
- Test questions.
- Instructions for using the Hewlett Packard, Sharp, Casio, and Texas Instruments graphic calculators.

The material in the Instructor's Manual may be reproduced for student use.

We acknowledge with sincere appreciation the encouraging and helpful comments of the reviewers of this fifth edition: Elsa M. Anderson, Oklahoma State University; Margaret Van Parys, Lakeland Community College (Ohio); E. James Peake, Iowa State University; and Jimmy L. Solomon, Mississippi State University.

We again extend a special thanks to our production supervisor Elisabeth Belfer for her professional help in the preparation of this edition.

We especially want to acknowledge the memory of Walter Hadel, our coauthor on previous editions. Walter was an inspirational coauthor and a good friend.

Irving Drooyan
Charles C. Carico

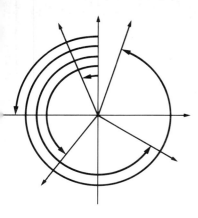

CONTENTS

APPENDIX

ESSENTIALS OF TRIGONOMETRY

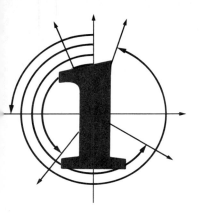

PROPERTIES OF RIGHT TRIANGLES

Historically, trigonometry developed primarily as a way to find the lengths of sides and the measures of angles of triangles for use in such practical areas as surveying and navigation. In more recent times, the tools of trigonometry have become important in describing phenomena that are periodic.

In this chapter we are concerned with some basic notions of trigonometry as they are applied to right triangles. In the following chapters we will apply extensions of these notions to obtain periodic functions that can be used in a variety of applied areas.

The appendix includes a review of prerequisite topics of algebra and geometry. References are made in the text to appropriate sections in the appendix wherever a review of the material would be helpful.

1.1 Trigonometric Ratios

Consider the right triangle in Figure 1.1 with legs a and b. From the Pythagorean theorem (see Appendix A.9) the length c of the hypotenuse is $\sqrt{a^2 + b^2}$. For any acute angle α,* we can form six ratios with the numbers a, b, and c. These ratios, called **trigonometric ratios,** are named as follows.

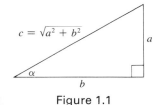

$$c = \sqrt{a^2 + b^2}$$

Figure 1.1

* See Appendix A.8 for a discussion of the names of angles.

1

Trigonometric Ratios:

$$\text{sine } \alpha = \frac{a}{c} \qquad\qquad \text{cosecant } \alpha = \frac{c}{a}$$

$$\text{cosine } \alpha = \frac{b}{c} \qquad\qquad \text{secant } \alpha = \frac{c}{b}$$

$$\text{tangent } \alpha = \frac{a}{b} \qquad\qquad \text{cotangent } \alpha = \frac{b}{a}$$

We can easily show that the trigonometric ratios are determined only by the measure of α and not by the lengths of the sides of the triangle.

Because the two right triangles in Figure 1.2 are similar (two angles, α and γ_1, of triangle AB_1C_1 are equal, respectively, to two angles, α and γ_2, of triangle AB_2C_2), the corresponding sides are proportional (see Appendix A.10). Hence,

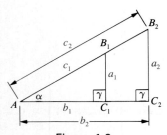

Figure 1.2

$$\frac{a_1}{c_1} = \frac{a_2}{c_2}, \qquad \frac{b_1}{c_1} = \frac{b_2}{c_2}, \qquad \text{and} \qquad \frac{a_1}{b_1} = \frac{a_2}{b_2}.$$

Similar equalities for the reciprocals of these ratios also hold. Hence, the six trigonometric ratios are determined *only* by the measure of α.

It is often helpful to view the trigonometric ratios in words as shown in the following definitions. The names of the ratios are commonly abbreviated.

For any acute angle α in a right triangle:

$$\sin \alpha = \frac{\text{length of side opposite } \alpha}{\text{length of hypotenuse}},$$

$$\cos \alpha = \frac{\text{length of side adjacent to } \alpha}{\text{length of hypotenuse}},$$

$$\tan \alpha = \frac{\text{length of side opposite } \alpha}{\text{length of side adjacent to } \alpha},$$

$$\csc \alpha = \frac{\text{length of hypotenuse}}{\text{length of side opposite } \alpha},$$

$$\sec \alpha = \frac{\text{length of hypotenuse}}{\text{length of side adjacent to } \alpha},$$

$$\cot \alpha = \frac{\text{length of side adjacent to } \alpha}{\text{length of side opposite } \alpha}.$$

Example In the figure, the side s opposite the right angle is the hypotenuse. Hence,

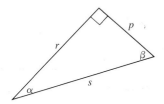

$$\sin \alpha = \frac{p}{s}, \qquad \cos \alpha = \frac{r}{s}, \qquad \tan \alpha = \frac{p}{r},$$

$$\csc \alpha = \frac{s}{p}, \qquad \sec \alpha = \frac{s}{r}, \qquad \cot \alpha = \frac{r}{p},$$

and

$$\sin \beta = \frac{r}{s}, \qquad \cos \beta = \frac{p}{s}, \qquad \tan \beta = \frac{r}{p},$$

$$\csc \beta = \frac{s}{r}, \qquad \sec \beta = \frac{s}{p}, \qquad \cot \beta = \frac{p}{r}.$$

Example Find the sine, cosine, and tangent of α and β to the nearest hundredth.*

Solution

$$\sin \alpha = \frac{5}{13} \approx 0.38, \qquad \cos \alpha = \frac{12}{13} \approx 0.92, \qquad \tan \alpha = \frac{5}{12} \approx 0.42$$

$$\sin \beta = \frac{12}{13} \approx 0.92, \qquad \cos \beta = \frac{5}{13} \approx 0.38, \qquad \tan \beta = \frac{12}{5} \approx 2.40$$

Example Find the six trigonometric ratios of α to the nearest hundredth.

Solution   Using the Pythagorean theorem,[†]

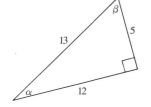

$$(2.1)^2 + (AC)^2 = (5.4)^2,$$

from which

$$AC = \sqrt{(5.4)^2 - (2.1)^2} \approx 4.97.$$

(*continued*)

---

* See Appendix B for a discussion of rounding-off procedures.
† See Appendix A.9 for a discussion of properties of triangles.

Hence, to the nearest hundredth,

$$\sin \alpha \approx \frac{2.1}{5.4} \approx 0.39, \qquad \cos \alpha \approx \frac{4.97}{5.4} \approx 0.92, \qquad \tan \alpha \approx \frac{2.1}{4.97} \approx 0.42;$$

$$\csc \alpha \approx \frac{5.4}{2.1} \approx 2.57, \qquad \sec \alpha \approx \frac{5.4}{4.97} \approx 1.09, \qquad \cot \alpha \approx \frac{4.97}{2.1} \approx 2.37.$$

Special care must be taken to determine trigonometric ratios of angles when there is more than one right triangle in a figure. For example, in Figure 1.3 there are three right triangles, $\triangle ABC$, $\triangle ABD$, and $\triangle BCD$, and four acute angles α, β, γ, and δ. There are six trigonometric ratios associated with each of these angles.

For example, in $\triangle ABC$,

$$\sin \alpha = \frac{n}{p+q}, \qquad \cos \alpha = \frac{m}{p+q}, \qquad \tan \alpha = \frac{n}{m},$$

and

$$\sin \delta = \frac{m}{p+q}, \qquad \cos \delta = \frac{n}{p+q}, \qquad \tan \delta = \frac{m}{n}.$$

Figure 1.3

You will be asked to find the trigonometric ratios of angles in the other two triangles in the exercises.

EXERCISE SET 1.1

A

■ *Find the six trigonometric ratios of each angle to the nearest hundredth.*

Example   Angle α

Solution   Using the Pythagorean theorem,

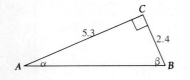

$$AB = \sqrt{(5.3)^2 + (2.4)^2} \approx 5.82.$$

Hence, to the nearest hundredth,

$$\sin \alpha \approx \frac{2.4}{5.82} \qquad \cos \alpha \approx \frac{5.3}{5.82} \qquad \tan \alpha \approx \frac{2.4}{5.3}$$

$$\approx 0.41, \qquad\qquad \approx 0.91, \qquad\qquad \approx 0.45,$$

$$\csc \alpha \approx \frac{5.82}{2.4} \qquad \sec \alpha \approx \frac{5.82}{5.3} \qquad \cot \alpha \approx \frac{5.3}{2.4}$$

$$\approx 2.43, \qquad\qquad \approx 1.10, \qquad\qquad \approx 2.21.$$

1. angle α **2.** angle β

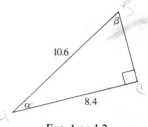

Exs. 1 and 2

3. $\angle A$ **4.** $\angle B$ **5.** $\angle CBA$ **6.** $\angle CAB$

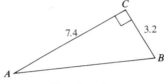

Exs. 3 and 4

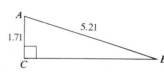

Exs. 5 and 6

■ *Find each trigonometric ratio in terms of m, n, p, q, r. Figure 1.3 is reproduced here.*

Examples

a. $\cos \gamma$ b. $\tan \beta$

Solutions

a. The hypotenuse of $\triangle BCD$ is n. Hence,

$$\cos \gamma = \frac{r}{n}.$$

b. The side opposite β is p and the side adjacent is r. Hence,

$$\tan \beta = \frac{p}{r}.$$

7. $\sin \beta$ **8.** $\cos \beta$ **9.** $\cot \beta$ **10.** $\csc \beta$

11. $\tan \gamma$ **12.** $\cot \gamma$ **13.** $\sec \gamma$ **14.** $\sin \gamma$

15. In $\triangle ABC$, $\sin \alpha$ **16.** In $\triangle ABC$, $\sin \delta$ **17.** In $\triangle ABD$, $\cos \alpha$

18. In $\triangle ABD$, $\tan \alpha$ **19.** In $\triangle BCD$, $\tan \delta$ **20.** In $\triangle BCD$, $\cos \delta$

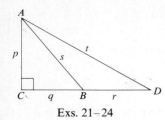

Exs. 21–24

■ *Find each trigonometric ratio in terms of p, q, r, s, and t.*

21. sin ∠CAB   **22.** tan ∠CBA   **23.** tan ∠ADC   **24.** cos ∠ADC

■ *Find the six trigonometric ratios of each angle to the nearest hundredth.*

25. ∠DAC   **26.** ∠ADC   **27.** ∠DBC   **28.** ∠BDC

29. ∠CDB   **30.** ∠CBD   **31.** ∠DAC   **32.** ∠ADC

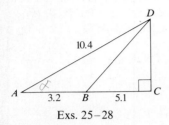

Exs. 25–28

B

■ *In Exercises 33–38, ∠C is the right angle in triangle ABC. Leave answers in radical form.*

33. If sin ∠A = $\frac{1}{2}$, find
csc ∠A, cos ∠B, and sec ∠A.

34. If cos ∠A = $\frac{2}{3}$, find
sec ∠A, sin ∠B, and csc ∠B.

35. If tan ∠A = 1, find
cot ∠A, cot ∠B, and sin ∠B.

36. If csc ∠B = 2, find
sin ∠B, sec ∠A, and csc ∠A.

37. If sec ∠B = 4, find
cos ∠B, csc ∠A, and cot ∠A.

38. If cot ∠A = $\frac{1}{3}$, find
tan ∠A, tan ∠B, and cot ∠B.

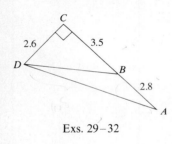

Exs. 29–32

39. Use the properties of an isosceles right triangle to find the exact values of the six trigonometric ratios of a 45° angle.

40. Use the properties of an equilateral triangle to find the exact values of the six trigonometric ratios of a 60° angle. *Hint: Construct the perpendicular from a vertex to the opposite side.*

1.2 Finding Trigonometric Ratios for Given Angles

In this section we will find the trigonometric ratios of an angle when we are given the measure of the angle. We will find *exact values* for the ratios of some special angles and use a calculator to find *decimal approximations* for the ratios for all other acute angles.

Two kinds of angle measure are commonly used. In this chapter we will use **degree measure** (see Appendix A.8) as we continue to work with triangles. **Radian measure,** which is used in advanced work in mathematics and the physical sciences, will be considered in the next chapter.

Degree Measure   Recall that each degree is divided into sixty minutes (60′). Each minute (1′) is also divided into sixty seconds (60″). However, the smallest unit in our work will be the minute.

Because angle measurements are sometimes given in degree-minute units and sometimes in decimal parts of a degree, it is often necessary to convert from one unit to the other. To convert from tenths of degrees to minutes, simply multiply the number of tenths by 60.

Examples   a. $0.4° = (0.4 \times 60)'$ b. $35.2° = 35° + (0.2 \times 60)'$

$= 24'$ $= 35° \ 12'$

Conversely, to convert from minutes to tenths of degrees, simply divide the number of minutes by 60.

Examples   a. $0° \ 36' = \left(\dfrac{36}{60}\right)°$ b. $75° \ 18' = 75° + \left(\dfrac{18}{60}\right)°$

$= 0.6°$ $= 75.3°$

Since the advent of calculators, the measurements involved in applications are usually given in decimal units. However, measurements are sometimes given in degree-minute units, and you should be familiar with the simple conversion methods shown in the examples above.

Ratios for 30°, 45°, and 60° Angles   First let us consider an equilateral triangle whose sides are each two units long. Recall from geometry (see Appendix A.9) that the measure of each angle is 60° and that an altitude of the triangle forms two congruent right triangles as shown in Figure 1.4a. As we noted in Section 1.1, the

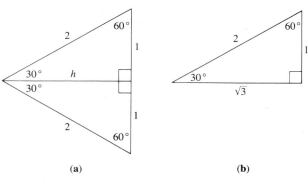

(a)   (b)

Figure 1.4

trigonometric ratios are determined by the measures of angles only. Hence, for simplicity, we used "2" as the length of each side of the equilateral triangle. From the Pythagorean theorem we can obtain the length of the altitude

$$h = \sqrt{2^2 - 1^2} = \sqrt{3}.$$

Hence, lengths of the sides of this triangle, and of course any $30° - 60° - 90°$ triangle, are proportional to the numbers 1, $\sqrt{3}$, and 2 as shown in Figure 1.4b.

Hence,

$$\sin 30° = \frac{1}{2}, \qquad\qquad \csc 30° = \frac{2}{1} = 2,$$

$$\cos 30° = \frac{\sqrt{3}}{2}, \qquad\qquad \sec 30° = \frac{2}{\sqrt{3}},$$

$$\tan 30° = \frac{1}{\sqrt{3}}, \qquad\qquad \cot 30° = \frac{\sqrt{3}}{1} = \sqrt{3},$$

and

$$\sin 60° = \frac{\sqrt{3}}{2}, \qquad\qquad \csc 60° = \frac{2}{\sqrt{3}},$$

$$\cos 60° = \frac{1}{2}, \qquad\qquad \sec 60° = \frac{2}{1} = 2,$$

$$\tan 60° = \frac{\sqrt{3}}{1} = \sqrt{3}, \qquad\qquad \cot 60° = \frac{1}{\sqrt{3}}.$$

Now let us consider a $45°$ right triangle whose sides opposite the $45°$ angles are one unit long. In this case, from the Pythagorean theorem, the hypotenuse is given by $\sqrt{1^2 + 1^2} = \sqrt{2}$. Hence, the lengths of the sides of this triangle, and of course any $45° - 45° - 90°$ triangle, are proportional to the numbers 1, 1, and $\sqrt{2}$, as shown in Figure 1.5. Hence,

$$\sin 45° = \frac{1}{\sqrt{2}}, \qquad\qquad \csc 45° = \frac{\sqrt{2}}{1} = \sqrt{2},$$

$$\cos 45° = \frac{1}{\sqrt{2}}, \qquad\qquad \sec 45° = \frac{\sqrt{2}}{1} = \sqrt{2},$$

$$\tan 45° = 1, \qquad\qquad \cot 45° = 1.$$

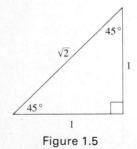

Figure 1.5

TABLE 1.1

α	$\sin \alpha$	$\cos \alpha$	$\tan \alpha$	$\csc \alpha$	$\sec \alpha$	$\cot \alpha$
30°	$\dfrac{1}{2}$	$\dfrac{\sqrt{3}}{2}$	$\dfrac{1}{\sqrt{3}}$	2	$\dfrac{2}{\sqrt{3}}$	$\sqrt{3}$
45°	$\dfrac{1}{\sqrt{2}}$	$\dfrac{1}{\sqrt{2}}$	1	$\sqrt{2}$	$\sqrt{2}$	1
60°	$\dfrac{\sqrt{3}}{2}$	$\dfrac{1}{2}$	$\sqrt{3}$	$\dfrac{2}{\sqrt{3}}$	2	$\dfrac{1}{\sqrt{3}}$

Although the function values for angles that are often used are available in Table 1.1, we strongly encourage you to construct a triangle (as we did to obtain these values) each time you want a particular trigonometric ratio for one of the special angles.

Note that some entries in Table 1.1 have a radical expression in the denominator. These forms are often more useful than the forms obtained it we had rationalized the denominators.

Calculators with trigonometric function value capability make it possible to obtain the table entries quite readily in decimal form. However, the definition of the trigonometric ratios will be more meaningful to you if you obtain these values in fractional form directly from the triangles in Figures 1.4 and 1.5 using radicals for exact values rather than obtaining decimal approximations. *In our work in this text we will express the trigonometric ratios for these special angles in such form.*

In the example below we use $\tan^2 30°$ for $(\tan 30°)^2$, $\sin^2 45°$ for $(\sin 45°)^2$, and $\tan^2 60°$ for $(\tan 60°)^2$. This placement of an exponent is commonly used for powers of all trigonometric ratios.

Examples   Compute

a. $(\csc 30°)(\cos 30°)(\tan^2 30°)$ b. $\sin^2 45° + \tan^2 60°$

Solutions   First construct appropriate triangles. The ratios can be read directly from the figures.

a.

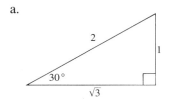

$(\csc 30°)(\cos 30°)(\tan^2 30)$

$$= 2\left(\frac{\sqrt{3}}{2}\right)\left(\frac{1}{\sqrt{3}}\right)^2$$

$$= \frac{\sqrt{3}}{3}$$

(continued)

b.

$$\sin^2 45° + \tan^2 60° = \left(\frac{1}{\sqrt{2}}\right)^2 + \left(\frac{\sqrt{3}}{1}\right)^2 = \frac{7}{2}$$

Trigonometric Ratios For Other Acute Angles   Exact values of trigonometric ratios for the special angles 30°, 45° and 60° were obtained using properties of geometry. *Approximations* for trigonometric ratios of other acute angles can be obtained by formulas that are introduced in the study of calculus. However, we can use a scientific calculator or specially prepared tables to obtain such approximations.

Using a Calculator   Finding trigonometric ratios by using a calculator is a simple process. Because the steps to be taken vary somewhat for different types of calculators, you may want to check the instruction booklet for your particular calculator if the steps shown in our examples do not apply to your calculator.

Calculators vary in the number of digits that are shown in the display. Unless otherwise specified, we will round off readings to four decimal places. Because we are using the degree measure of angles at this time, set the Degree/Radian switch on your calculator on DEG for the work in this chapter.

Examples
   a. cos 15.3°; calculator key stroke:

$$15.3 \boxed{\text{COS}} \approx 0.964557$$

Rounding off the result to four decimal places gives

$$\cos 15.3° \approx 0.9646.$$

   b. sin 21° 42′ = sin 21.7°

$$21.7 \boxed{\text{SIN}} \approx 0.369747$$

Rounding off the result to four decimal places gives

$$\sin 21° \ 42' \approx 0.3697.$$

c. tan 63.7°

$$63.7 \ \boxed{\text{TAN}} \approx 2.023346$$

Rounding off the result to four decimal places gives

$$\tan 63.7° \approx 2.0233.$$

Note that in Example b above it was necessary first to express the angle in decimal form before using a calculator.

$$42' = \left(\frac{42}{60}\right)° = 0.7°.$$

Using Calculators for csc α, sec α, and cot α

Values for sin α, cos α, and tan α can be obtained directly on scientific calculators. Values for sec α, csc α, and cot α can be obtained by using certain reciprocal relationships. Note that from the definitions of the trigonometric ratios and Figure 1.6,

$$\sin \alpha = \frac{a}{c} \qquad \text{and} \qquad \csc \alpha = \frac{c}{a}.$$

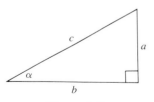

Figure 1.6

Hence, these ratios are reciprocals of each other and

$$\csc \alpha = \frac{1}{\sin \alpha}.$$

Similarly, sec α is the reciprocal of cos α and cot α is the reciprocal of tan α. Hence,

$$\sec \alpha = \frac{1}{\cos \alpha} \qquad \text{and} \qquad \cot \alpha = \frac{1}{\tan \alpha}.$$

Note that since we are obtaining trigonometric ratios of acute angles of a triangle, sin α, cos α, and tan α cannot equal zero, and hence csc α, sec α, and tan α are defined for all acute angles α.

Examples

a. $\sec 22.6° = \dfrac{1}{\cos 22.6°}$

$$22.6 \ \boxed{\text{COS}} \ \boxed{1/x} \approx 1.083177 \qquad \text{(continued)}$$

Rounding off to four places gives

$$\sec 22.6° \approx 1.0832.$$

b. $\cot 14° \ 18' = \cot 14.3° = \dfrac{1}{\tan 14.3°}$.

$$14.3 \ \boxed{\text{TAN}} \ \boxed{1/x} \approx 3.923156$$

Rounding off to four places gives

$$\cot 14° \ 18' \approx 3.9232.$$

Using Tables   Before the advent of scientific calculators, approximations for trig onometric ratios were obtained from extensive tables prepared for th purpose. Although such tables have lost their primary purpose, you ma wish to examine the tables in Appendix C that explicitly list such value to get an overview of the behavior of the trigonometric ratios.

EXERCISE SET 1.2

A

■ *Find the exact value of each expression.*

Example $\sin 45° + (\cos^2 30°)(\tan 60°)$

Solution   Sketch the appropriate triangles.

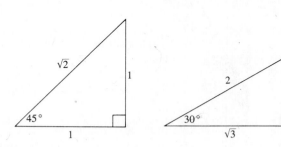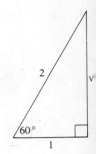

Hence,

$$\sin 45° + (\cos^2 30°)(\tan 60°) = \frac{1}{\sqrt{2}} + \left(\frac{3}{4}\right)\left(\frac{\sqrt{3}}{1}\right)$$

$$= \frac{1(4)}{\sqrt{2}(4)} + \frac{3\sqrt{3}(\sqrt{2})}{4(\sqrt{2})} = \frac{4 + 3\sqrt{6}}{4\sqrt{2}}.$$

1. $\sin^2 30°$ 2. $\tan^2 45°$

3. $(\cos 30°)(\tan^2 60°)$ 4. $(\sin 45°)(\cos^2 60°)$

5. $\sec 60° - \cos^2 45°$ 6. $\csc 30° - \tan^2 30°$

7. $\tan^2 45° + \cot^2 60°$ 8. $\sec^2 30° + \sin^2 45°$

9. $\cos^2 60° - (\sec 30°)(\tan 45°)$ 10. $\sin^2 45° + (\tan 30°)(\cos 45°)$

11. $4 \sin^2 45° + 2 \cos 30° (\sec 30°)$

12. $2 \tan^2 60° + 4 \sin 30° (\csc 30°)$

13. $\dfrac{1 + \tan^2 30°}{\sec^2 30°}$ 14. $\dfrac{1 - \cos^2 45°}{\sin^2 45°}$

15. $\dfrac{\sin 60° \pm \cos^2 45°}{\cos 30°}$ 16. $\dfrac{\tan^2 30° + \sec^2 45°}{\tan 60°}$

■ *Find an approximation to four decimal places.*

Examples   a. cos 63.7°   b. cot 17° 6′

Solutions
  a. cos 63.7°

$$63.7 \boxed{\text{COS}} \approx 0.443071$$

Hence, to four decimal places

$$\cos 63.7° \approx 0.4431.$$

  b. $\cot 17° 6' = \cot 17.1° = \dfrac{1}{\tan 17.1}$

$$17.1 \boxed{\text{TAN}} \boxed{1/\text{X}} \approx 3.250551$$

Hence, to four decimal places

$$\cot 17° 6' \approx 3.2506.$$

17. sin 16.4°   18. cos 21.5°   19. tan 69.2°   20. sin 82.6°

21. cos 18° 48′   22. sin 27° 6′   23. tan 6° 12′   24. cos 63° 30′

25. cot 14.3°   26. sec 79.6°   27. csc 12° 24′   28. cot 62° 12′

29. sec 64° 48′   30. csc 14° 54′

B

■ *Complete each table. Show values to the nearest hundredth.*

31.

α	10°	5°	1°	0.1°
sin α				
csc α				

Make a conjecture about sin α and csc α as α approaches 0°.

32.

α	10°	5°	1°	0.1°
cos α				
sec α				

Make a conjecture about cos α and sec α as α approaches 0°.

33.

α	10°	5°	1°	0.1°
tan α				
cot α				

Make a conjecture about tan α and cot α as α approaches 0°.

34.

α	80°	85°	89°	89.9°
sin α				
csc α				

Make a conjecture about sin α and csc α as α approaches 90°.

35.

α	80°	85°	89°	89.9°
cos α				
sec α				

Make a conjecture about cos α and sec α as α approaches 90°.

36.

α	80°	85°	89°	89.9°
tan α				
cot α				

Make a conjecture about tan α and cot α as α approaches 90°.

1.3 Finding Measures of Acute Angles

If we are given a value for a trigonometric ratio listed in Table 1.1, we can sketch a triangle or use Table 1.1 to find the acute angle associated with the given ratio. For example, if sin $\alpha = \sqrt{3}/2$ we can sketch a right triangle (see Figure 1.7) where the side opposite α is $\sqrt{3}$ and the hypotenuse is 2 to note that $\alpha = 60°$. Alternatively, we can use Table 1.1 to obtain the same result.

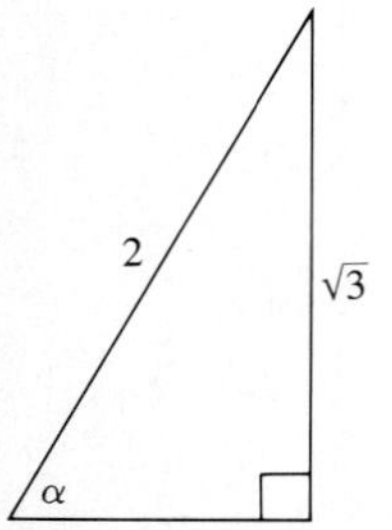

Figure 1.7

Inverse Notation

A special notation called **inverse notation** is used to express an angle α explicitly in terms of a corresponding trigonometric ratio. For example,

$$\sin \alpha = \frac{\sqrt{3}}{2}$$

can be expressed equivalently as

$$\alpha = \text{Sin}^{-1} \frac{\sqrt{3}}{2} \qquad \text{or} \qquad \alpha = \text{Arcsin} \frac{\sqrt{3}}{2},$$

which are both read as "α equals Arcsin $\sqrt{3}/2$." The phrase "inverse sine of $\sqrt{3}/2$" is also used in this context. In each case we can simply view each of the expressions

$$\text{Sin}^{-1} \frac{\sqrt{3}}{2} \quad \text{and} \quad \text{Arcsin} \frac{\sqrt{3}}{2}$$

as the *acute angle* whose sine is $\sqrt{3}/2$. Hence, from the example above,

$$\alpha = \text{Sin}^{-1} \frac{\sqrt{3}}{2} = 60°.$$

Note that a capital letter is used for the first letter in each symbol, Sin^{-1} and Arcsin.

Inverse notation is used with each of the six trigonometric ratios. Hence, using the symbol **trig α** to represent any one of these ratios, we can define $\text{Trig}^{-1} \, y$ is as follows.

Trig^{-1} y or **Arctrig y** names an acute angle α such that

$$\text{trig } \alpha = y \quad (y > 0).*$$

Inverse notation is applicable to all trigonometric ratios, including those that are given in decimal form.

Examples   Express using inverse notation.

a. $\cos \alpha = \dfrac{1}{2}$ b. $\tan \alpha = 1.4882$

Solutions

a. $\alpha = \text{Cos}^{-1} \dfrac{1}{2}$ or $\alpha = \text{Arccos} \dfrac{1}{2}$

b. $\alpha = \text{Tan}^{-1} 1.4882$ or $\alpha = \text{Arctan } 1.4882$

It may happen that a problem involves an equation of the form $\alpha = \text{Trig}^{-1} \, y$. In such a case, it is often helpful to rewrite the equation as $\text{trig } \alpha = y$, since this form sometimes provides more insight regarding the relationship.

Examples   Expresss without using inverse notation.

a. $\alpha = \text{Tan}^{-1} \dfrac{1}{\sqrt{3}}$ b. $\alpha = \text{Arccos } 0.4732$

---

* In the following chapters we will extend this definition to apply to all angles and $y \le 0$.

Solutions

a. $\tan \alpha = \dfrac{1}{\sqrt{3}}$ b. $\cos \alpha = 0.4732$

We can always find the *exact value* for the measure of an acute angle if the given trigonometric ratio is one of the entries in Table 1.1.

Examples   Find the exact value for the measure of angle α.

a. $\cos \alpha = \dfrac{1}{2}$ b. $\cot \alpha = 1$

Solutions   Sketch the appropriate right triangles (if necessary).

a. 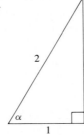   b.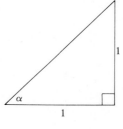

$\alpha = \cos^{-1} \dfrac{1}{2} = 60°$ $\alpha = \cot^{-1} 1 = 45°$

We can find *approximations* for the measures of acute angles from trigonometric ratios $\sin \alpha$, $\cos \alpha$, and $\tan \alpha$ given in decimal form.

Examples   Find an approximation (to the nearest tenth of a degree) for the measure of each angle α.

a. $\sin \alpha = 0.4321$ b. $\tan \alpha = 2.4132$

Solutions

a. $\alpha = \mathrm{Sin}^{-1}\, 0.4321.$ b. $\alpha = \mathrm{Tan}^{-1}\, 2.4132.$

$0.4321\ \boxed{\text{INV}}\ \boxed{\text{SIN}} \approx 25.60091°$ $2.4132\ \boxed{\text{INV}}\ \boxed{\text{TAN}} \approx 67.49149°$

To the nearest tenth, $\alpha \approx 25.6°$.   To the nearest tenth $\alpha \approx 67.5°$.

Since calculators are not programmed for the cosecant, secant, and cotangent ratios, we have to use the reciprocals of these ratios as shown in the following example.

Example   Find an approximation to the nearest tenth for the measure of α if csc $\alpha = 2.1321$.

Solution

$$\sin \alpha = \frac{1}{\csc \alpha} = \frac{1}{2.1321};$$

$$\alpha = \mathrm{Sin}^{-1} \frac{1}{2.1321}.$$

$$2.1321 \boxed{1/\mathrm{x}} \boxed{\mathrm{INV}} \boxed{\mathrm{SIN}} \approx 27.97078°$$

To the nearest tenth, $\alpha \approx 28.0°$.

Given any two sides of a right triangle, we can now find the measures (or approximations) of the acute angles of the triangle. Using the fact that the sum of the measures of a triangle equals 180°, and hence the sum of the measures of the acute angles in a right triangle equals 90°, will simplify our work.

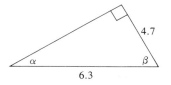

Example   Find an approximation to the nearest tenth for the measures of α and β in the figure.

Solution   Since the side opposite α and the hypotenuse are known, we use the sine relationship to obtain

$$\sin \alpha = \frac{4.7}{6.3},$$

from which

$$\alpha = \mathrm{Sin}^{-1} \frac{4.7}{6.3}.$$

$$4.7 \boxed{\div} 6.3 \boxed{=} \boxed{\mathrm{INV}} \boxed{\mathrm{SIN}} \approx 48.24780°$$

Hence, to the nearest tenth, $\alpha \approx 48.2°$. Since the sum of α and β equals 90°,

$$\beta = 90° - \alpha \approx 90° - 48.2° = 41.8°.$$

In any chain of statements involving both the symbols $=$ and $\approx$ as in the example above, it should be understood that either relationship is valid only for the expressions on either side of the particular symbol.

EXERCISE SET 1.3

A

▪ *Express each of the following using inverse notation.*

Examples

a. $\sin \alpha = \dfrac{\sqrt{3}}{2}$ b. $\sec \beta = 2.4312$

Solutions

a. $\alpha = \text{Sin}^{-1} \dfrac{\sqrt{3}}{2}$ or $\alpha = \text{Arcsin} \dfrac{\sqrt{3}}{2}$

b. $\beta = \text{Sec}^{-1} 2.4312$ or $\beta = \text{Arcsec} \, 2.4312$

1. $\tan \alpha = \dfrac{1}{\sqrt{3}}$ **2.** $\cos \alpha = \dfrac{1}{\sqrt{2}}$ **3.** $\csc \beta = 3.1421.$

4. $\cot \beta = 0.4721$ **5.** $\sec \gamma = 2.6312$ **6.** $\sin \gamma = 0.7172$

▪ *Express each of the following without using inverse notation.*

Examples   a. $\alpha = \text{Sec}^{-1} \sqrt{2}$ b. $\beta = \text{Arctan} \, 2.1413$

Solutions   a. $\sec \alpha = \sqrt{2}$ b. $\tan \beta = 2.1413$

7. $\alpha = \text{Sin}^{-1} \dfrac{1}{\sqrt{2}}$ **8.** $\alpha = \text{Cot}^{-1} \sqrt{3}$ **9.** $\beta = \text{Cos}^{-1} 0.62$

10. $\beta = \text{Csc}^{-1} 1.7431$ **11.** $\gamma = \text{Arcsec} \, 2.6060$ **12.** $\gamma = \text{Arctan} \, 0.73$

▪ *Find the exact value for the measure of angle α.*

Examples

a. $\sec \alpha = \sqrt{2}$ b. $\tan \gamma = \sqrt{3}$

Solutions   *Sketch the appropriate right triangles (if necessary).*

a. $\sec \alpha = \sqrt{2};$

$\alpha = \text{Sec}^{-1} \sqrt{2} = 45°$

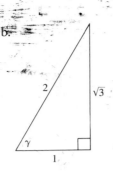

$$\tan \gamma = \sqrt{3};$$

$$\gamma = \text{Tan}^{-1}\sqrt{3} = 60°$$

13. $\sin \alpha = \dfrac{1}{2}$ **14.** $\sec \alpha = \sqrt{2}$ **15.** $\cot \beta = \dfrac{1}{\sqrt{3}}$

16. $\cos \beta = \dfrac{1}{2}$ **17.** $\csc \gamma = 2$ **18.** $\tan \gamma = 1$

▪ *Use a calculator to find an approximation (to the nearest tenth of a degree)
for the measure of each angle.*

Examples

a. $\cos \alpha = 0.6214$ b. $\cot \beta = 1.4324$

Solutions

   a. $\alpha = \text{Cos}^{-1}\, 0.6214$.

$$0.6214 \boxed{\text{INV}}\ \boxed{\text{COS}} \approx 51.58156°$$

To the nearest tenth, $\alpha \approx 51.6°$.

   b. $\tan \beta = \dfrac{1}{\cot \beta} = \dfrac{1}{1.4324}$; $\beta = \text{Tan}^{-1}\dfrac{1}{1.4324}$.

$$1.4324 \boxed{1/\text{x}}\ \boxed{\text{INV}}\ \boxed{\text{TAN}} \approx 34.92001°$$

To the nearest tenth, $\beta = 34.9°$.

19. $\sin \alpha = 0.3162$ **20.** $\tan \alpha = 0.7411$

21. $\tan \beta = 3.1462$ **22.** $\cos \beta = 0.2416$

23. $\csc \gamma = 1.6421$ **24.** $\cot \gamma = 2.4301$

25. $\cot \alpha = 0.6112$ **26.** $\sec \alpha = 3.0211$

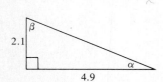

■ *Find approximations (to the nearest tenth) for the measures of the speci-fied acute angles in each figure.*

Example

Solution   Since the sides opposite α and adjacent to α are known, we use the tangent relationship to obtain

$$\tan \alpha = \frac{2.1}{4.9},$$

from which

$$\alpha = \text{Tan}^{-1} \frac{2.1}{4.9}.$$

$$2.1 \boxed{\div} 4.9 \boxed{=} \boxed{\text{INV}} \boxed{\text{TAN}} \approx 23.19859°$$

To the nearest tenth, $\alpha \approx 23.2°$. Since the sum of α and β equals $90°$

$$\beta = 90° - \alpha \approx 90° - 23.2° = 66.8°.$$

27.

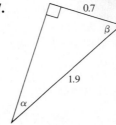

28.

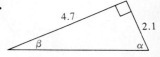

29.

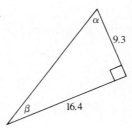

30.

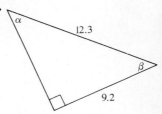

31.

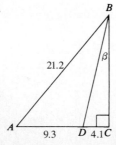

32.

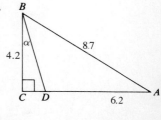

■ *Given the specified trigonometric ratio for an angle α, find approximations (to four decimal places) for the other five trigonometric ratios.*

33. $\sin \alpha = 0.4231$

34. $\cos \alpha = 0.6149$

35. $\sec \alpha = 1.7942$

36. $\cot \alpha = 2.4102$

■ *Given the specified trigonometric ratio for an angle γ, find exact values for the other five trigonometric ratios.*

37. $\cos \gamma = \dfrac{1}{\sqrt{2}}$

38. $\tan \gamma = \dfrac{1}{\sqrt{3}}$

39. $\csc \gamma = 2$

40. $\sec \gamma = 2$

1.4 Applications

In Section 1.2 we obtained trigonometric ratios when we knew the measures of two sides of a right triangle. In Section 1.3 we obtained the measures of the acute angles of a right triangle when we knew a trigonometric ratio of one acute angle. These "tools" will enable us to solve for all missing measures of a right triangle when the triangle is determined by some given information. Then we will be able to solve a variety of applied problems whose geometric representation involves a right triangle.

The computational work in solving triangles is usually facilitated by using a calculator.

Suggestions on Using Calculators   In Section 1.3 we displayed calculator key-stroke sequences that were used to find trigonometric ratios and the measure of an acute angle from an associated ratio. We will not continue to show key-stroke sequences because applied problems often involve complicated combined operations and the sequences may vary with different calculators.

As you use your calculator, you will eventually be able to perform many complicated combined operations involving trigonometric ratios by using a single chain of key strokes. Until you become proficient in performing combined operations, however, you can simply use your calculator in a sequence of steps:

1. Obtain approximations for trigonometric ratios.

2. Write these values in expressions that involve one or more computations.

3. Complete the computations using the simple function keys for the basic operations.

The approximations that you obtain using this procedure may differ slightly from the values you would obtain if you used a single chain of key strokes.

Solution of Triangles   Right triangles can be solved (i.e., the measures of all parts can be found) by using the trigonometric ratios and the fact that *the sum of the measures of the angles of any triangle equals* 180°.

Example   One angle of a right triangle measures 34° and the side opposite this angle has a length of 12 inches. Solve the triangle.

Solution   Sketch the right triangle, showing the given information. First, from the fact that $\alpha + \beta + \gamma = 180°$ and $\gamma = 90°$.

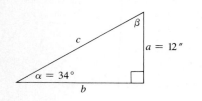

$$\beta = 90° - \alpha$$
$$= 90° - 34° = 56°.$$

By definition,

$$\tan 34° = \frac{a}{b} = \frac{12}{b} \quad \text{or} \quad b = \frac{12}{\tan 34°}$$

and

$$\sin 34° = \frac{a}{c} = \frac{12}{c} \quad \text{or} \quad c = \frac{12}{\sin 34°}.$$

Thus,

$$b \approx \frac{12}{0.6745} \approx 17.8 \quad \text{and} \quad c \approx \frac{12}{0.5592} \approx 21.5.$$

Hence, the measure of β is 56°, $b \approx 17.8$ in., and $c \approx 21.5$ in.

In the above example we have shown values for tan 34° and sin 34° to four decimal places as intermediate steps in the computation. The final results have been rounded off to three significant digits.* We will continue to show function values in intermediate steps even though as soon as you develop calculator proficiency, you will be able to make a single key-stroke sequence to obtain most results.

Example   The hypotenuse of a right triangle is 14.7 centimeters long and the length of a side is 6.8 centimeters. Solve the triangle.

Solution   Sketch the right triangle showing the given information. From the two given dimensions we could first obtain the measure of using the ratios

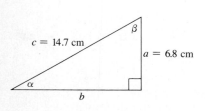

$$\sin \alpha = \frac{6.8}{14.7} \quad \text{or} \quad \csc \alpha = \frac{14.7}{6.8}.$$

---

* See Appendix B for a discussion of significant digits.

Alternatively, we could first obtain the measure of β using the ratios

$$\cos \beta = \frac{6.8}{14.7} \qquad \text{or} \qquad \sec \beta = \frac{14.7}{6.8}.$$

We will use

$$\sin \alpha = \frac{6.8}{14.7} \approx 0.4626,$$

from which

$$\alpha \approx \text{Sin}^{-1}\, 0.4626 \approx 27.6°.$$

Then we have that

$$\beta \approx 90° - 27.6° = 62.4°.$$

We can find an approximation for b by using the Pythagorean theorem:

$$b^2 = \sqrt{(14.7)^2 - (6.8)^2},$$
$$b = \sqrt{216.09 - 46.24} \approx 13.0.$$

Alternatively, we can find b by using the results we obtained for the measure of α or β. If we use $\alpha \approx 27.6°$, we can set up an equation using a variety of trigonometric ratios. One ratio is

$$\cos \alpha = \frac{b}{14.7}.$$

from which

$$b = 14.7 \cos \alpha \approx 14.7 \cos 27.6°$$

$$\approx 14.7(0.8862) \approx 13.0.$$

Hence, the length b is approximately 13.0 cm.

Accuracy of Results   In the foregoing example the lengths b and c were expressed by an approximation using three significant digits. Unless otherwise stated, we will continue to specify lengths of line segments to this degree of accuracy when solutions are given in decimal notation, and we will specify measures of angles to the *nearest tenth of a degree or nearest six minutes*. Furthermore, we will assume that all data have similar accuracy. Thus $a = 12$ in the above example implies that $a = 12.0$, and $\alpha = 34$ implies that $\alpha = 34.0 = 34° \, 00'$.

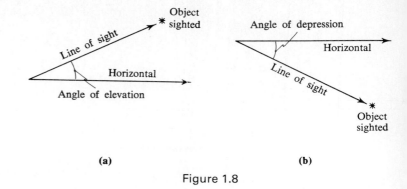

(a)   **(b)**

Figure 1.8

Special Names for Angles   Certain angles are given special names in the fields of surveying and navigation. For example, the angle formed by a horizontal ray and an observer's "line of sight" to any object above the horizontal is called an **angle of elevation** (Figure 1.8a). The angle formed by a horizontal ray and an observer's "line of sight" to any object below the horizontal is called an **angle of depression** (Figure 1.8b).

When solving problems that are expressed in words, it is usually helpful to first sketch a figure including the appropriate measures that are given.

Example On level ground at point A the measure of the angle of elevation of the top of a tower is $57°\ 24'$. The distance from A to the base of the tower is 50 meters. How high is the tower?

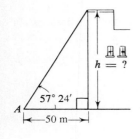

Solution Sketch a figure. By definition,

$$\tan 57°\ 24' = \frac{h}{50},$$

where h represents the height of the tower. Equivalently,

$$h = 50 \tan 57°\ 24' = 50 \tan 57.4°$$

$$\approx 50(1.5637) \approx 78.2.$$

Thus, the height of the tower is approximately 78.2 meters.

Example As a tanker is steaming away from the shore, from an observation point 30 feet above the top of a 330-foot cliff, the angles of depression of the bow and the stern are 22.4° and 16.6°, respectively. Find the length of the tanker.

Solution   Sketch a figure, since $d = AC - BC$, we will solve $\triangle ADC$ to find AC and $\triangle BDC$ to find BC. Since

$$\angle ADC = 90° - 16.6° = 73.4°,$$

$$\tan 73.4° = \frac{AC}{360},$$

from which

$$AC = 360 \tan 73.4° \approx 1207.60.$$

Since

$$\angle BDC = 90° - 22.4° = 67.6°,$$

$$\tan 67.6° = \frac{BC}{360},$$

from which

$$BC = 360 \tan 67.6° \approx 873.43.$$

Hence,

$$d = AC - BC$$

$$\approx 1207.60 - 873.43 = 334.17,$$

and the length of the tanker is approximately 334 feet.

EXERCISE SET 1.4

A

■ *The following measurements are lengths of sides and/or measures of angles of a right triangle, where in each case, $\gamma = 90°$.*

Examples

a. $\alpha = 23.4°$; find β.   **b.** $a = 6.3$ and $\beta = 47°$; find b.

c. $a = 5.31$ and $c = 7.27$; find α.   **d.** $a = 6.2$ and $b = 4.3$; find c.

Solutions   A sketch for each case is helpful.

  a. Since α and β are complementary angles, $\beta = 90° - 23.4° = 66.6°$.

  b. b is related to a and β by the tangent or cotangent functions. Using the tangent, we have

$$\tan 47° = \frac{b}{6.3},$$

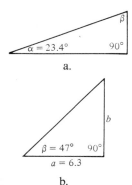
a.

b.

from which

$$b = 6.3 \tan 47° \approx 6.76.$$

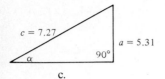

c.

c. α is related to a and c by the sine or cosecant functions. Using th
sine, we have

$$\sin \alpha = \frac{a}{c} = \frac{5.31}{7.27},$$

from which

$$\alpha = \text{Arcsin } \frac{5.31}{7.27} \approx 46.9°.$$

d. From the Pythagorean theorem,

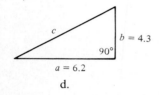

d.

$$c^2 = (6.2)^2 + (4.3)^2.$$

Hence,

$$c = \sqrt{(6.2)^2 + (4.3)^2} \approx 7.55.$$

1. $\beta = 42°$; find α.   2. $\alpha = 16.1°$; find β.

3. $b = 4.2$ and $\alpha = 23°$; find a.   4. $a = 21.3$ and $\alpha = 24°$; find

5. $b = 2.9$ and $c = 3.5$; find β.   6. $a = 4.9$ and $c = 7.3$; find α.

7. $a = 7$ and $\beta = 9.1°$; find c.   8. $b = 6$ and $\alpha = 16.7°$; find a

9. $a = 4.2$ and $b = 6.3$; find c.   10. $a = 7.3$ and $b = 2.1$; find c.

11. $a = 0.6$ and $c = 2.4$; find b.   12. $b = 0.4$ and $c = 1.3$; find a.

■ *Solve each right triangle. In each case, $\gamma = 90°$.*

Example $a = 4.1$ and $c = 7.9$

Solution   A sketch is helpful. In this case, we want to solve for α,
and b. Sine or cosecant functions relate the two known quantities
and c with α. Using the sine, we have

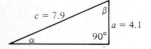

$$\sin \alpha = \frac{4.1}{7.9},$$

from which

$$\alpha = \text{Arcsin } \frac{4.1}{7.9} \approx 31.3°.$$

Then,

$$\beta = 90° - 31.3° = 58.7°.$$

The length b can now be obtained by using any one of several different trigonometric ratios (or the Pythagorean theorem). Using cos α, we have

$$\cos 31.3° = \frac{b}{7.9},$$

from which

$$b = 7.9 \cos 31.3° \approx 6.75.$$

13. $a = 2$ and $b = 3$ **14.** $c = 41$ and $b = 9$

15. $c = 17$ and $a = 15$ **16.** $c = 14$ and $\alpha = 17°$

17. $c = 23.5$ and $\beta = 51°$ **18.** $a = 6.1$ and $\alpha = 73°$

19. $a = 11$ and $\beta = 22.2°$ **20.** $b = 5$ and $\alpha = 64.7°$

21. $b = 15$ and $\alpha = 50° \ 18'$ **22.** $c = 20$ and $\beta = 27° \ 30'$

23. $b = 30$ and $\beta = 41° \ 36'$ **24.** $a = 32$ and $\beta = 75° \ 6'$

■ *Solve.*

25. The measure of an angle of elevation from a surveyor's position (ignoring the surveyor's height) to the top of a hill is $37° \ 24'$. The top of the hill is 275 meters above a level line through the surveyor's position. How far is the surveyor from a point directly below the top of the hill?

26. From the top of a vertical cliff, 80 meters above the surface of the ocean, the measure of the angle of depression to a marker on the surface of the ocean is $18° \ 12'$. How far is the marker from the foot of the cliff?

27. A 40-foot ladder is used to reach the top of a 24-foot wall. If the ladder extends 10 feet past the top of the wall, find the measure of the angle that the ladder forms with the horizontal ground.

28. The lengths of the sides of an isosceles triangle are 17, 17, and 30 centimeters. Find the measures of the angles in the triangle. *Hint:* Use the altitude of the triangle.

29. A ladder 30 decimeters long is leaning against the side of a building. If the ladder makes an angle of $25° \ 18'$ with the side of the building, how far up from the ground does the ladder touch the building?

30. A tower 120 feet high casts a shadow 60 feet long. Find the angle of elevation of the sun.

31. A railroad track makes an angle of 4° 48′ with the horizontal. How far must a train go up the track to gain 20 meters in altitude?

32. A rectangle is 20 centimeters long and 14 centimeters wide. Find the angles that a diagonal makes with the sides.

33. A parachutist is descending vertically. How far does the parachutist fall as the angle of elevation from A changes from 50° to 30°?

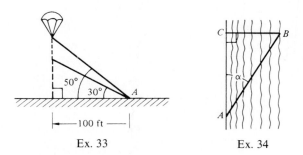

Ex. 33   Ex. 34

34. A surveyor measures AC as 35.3 meters and α as 57.2°. How wide is the river?

35. An observer at P notes that the angle of elevation to the top of mountain is 16.5°. If the mountain is 3500 feet high, how far is from the top of the mountain?

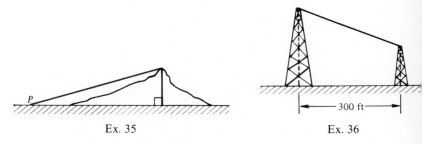

Ex. 35   Ex. 36

36. In the figure, the angle of depression from the top of one tower to the top of the other is 20°. If the height of the taller tower is 150 feet, how high is the shorter?

B

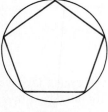

Ex. 37

37. A regular pentagon is inscribed in a circle of radius 5 centimeter Find the length of a side of the pentagon.

38. The angle of elevation of the top of the tower is 62.3° from A and 37.1° from B. If AB is 60 feet, find the height of the tower.

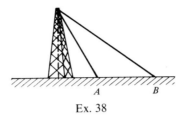

Ex. 38

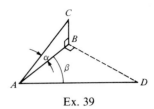

Ex. 39

39. If $\alpha = 30°$, $\beta = 60°$, and $BC = 2$, find AD.

40. Prove that when a simple pendulum of length l is inclined at an angle α with the vertical, the bob is at a height h above its lowest position where $h = l(1 - \cos \alpha)$.

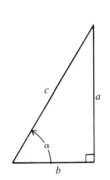

Ex. 40

Chapter Summary

[1.1] The six trigonometric ratios of an acute angle α may be expressed in terms of the sides of a right triangle, where a is the length of the side opposite α, b is the length of the side adjacent to α, and c is the length of the hypotenuse:

$$\sin \alpha = \frac{a}{c}, \qquad \csc \alpha = \frac{c}{a},$$

$$\cos \alpha = \frac{b}{c}, \qquad \sec \alpha = \frac{c}{b},$$

$$\tan \alpha = \frac{a}{b}, \qquad \cot \alpha = \frac{b}{a}.$$

[1.2] Exact trigonometric ratios for 30°, 45°, and 60° angles can be obtained from geometric considerations of right triangles. A calculator can be used to find decimal approximations for ratios of all acute angles. Alternatively, tables in Appendix C can be used to obtain such ratios.

[1.3] A calculator or the tables can also be used to find a positive acute angle α, where trig α is given. The **inverse notation** Trig^{-1} names an acute angle α such that trig $\alpha = y$.

[1.4] Right triangles can be solved by using the trigonometric ratios and the fact that the sum of the measures of the angles of any triangle equals 180°.

Review Exercises

A

[1.1] ▪ *Find the six trigonometric ratios of each angle to the nearest tenth.*

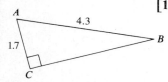

Exs. 1 and 2

1. ∠A   **2.** ∠B   **3.** angle α   **4.** angle β

5. ∠CAD   **6.** ∠CDA   **7.** ∠CBD   **8.** ∠CDB

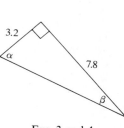

Exs. 3 and 4

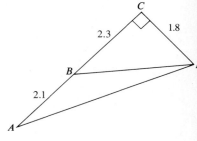

Exs. 5–8

[1.2] ▪ *Find the exact value of each expression.*

9. $\cos 60° - \tan^2 30°$ **10.** $\sin 45° + (\cos 30°)(\cot 45°)$

11. $\dfrac{\sec^2 45° - \tan 30°}{\cos 60°}$ **12.** $\dfrac{\csc^2 30° + \cos 45°}{\sin 60°}$

▪ *Find an approximation to four decimal places.*

13. $\cos 12.6°$ **14.** $\tan 34.6°$ **15.** $\sin 68° \; 24'$

16. $\cos 44° \; 36'$ **17.** $\sec 12° \; 48'$ **18.** $\cot 34° \; 12'$

[1.3] ▪ **a.** *Express each of the following using inverse notation.* **b.** *Find the exact value of* α.

19. $\cos \alpha = \dfrac{1}{2}$ **20.** $\tan \alpha = \sqrt{3}$

21. $\csc \alpha = 2$ **22.** $\cot \alpha = 1$

▪ **a.** *Express each of the following using inverse notation.* **b.** *Find an approximation* (*to the nearest tenth of a degree*) *for* α.

23. $\sin \alpha = 0.3124$ **24.** $\tan \alpha = 1.6219$

25. $\cot \alpha = 1.7213$ **26.** $\sec \alpha = 2.1462$

■ *Find approximations (to the nearest tenth of a degree) for the measures of the specified acute angles.*

27. $\angle CAB$ **28.** $\angle CBD$ **29.** angle β **30.** angle α

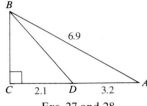

Exs. 27 and 28

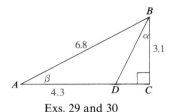

Exs. 29 and 30

[1.4] ■ *Solve each right triangle. In each case, $\gamma = 90°$.*

31. $c = 14,\ \beta = 51°$ **32.** $c = 6.7,\ a = 5.4$

33. $b = 6.8,\ \beta = 46.2°$ **34.** $c = 162,\ \alpha = 12.4°$

35. $a = 12.9,\ \alpha = 32°\ 36'$ **36.** $b = 9.72,\ \alpha = 71°\ 18'$

■ *Solve*

37. At a point 160 feet from the base of a building, the angle of elevation to the top of the building is 64.8°. Find the height of the building to the nearest tenth of a foot.

38. A remote-controlled target plane is launched from a pad on the ground at an angle α so as to travel along a straight line at an average speed of 60 kilometers per minute. If the plane is to be 8.4 kilometers directly above a certain point on the ground 30 seconds after launching, find the measure of α to the nearest tenth of a degree.

39. At an altitude of 4000 feet, the engine of an airplane fails. What glide angle α (to the nearest tenth of a degree) should the pilot use to reach an airfield at a point 3.5 miles away?   *Hint:* 1 mile = 5280 feet.

40. A burglar alarm mounted on a wall detects motion within an angle of 68°. Find the length (to the nearest tenth of a foot) of the interval protected on the property line 42 feet from the alarm. *Hint:* construct two right triangles.

TRIGONOMETRIC AND CIRCULAR FUNCTIONS

In Chapter 1 we defined the trigonometric ratios associated with acute angles measured in degrees. In this chapter we extend the definitions of these ratios for all angles. We will also consider another unit of measure of angles, called *radian measure,* which is widely used in mathematics and the physical sciences.

2.1 General Angle; Radian Measure

In Chapter 1 we simply viewed an angle to be two rays with a common end point. Let us now consider an angle as being formed by a *notation* of one of the sides. In Figure 2.1 consider $\angle AOB$ and visualize it as being formed by rotating side $\overrightarrow{OA}$, with O as the pivotal point, to side $\overrightarrow{OB}$; side $\overrightarrow{OA}$ is called the **initial side,** and $\overrightarrow{OB}$ is called the **terminal side** (see Figure 2.1a). Angles viewed in this way, for which rotation is considered, are called **directed angles.** Furthermore, we will specify that an angle is i

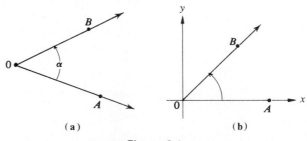

(a)   (b)

Figure 2.1

standard position if its initial side lies in the positive x-axis and its vertex is the origin (see Figure 2.1b).

The terminal side of an angle in standard position may lie in any quadrant or on any axis. The angle is said to be in that quadrant in which the terminal side lies. If the terminal side lies on an axis, the angle is called a **quadrantal angle.** For example, in Figure 2.2, α is in Quadrant I, β is in Quadrant II, γ is in Quadrant III, δ is in Quadrant IV, and θ is a quadrantal angle.

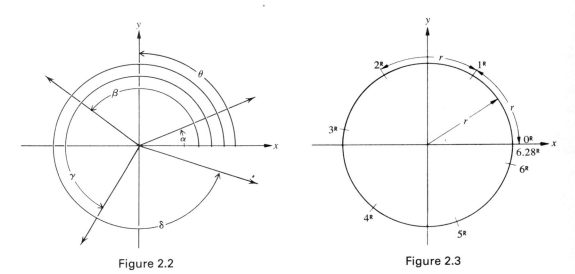

Figure 2.2   Figure 2.3

Radian Measure   In radian measure, a circle is divided into arcs, each having a length equal to the radius of the circle, as shown in Figure 2.3. The circumference of a circle with radius r is $2\pi r$;* hence, the circle will include 2π, or *approximately* 6.28, of these arcs.

The number of such arcs intercepted by a central angle is the measure of the angle in radian units. For this measure we use the notation α^R.

Now if the measure in radians of the angle associated with the entire circle is $2\pi^R$, then the measure of the angle associated with a semicircle is π^R. Measures of other angles that are rational number multiples of π^R as $(\pi/2)^R$, $(\pi/4)^R$, $(\pi/6)^R$, and $(\pi/3)^R$, are shown in Figure 2.4 on page 34.

Positive numbers are assigned for the measures of angles in which the rotation from the initial to the terminal side is in a counterclockwise

---

* π is an irrational number that *cannot* be expressed exactly as a decimal. Accurate to seven places $\pi = 3.1415927$.

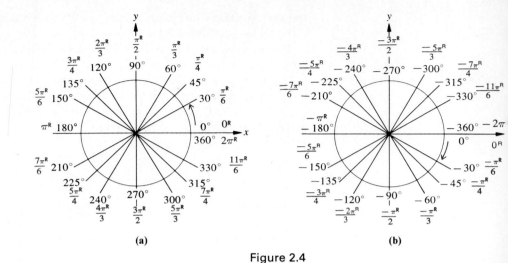

Figure 2.4

direction, and negative numbers are assigned for the measure if the rotation is in a clockwise direction. In Figure 2.4a, the measuring circle is scaled in degree units and radian units for positive measures that are often used. In Figure 2.4b, the measuring circle is scaled for similar negative units.

For example, in Figure 2.5a,

$$\alpha = 150° \qquad \text{and} \qquad \alpha = \frac{5\pi^R}{6},$$

and in Figure 2.5b,

$$\alpha = -60° \qquad \text{and} \qquad \alpha = -\frac{\pi^R}{3}.$$

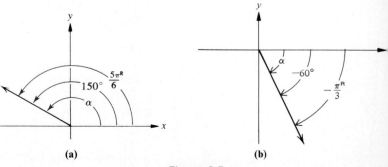

Figure 2.5

As is customary, we will use the terms "least positive angle" to mean that angle of a set of angles which has the least positive measure and "negative angle" to mean an angle with a negative measure.

Degree–Radian Conversion Formulas

The measure of an angle in radians can be related to the measure of the angle in degrees. Since the ratio of the measure of a given central angle to the measure of the central angle intercepting the entire circle is the same for any unit of measurement, we see that for degree measure and radian measure in particular

$$\frac{\alpha^\circ}{360} = \frac{\alpha^R}{2\pi}.$$

This equation is sometimes written in either of the following forms.

$$\text{I.} \quad \alpha^\circ = \frac{180}{\pi}\alpha^R \qquad \text{II.} \quad \alpha^R = \frac{\pi}{180}\alpha^\circ$$

Examples   Find the degree measure (to the nearest tenth of a degree) of each angle α with the given radian measure.

a. $\alpha = \dfrac{3\pi}{5}^R$

b. $\alpha = 0.81^R$

Solutions

a. $\alpha^\circ = \left(\dfrac{180}{\pi} \cdot \dfrac{3\pi}{5}\right)^\circ = 108.0^\circ$

b. $\alpha = \left(\dfrac{180}{\pi} \cdot 0.81\right)^\circ \approx 46.4^\circ$

Examples   Find the radian measure (to the nearest hundredth of a radian) of each angle α with the given degree measure.

a. $\alpha = 60^\circ$

b. $\alpha = 330^\circ$

Solutions

a. $\alpha^R = \left(\dfrac{\pi}{180} \cdot 60\right)^R$

$= \dfrac{\pi}{3}^R \approx 1.05^R$

b. $\alpha^R = \left(\dfrac{\pi}{180} \cdot 330\right)^R$

$= \dfrac{11\pi}{6}^R \approx 5.76^R$

Many exercises in this text, like the examples above, require some approximation for the value of π. Our results in the answers to the examples and exercises in this text were obtained using a calculator value for π that contained more than two decimal places. If you use an approximation for π with less accuracy, such as 3.1 or 3.14, your result may differ slightly.

From conversion formulas I and II on page 35, we observe that the measure in degrees of an angle α with a measure of 1^R is given by

$$\alpha^\circ = \left(\frac{180}{\pi} \cdot 1\right)^\circ \approx 57.30^\circ$$

and that the measure in radians of an angle α with a measure of 1° is given by

$$\alpha^R = \left(\frac{\pi}{180} \cdot 1\right)^R \approx 0.01745^R.$$

Coterminal Angles

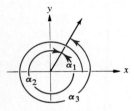

Figure 2.6

If an angle is viewed as being formed by rotation of the initial side to the terminal side, then it is possible to have an angle whose initial side is rotated more than 360°. For example, in Figure 2.6, $\alpha_3 > 360°$. As indicated in the figure, *angles may have the same initial side and the same terminal side but different measures.* Such angles are called **coterminal angles.** Their measures can be positive or negative. In Figure 2.6, α_1, α_2, and α_3 are coterminal. Angles α_1 and α_3 are positive and α_2 is negative.

Examples   Find the least positive angle that is coterminal with each of the following.

a. 6.59^R b. 520°

Solutions

a. Sketch the angle, showing its initial and terminal sides. Since the angle generated in one revolution is $2\pi^R \approx 6.28^R$ to the nearest hundredth, the least positive angle having the same initial and same terminal sides is approximately $(6.59 - 6.28)^R = 0.31^R$.

b. Sketch the angle. In this case, the least positive angle having the same initial and terminal sides is $(520 - 360)^\circ = 160^\circ$.

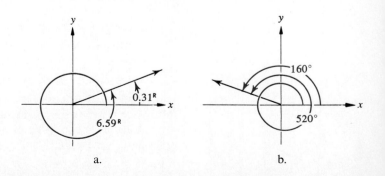

a.   b.

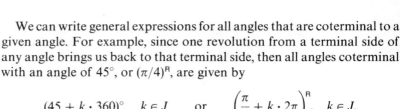

We can write general expressions for all angles that are coterminal to a given angle. For example, since one revolution from a terminal side of any angle brings us back to that terminal side, then all angles coterminal with an angle of 45°, or $(\pi/4)^R$, are given by

$$(45 + k \cdot 360)°, \quad k \in J \qquad \text{or} \qquad \left(\frac{\pi}{4} + k \cdot 2\pi\right)^R, \quad k \in J,$$

where the letter J is used to name the set of integers and the symbol $\in$ is read "is an element of." Note that

$$\text{for } k = 0, \qquad \alpha_1 = (45 + 0)° = 45° \qquad \text{or} \quad \left(\frac{\pi}{4} + 0\right)^R = \frac{\pi}{4}^R;$$

$$\text{for } k = 1, \qquad \alpha_2 = (45 + 360)° = 405° \qquad \text{or} \quad \left(\frac{\pi}{4} + 2\pi\right)^R = \frac{9\pi}{4}^R;$$

$$\text{for } k = -1, \quad \alpha_3 = (45 - 360)° = -315° \quad \text{or} \quad \left(\frac{\pi}{4} - 2\pi\right)^R = -\frac{7\pi}{4}^R.$$

Arc Length Because α^R indicates how many times the length of the radius r of the circle has been used as a unit length along the arc of a circle, the length s of an intercepted arc is given by

$$s = r \cdot \alpha^R. \tag{1}$$

The length s is a directed length; it is a positive or negative real number as α^R is a positive or negative real number, respectively. If the measure of an angle is given in degrees, the measure must first be changed to radian measure before Equation (1) can be applied.

Examples On each circle with the given radius, find the length (to the nearest tenth) of the arc intercepted by the given angle.

a. $r = 4$ cm; $\alpha = 2^R$ \qquad\qquad b. $r = 6$ cm; $\alpha = -150°$

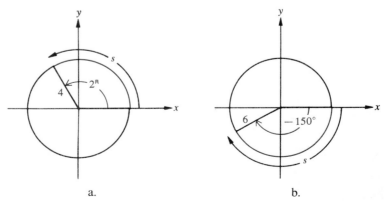

a. \qquad\qquad\qquad\qquad\qquad b.

(continued)

Solutions

a. From Equation (1) above, $s = 4 \cdot 2 = 8.0$. Therefore, the arc is 8 centimeters in length.

b. From Equation II (page 35), the related radian measure of $-150°$ is

$$\alpha = \left[\frac{\pi}{180}(-150) \right]^R = -\frac{5\pi}{6}^R.$$

From Equation (1) above,

$$s = 6\left(-\frac{5\pi}{6} \right) = -5\pi \approx -15.7079.$$

The arc length to the nearest tenth is -15.7 centimeters.

Trigonometric Ratios of Angles with Radian Measure

In Chapter 1 we defined the trigonometric ratios for acute angles with degree measure. We can now extend the definitions to apply to angles with radian measure.

Note that

$$30° = \left(30 \cdot \frac{\pi}{180} \right)^R = \frac{\pi}{6}^R,$$

$$45° = \left(45 \cdot \frac{\pi}{180} \right)^R = \frac{\pi}{4}^R,$$

and

$$60° = \left(60 \cdot \frac{\pi}{180} \right) = \frac{\pi}{3}^R.$$

For convenient reference, the *exact* trigonometric ratios of these special angles are listed again in Table 2.1 with the corresponding radian measures of the angles.

TABLE 2.1

$\alpha°$	α^R	$\sin \alpha$	$\cos \alpha$	$\tan \alpha$	$\csc \alpha$	$\sec \alpha$	$\cot \alpha$
$30°$	$\frac{\pi}{6}^R$	$\frac{1}{2}$	$\frac{\sqrt{3}}{2}$	$\frac{1}{\sqrt{3}}$	2	$\frac{2}{\sqrt{3}}$	$\sqrt{3}$
$45°$	$\frac{\pi}{4}^R$	$\frac{1}{\sqrt{2}}$	$\frac{1}{\sqrt{2}}$	1	$\sqrt{2}$	$\sqrt{2}$	1
$60°$	$\frac{\pi}{3}^R$	$\frac{\sqrt{3}}{2}$	$\frac{1}{2}$	$\sqrt{3}$	$\frac{2}{\sqrt{3}}$	2	$\frac{1}{\sqrt{3}}$

EXERCISE SET 2.1

A

▪ *Complete the following degree–radian conversion wheels for special angles without referring to Figure 2.4. (These wheels are a handy reference.)*

1.

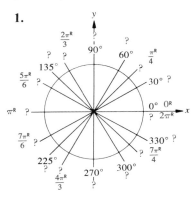

2.

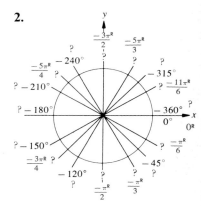

▪ *Find the degree measure (to the nearest tenth of a degree) of an angle α with the given radian measure. Specify the quadrant in which the terminal side of α lies.*

Examples   a. $\dfrac{6\pi^R}{5}$ b. -1.24^R

Solutions   a. $= \left(\dfrac{180}{\pi} \cdot \dfrac{6\pi}{5}\right)^{\circ}$ b. $\alpha^{\circ} = \left[\dfrac{180}{\pi}(-1.24)\right]^{\circ}$

$= 216.0°;$ Quad. III $\approx -71.0°;$ Quad. IV

3. $\dfrac{\pi^R}{3}$ **4.** $\dfrac{2\pi^R}{3}$ **5.** $\dfrac{14\pi^R}{3}$ **6.** $\dfrac{15\pi^R}{4}$

7. $\dfrac{-8\pi^R}{5}$ **8.** $\dfrac{-11\pi^R}{8}$ **9.** $\dfrac{-18\pi^R}{8}$ **10.** $\dfrac{-12\pi^R}{5}$

11. 0.18^R **12.** 0.94^R **13.** 5.16^R **14.** 4.55^R

15. -0.27^R **16.** -0.86^R **17.** -4.04^R **18.** -3.61^R

▪ *Find the radian measure (to the nearest hundredth) of an angle α with the given degree measure. Specify the quadrant in which the terminal side of α lies.*

Examples   a. $118°$ b. $-213°$

(continued)

Solutions   a. $\alpha^R = \left(\dfrac{\pi}{180} \cdot 118\right)^R$ b. $\alpha^R = \left[\dfrac{\pi}{180} \cdot (-213)\right]^R$

$\approx 2.06^R$;   Quad. II $\approx -3.72^R$;   Quad. II

19. $30°$ **20.** $45°$ **21.** $75°$ **22.** $125°$

23. $143°$ **24.** $137°$ **25.** $402°$ **26.** $560°$

27. $-37°$ **28.** $-51°$ **29.** $-80°$ **30.** $-130°$

31. $-240°$ **32.** $-120°$ **33.** $-510°$ **34.** $-400°$

■ *Find the least positive angle (to the nearest hundredth) that is coterminal with each of the following. Sketch each angle.*

Examples   a. 7.34^R b. $-620°$

Solutions   a. $[7.34 + (-1) \cdot 2\pi]^R$ b. $(-620 + 2 \cdot 360)°$

$\approx (7.34 - 6.28)^R$ $= (-620 + 720)°$

$= 1.06^R$ $= 100°$

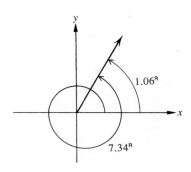 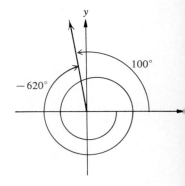

35. 6.72^R **36.** 8.91^R **37.** $379°$ **38.** $423°$

39. -11.11^R **40.** -13.48^R **41.** $-792°$ **42.** $-863°$

■ *Write a general expression for all angles coterminal with each of the following.*

Examples   a. $16°$ b. $\dfrac{\pi}{2}^R$

Solutions   a. $(16 + k \cdot 360)°$, $k \in J$ b. $\left(\dfrac{\pi}{2} + k \cdot 2\pi\right)^R$, $k \in J$

43. $21°$ **44.** $128°$ **45.** $242°$ **46.** $310°$

47. $\dfrac{\pi}{6}^{R}$ **48.** $\dfrac{\pi}{3}^{R}$ **49.** $\dfrac{5\pi}{3}^{R}$ **50.** $\dfrac{7\pi}{6}^{R}$

■ *On each circle with the given radius, find the length (to the nearest tenth of a unit) of the arc intercepted by each angle with given measure.*

Example $r = 9$ centimeters; $240°$

Solution   The related radian measure of $240°$ is

$$m^{R}(\alpha) = \left(\frac{\pi}{180} \cdot 240\right)^{R}$$

$$= \frac{4\pi}{3}^{R}.$$

Thus,

$$s = r \cdot m^{R}(\alpha) = 9 \cdot \frac{4\pi}{3}$$

$$= 12\pi \approx 37.7,$$

and the length of the intercepted arc is approximately 37.7 centimeters.

51. $r = 2$ meters; 3.1^{R} **52.** $r = 30$ centimeters; 1.5^{R}

53. $r = \pi$ decimeters; $150°$ **54.** $r = 2\pi$ meters; $210°$

55. $r = 3$ feet; $-135°$ **56.** $r = 20$ inches; $-240°$

■ *Find the measure (to the nearest tenth of a degree) of the central angle that intercepts each arc on a circle with given radius.*

Example $s = 15$ centimeters; $r = 5$ centimeters

Solution   Substituting 5 for r and 15 for s in $s = r \cdot m^{R}(\alpha)$ yields

$$15 = 5 \cdot m^{R}(\alpha).$$

Thus,

$$m^{R}(\alpha) = \frac{15^{R}}{5} = 3^{R},$$

and

$$m°(\alpha) = \left(\frac{180°}{\pi} \cdot 3\right)° \approx 171.9°.$$

57. $s = 31$ decimeters; $r = 14$ decimeters

58. $s = 2.5$ meters; $r = 2.5$ meters

59. $s = 4.9$ centimeters; $r = 1$ centimeter

60. $s = 14.20$ inches; $r = 1.27$ inches

■ *Evaluate each expression (exact value).*

Example $\sin \dfrac{\pi}{4}^{\text{R}} + \left(\cos^2 \dfrac{\pi}{6}^{\text{R}} \right)\left(\tan \dfrac{\pi}{3}^{\text{R}} \right)$

Solution   By sketching the appropriate triangles, we find

$$\sin \frac{\pi}{4}^{\text{R}} = \frac{1}{\sqrt{2}}, \qquad \cos^2 \frac{\pi}{6}^{\text{R}} = \left(\frac{\sqrt{3}}{2} \right)^2 = \frac{3}{4}, \qquad \text{and} \qquad \tan \frac{\pi}{3}^{\text{R}} = \sqrt{3}.$$

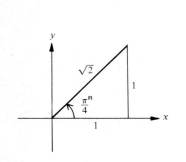
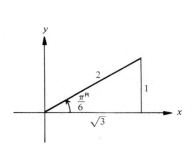
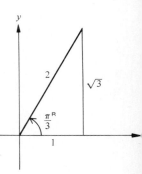

Hence,

$$\sin \frac{\pi}{4}^{\text{R}} + \left(\cos^2 \frac{\pi}{6}^{\text{R}} \right)\left(\tan \frac{\pi}{3}^{\text{R}} \right) = \frac{1}{\sqrt{2}} + \left(\frac{3}{4} \right)\left(\frac{\sqrt{3}}{1} \right)$$

$$= \frac{1(4)}{\sqrt{2}(4)} + \frac{3\sqrt{3}(\sqrt{2})}{4(\sqrt{2})}$$

$$= \frac{4 + 3\sqrt{6}}{4\sqrt{2}}.$$

61. $\left(\sin^2 \dfrac{\pi}{6}^{\text{R}} \right)\left(\cot^2 \dfrac{\pi}{3} \right)^{\text{R}}$

62. $\left(\tan^2 \dfrac{\pi}{6}^{\text{R}} \right)\left(\cos^2 \dfrac{\pi}{4} \right)$

63. $\tan^2 \dfrac{\pi}{4}^{\text{R}} - \cot^2 \dfrac{\pi}{6}^{\text{R}}$

64. $\tan^2 \dfrac{\pi}{3}^{\text{R}} - \cot^2 \dfrac{\pi}{6}^{\text{R}}$

65. $\left(\cos \dfrac{\pi}{3}^{\text{R}} \right)\left(\sin \dfrac{\pi}{6}^{\text{R}} \right) + \tan \dfrac{\pi}{4}^{\text{R}}$

66. $\cot \dfrac{\pi}{4}^{\text{R}} + \left(\tan \dfrac{\pi}{3}^{\text{R}} \right)\left(\cot \dfrac{\pi}{3}^{\text{R}} \right)$

■ *Find an approximation for each trigonometric ratio to four decimal places.*

Examples   a. $\tan 1.3^R$ b. $\sec 0.94^R$

Solutions   Set the calculator on radian mode (RAD).

a. $\tan 1.3^R \approx 3.6021$ b. $\sec 0.94^R = \dfrac{1}{\cos 0.94} \approx 1.6955$

67. $\sin 0.6^R$ **68.** $\cos 1.4^R$ **69.** $\cot 1.3^R$
70. $\csc 1.12^R$ **71.** $\sec 0.92^R$ **72.** $\cot 0.44^R$

■ *Find an approximation (to the nearest tenth of a radian) for the measure of each acute angle α.*

Examples   a. $\cos \alpha = 0.4231$ b. $\csc \alpha = 1.4216$

Solutions   Set the calculator in radian mode (RAD).

a. $\alpha = \text{Cos}^{-1} 0.4231$
    $\approx 1.1^R$

b. $\sin \alpha = \dfrac{1}{\csc \alpha} = \dfrac{1}{1.4216}$

    $\alpha = \text{Sin}^{-1} \dfrac{1}{1.4216}$

    $\approx 0.8^R$

73. $\sin \alpha = 0.6921$ **74.** $\tan \alpha = 1.4392$
75. $\cot \alpha = 0.8421$ **76.** $\sec \alpha = 2.6942$

B

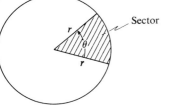

Ex. 77

77. Show that the area $\mathscr{A}$ of the sector of a circle (see figure) with central angle θ is given by $\mathscr{A} = \frac{1}{2}r^2\theta^R$.

78. Use the formula in Exercise 77 to find the area of a sector of a circle whose radius has a length of 1 meter, where the sector has a central angle with measure of $148°$.

79. If a bicycle wheel has a diameter of 27 inches, find the radian measure of the angle through which the wheel turns as the bicycle travels a distance of 1 mile.

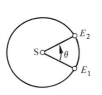

Ex. 80

80. Assuming that the Earth travels in a circular orbit about the sun as in the figure, what would be the measure of θ during one week's travel? If the Earth is 93,000,000 miles from the sun, how far does it travel in its orbit in one week?

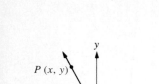

(a)

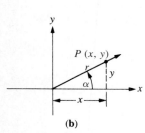

(b)

Figure 2.7

2.2 Trigonometric Ratios of All Angles

In our earlier work the trigonometric ratios were defined for acut
angles in terms of the sides of a right triangle. We now extend thes
definitions for all angles (see Figure 2.7).

For any point $P(x, y)$ on the terminal side of an angle α in standard
position, with $r = \sqrt{x^2 + y^2}$,

$$\sin \alpha = \frac{y}{r}, \qquad\qquad \csc \alpha = \frac{r}{y} \quad (y \neq 0),$$

$$\cos \alpha = \frac{x}{r}, \qquad\qquad \sec \alpha = \frac{r}{x} \quad (x \neq 0),$$

$$\tan \alpha = \frac{y}{x} \quad (x \neq 0), \qquad \cot \alpha = \frac{x}{y} \quad (y \neq 0).$$

Note that if α is an acute angle (see Figure 2.7b), the trigonometri
ratios are the same as those given on page 2 for the angles of a righ
triangle.

Examples

a.

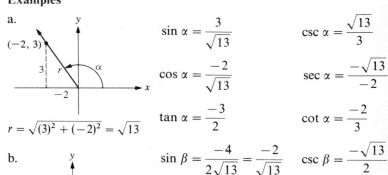

$$\sin \alpha = \frac{3}{\sqrt{13}} \qquad\qquad \csc \alpha = \frac{\sqrt{13}}{3}$$

$$\cos \alpha = \frac{-2}{\sqrt{13}} \qquad\qquad \sec \alpha = \frac{-\sqrt{13}}{-2}$$

$$\tan \alpha = \frac{-3}{2} \qquad\qquad \cot \alpha = \frac{-2}{3}$$

$$r = \sqrt{(3)^2 + (-2)^2} = \sqrt{13}$$

b.

$$\sin \beta = \frac{-4}{2\sqrt{13}} = \frac{-2}{\sqrt{13}} \qquad \csc \beta = \frac{-\sqrt{13}}{2}$$

$$\cos \beta = \frac{-6}{2\sqrt{13}} = \frac{-3}{\sqrt{13}} \qquad \sec \beta = \frac{-\sqrt{13}}{3}$$

$$\tan \beta = \frac{-4}{-6} = \frac{2}{3} \qquad \cot \beta = \frac{3}{2}$$

$$r = \sqrt{(-6)^2 + (-4)^2} = 2\sqrt{13}$$

Trigonometric Ratios for
Quadrantal Angles

As we have noted, the terminal side of a quadrantal angle coincide
with a coordinate axis, and one of the coordinates of any point on th
terminal side is zero. Hence, some of the trigonometric ratios fo

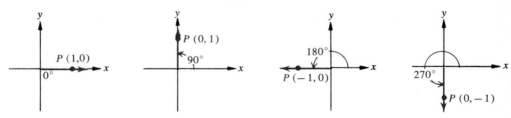

Figure 2.8

quadrantal angles are not defined. For convenience in finding the trigonometric ratios that are defined, we select the point on the terminal side with the nonzero coordinate 1 or -1, as shown in Figure 2.8. For $0°$, $90°$, $180°$, and $270°$,

$$r = \sqrt{x^2 + y^2} = \sqrt{1^2} = 1.$$

For example,

$$\sin 0° = \frac{y}{r} = \frac{0}{1} = 0, \qquad \cos 0° = \frac{x}{r} = \frac{1}{1} = 1, \qquad \text{etc.}$$

You should verify each of the trigonometric ratios of the quadrantal angles listed in Table 2.2.

TABLE 2.2

$\alpha°$	α^R	$\sin \alpha$	$\cos \alpha$	$\tan \alpha$	$\csc \alpha$	$\sec \alpha$	$\cot \alpha$
$0°$	0^R	0	1	0	undef.	1	undef.
$90°$	$\dfrac{\pi}{2}^R$	1	0	undef.	1	undef.	0
$180°$	π^R	0	-1	0	undef.	-1	undef.
$270°$	$\dfrac{3\pi}{2}^R$	-1	0	undef.	-1	undef.	0

In particular, note that $\tan \alpha$ and $\sec \alpha$ are *undefined* if $x = 0$ (the terminal side of α is on the y-axis); that is,

$$\alpha \neq 90° \text{ or } 270°.$$

Furthermore, $\cot \alpha$ and $\csc \alpha$ are *undefined* if $y = 0$ (the terminal side of α is on the x-axis); that is,

$$\alpha \neq 0° \text{ or } 180°.$$

In order to find the trigonometric ratios associated with angles α, such that $\alpha > 90°$ or $\alpha < 0°$ that are not quadrantal angles, we will first introduce several notions that will be helpful.

Signs of Trigonometric Ratios

For an angle α in standard position and x and y the coordinates of point on the terminal side, $r = \sqrt{x^2 + y^2}$ is always positive. Thus, th sign of a trigonometric ratio depends on the values of x and y, and henc on the quadrant in which the terminal side of the angle lies.

Example If the terminal side of an angle α in standard position lies i the second quadrant, the x-coordinate of a point on the terminal side i negative and the y-coordinate is positive. Hence,

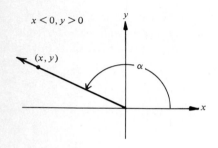

$x < 0, y > 0$

(x, y)

$$\sin \alpha \left(\text{equal to } \frac{y}{r} \right) \text{ is positive,}$$

$$\cos \alpha \left(\text{equal to } \frac{x}{r} \right) \text{ is negative, and}$$

$$\tan \alpha \left(\text{equal to } \frac{y}{x} \right) \text{ is negative.}$$

The signs of the trigonometric ratios of angles in the four quadrants ar shown in Table 2.3

TABLE 2.3

Quadrant	I $x, y > 0$	II $x < 0, y > 0$	III $x, y < 0$	IV $x > 0, y < 0$
sin α or csc α	+	+	−	−
cos α or sec α	+	−	−	+
tan α or cot α	+	−	+	−

Reference Angles

*The **reference angle** of any nonquadrantal angle α in standard position is the positive acute angle $\tilde{\alpha}$ whose sides are the terminal side of α and a ray of the x-axis.*

The reference angles for angles in each of the four quadrants are shown in Figure 2.9.

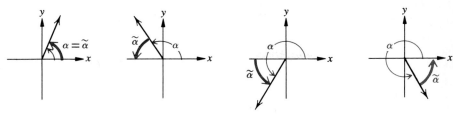

Figure 2.9

Examples   Sketch $\tilde{\alpha}$ for each angle and find $\tilde{\alpha}°$.

a. $\alpha = 135°$ b. $\alpha = 240°$

Solutions

a. $\tilde{\alpha} = (180 - 135)°$ b. $\tilde{\alpha} = (240 - 180)°$

      $= 45°$ $= 60°$

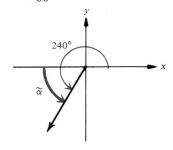

Using Reference Angles

Using geometric considerations, we can establish that

The absolute values of the trigonometric ratios of any angle α are equal, respectively, to the values of the corresponding trigonometric ratios for $\tilde{\alpha}$.

For the case in which α is in Quadrant II, as shown in Figure 2.10 on page 48, we can construct

1. $\overline{PA}$ perpendicular to the x-axis for any point P on the terminal side of α.
2. Semicircle with radius $\overline{OP}$.
3. Radius $\overline{OB}$, making an angle $\tilde{\alpha}$ with the x-axis.
4. $\overline{BC}$ perpendicular to the x-axis.

Because the hypotenuse and an acute angle of triangle POA are congruent, respectively, to the hypotenuse and an acute angle of triangle BOC, the triangles are congruent, and hence $PA = BC$ and

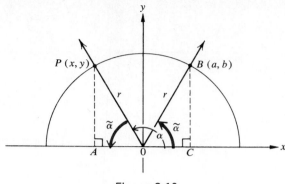

Figure 2.10

$OA = OC$. Thus, $x = -a$ and $y = b$. We then have

$$|\sin \alpha| = \left|\frac{y}{r}\right| = \frac{b}{r} = \sin \tilde{\alpha}, \qquad |\csc \alpha| = \left|\frac{r}{y}\right| = \frac{r}{b} = \csc \tilde{\alpha},$$

$$|\cos \alpha| = \left|\frac{x}{r}\right| = \frac{a}{r} = \cos \tilde{\alpha}, \qquad |\sec \alpha| = \left|\frac{r}{x}\right| = \frac{r}{a} = \sec \tilde{\alpha},$$

$$|\tan \alpha| = \left|\frac{y}{x}\right| = \frac{b}{a} = \tan \tilde{\alpha}, \qquad |\cot \alpha| = \left|\frac{x}{y}\right| = \frac{a}{b} = \cot \tilde{\alpha}.$$

The above relationships hold for an angle in any quadrant, as can b‍
shown by using congruent triangles as above. Using the symbol "trig α‍
to represent any trigonometric ratio, we have that

$$|\text{trig } \alpha| = \text{trig } \tilde{\alpha},$$

for all α for which trig α is defined.

The notion of a reference angle enables us to find *exact* trigonometri‍
ratios of angles with measures that are integral multiples of 30° and 45‍
[or $(\pi/6)^R$ and $(\pi/4)^R$], and *approximations* for all other nonquadranta‍
angles. For such angles we proceed as follows.

1. Find trig $\tilde{\alpha}$.
2. Prefix a negative sign if the original ratio is negative; use Table 2.‍
   as necessary.

Exact Trigonometric
Ratios for Integral
Multiples of 30° and 45°

Example Find tan 210°.

Solution A sketch showing the reference angle is helpful. Observe tha‍

$$\tilde{\alpha} = (210 - 180)° = 30°.$$

Since α is in Quadrant III, tan 210° is positive (see Table 2.3). Therefore,

$$\tan 210° = \tan \tilde{\alpha} = \tan 30° = \frac{1}{\sqrt{3}}.$$

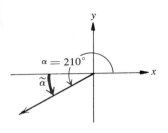

Example Find $\cos \dfrac{3\pi}{4}^{R}$.

Solution A sketch showing the reference angle is helpful. Observe that

$$\tilde{\alpha} = \left(\pi - \frac{3\pi}{4}\right)^{R} = \frac{\pi}{4}^{R}.$$

Since α is in Quadrant II, $\cos (3\pi/4)^{R}$ is negative (see Table 2.3). Therefore,

$$\cos \frac{3\pi}{4}^{R} = -\cos \tilde{\alpha}$$

$$= -\cos \frac{\pi}{4} = -\frac{1}{\sqrt{2}}.$$

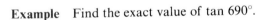

Trigonometric ratios of angles that are greater than 360° and that are integral multiples of 30° or 45° can also be obtained by using a reference angle.

Example Find the exact value of tan 690°.

Solution Since $690° - 360° = 330°,$ the reference angle

$$\tilde{\alpha} = 360° - 330° = 30°.$$

Because 690° is in Quadrant IV, we have

$$\tan 690° = -\tan 30°$$

$$= -\frac{1}{\sqrt{3}}.$$

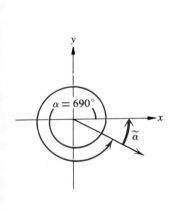

Example Find the exact value of $\cos \dfrac{16\pi}{3}^{R}$.

Solution Since

$$\frac{16\pi}{3}^{R} - 4\pi^{R} = \frac{16\pi}{3}^{R} - \frac{12\pi}{3}^{R} = \frac{4\pi}{3}^{R},$$

(*continued*)

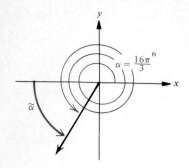

the reference angle

$$\tilde{\alpha} = \frac{4\pi^{\text{R}}}{3} - \pi^{\text{R}} = \frac{\pi^{\text{R}}}{3}.$$

Because $(16\pi/3)^{\text{R}}$ is in Quadrant III, we have

$$\cos\frac{16\pi^{\text{R}}}{3} = -\cos\frac{\pi^{\text{R}}}{3} = -\frac{1}{2}.$$

Trigonometric ratios for angles with negative measures can also be found using similar procedures.

Example   Find the exact value of $\sin(-45°)$.

Solution   The reference angle is

$$\tilde{\alpha} = 45°.$$

Because $-45°$ is in Quadrant IV, we have

$$\sin(-45°) = -\sin 45° = -\frac{1}{\sqrt{2}}.$$

Using a Calculator   Calculators are programmed to give approximations for $\sin\alpha$, $\cos\alpha$ and $\tan\alpha$ directly for all values of α.

Examples
a. Set Degree/Radian switch on RAD.

$$\sin 5.84^{\text{R}} \approx -0.4288.$$

b. Set Degree/Radian switch on DEG.

$$\cos 284.3° \approx 0.2470.$$

Again, values for $\sec\alpha$, $\csc\alpha$, and $\cot\alpha$ can be obtained on a calculator by using the reciprocal relationships.

Example   Set Degree/Radian switch on DEG.

$$\sec 163° \; 18' = \sec 163.3°$$

$$= \frac{1}{\cos 163.3°} \approx -1.0440.$$

As we have previously noted, we will only use a calculator for trigonometric ratios of angles that are not integral multiples of 30° and 45°.

Using Tables   The use of tables to obtain trigonometric ratios of angles outside the interval 0^R to $(\pi/2)^R$ is discussed in Appendix C. Reference angles in conjunction with tables are used for such values, and 3.14 is used as an approximation for π when the angle is given in radian measure. Hence, the approximations obtained are less accurate than the approximations obtained by using a calculator programmed for a closer approximation to π.

EXERCISE 2.2

■ *Find the six trigonometric ratios of an angle α (positive or negative) whose terminal side contains the given point.*

Examples   a. $(-3, -6)$ b. $(0, -1)$

Solutions

a.

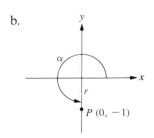

$$r = \sqrt{(-3)^2 + (-6)^2} = 3\sqrt{5}$$

$$\sin \alpha = \frac{-6}{3\sqrt{5}} = \frac{-2}{\sqrt{5}} \qquad \csc \alpha = \frac{-\sqrt{5}}{2}$$

$$\cos \alpha = \frac{-3}{3\sqrt{5}} = \frac{-1}{\sqrt{5}} \qquad \sec \alpha = -\sqrt{5}$$

$$\tan \alpha = \frac{-6}{-3} = 2 \qquad \cot \alpha = \frac{1}{2}$$

b.

$$r = \sqrt{0^2 + (-1)^2} = 1$$

$$\sin \alpha = \frac{-1}{1} = -1 \qquad \csc \alpha = -1$$

$$\cos \alpha = \frac{0}{1} = 0 \qquad \sec \alpha \text{ (undef.)}$$

$$\tan \alpha \text{ (undef.)} \qquad \cot \alpha = \frac{0}{-1} = 0$$

1. $(5, 2)$ **2.** $(-3, 7)$ **3.** $(-2, -3)$ **4.** $(4, -1)$

5. $(3, 0)$ **6.** $(0, 2)$ **7.** $(-2, 0)$ **8.** $(0, -3)$

■ *Determine the quadrant in which the terminal side of each angle lies.*

Example $\sin \alpha < 0$ and $\tan \alpha > 0$

Solution   For $\sin \alpha < 0$, the terminal side of α lies in Quadrant III
IV, and for $\tan \alpha > 0$, the terminal side of α lies in Quadrant I or
Therefore, the terminal side of α must lie in Quadrant III.

9. $\sin \alpha < 0$ and $\cos \alpha > 0$ **10.** $\sin \alpha < 0$ and $\tan \alpha < 0$

11. $\sin \alpha > 0$ and $\sec \alpha > 0$ **12.** $\sin \alpha > 0$ and $\cot \alpha < 0$

13. $\cos \alpha < 0$ and $\tan \alpha > 0$ **14.** $\cos \alpha < 0$ and $\csc \alpha < 0$

15. $\cos \alpha < 0$ and $\cot \alpha < 0$ **16.** $\tan \alpha > 0$ and $\csc \alpha > 0$

■ *Find $\tilde{\alpha}$ in the same units in which each α is given.*

Examples   a. $150°$ b. $\dfrac{11\pi^{R}}{6}$

Solutions   a. $\tilde{\alpha} = 180° - 150°$ b. $\tilde{\alpha} = 2\pi^{R} - \dfrac{11\pi^{R}}{6} = \dfrac{\pi^{R}}{6}$

                $= 30°$

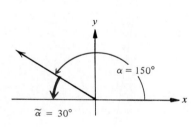 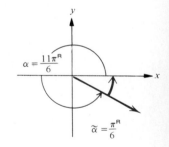

17. $120°$ **18.** $210°$ **19.** $225°$ **20.** $315°$

21. $\dfrac{5\pi^{R}}{4}$ **22.** $\dfrac{7\pi^{R}}{4}$ **23.** $\dfrac{5\pi^{R}}{3}$ **24.** $\dfrac{4\pi^{R}}{3}$

■ *Find the exact value of each trigonometric ratio if it exists.*

Examples

a. $\sin 240°$ b. $\tan \dfrac{7\pi^{R}}{4}$ c. $\tan 180°$

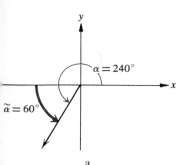

a.

Solutions

a.   The reference angle

$$\tilde{\alpha} = (240 - 180)° = 60°.$$

Since 240° is in Quadrant III, sin 240° is negative. Therefore,

$$\sin 240° = -\sin 60° = -\frac{\sqrt{3}}{2}.$$

b. The reference angle

$$\tilde{\alpha} = 2\pi^R - \frac{7\pi^R}{4} = \frac{\pi^R}{4}.$$

Since $(7\pi/4)^R$ is in Quadrant IV, $\tan(7\pi/4)^R$ is negative. Therefore,

$$\tan\frac{7\pi^R}{4} = -\tan\frac{\pi^R}{4} = -1.$$

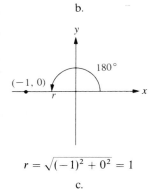

b.

c. 180° is the measure of a quadrantal angle.

$$\tan 180° = \frac{y}{x} = \frac{0}{-1} = 0.$$

25. tan 225°	**26.** sin 315°	**27.** cos 120°
28. sec 330°	**29.** cot 225°	**30.** csc 150°
31. $\sin\dfrac{5\pi^R}{4}$	**32.** $\cos\dfrac{3\pi^R}{4}$	**33.** $\tan\dfrac{2\pi^R}{3}$
34. $\cot\dfrac{5\pi^R}{3}$	**35.** $\sec\dfrac{7\pi^R}{6}$	**36.** $\csc\dfrac{5\pi^R}{6}$
37. sin 270°	**38.** cos 180°	**39.** cot 180°
40. sec 90°	**41.** $\tan 0^R$	**42.** $\csc\dfrac{\pi^R}{2}$

■ *Find the exact value of each trigonometric ratio.*

Example   tan 480°

(*continued*)

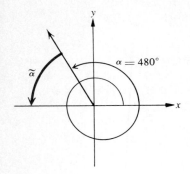

Solution   Since $480° - 360° = 120°$, the reference angle

$$\tilde{\alpha} = 180° - 120° = 60°.$$

Because 480° is in Quadrant II, tan 480° is negative. Hence,

$$\tan 480° = -\tan 60° = -\sqrt{3}.$$

43. $\tan 420°$ **44.** $\sin 570°$ **45.** $\cos 6\tilde{}5°$

46. $\sec 510°$ **47.** $\cot 930°$ **48.** $\tan 870°$

49. $\sin 4\pi^R$ **50.** $\tan 5\pi^R$ **51.** $\cos \dfrac{11\pi}{4}^R$

52. $\sec \dfrac{8\pi}{3}^R$ **53.** $\tan \dfrac{10\pi}{3}^R$ **54.** $\csc \dfrac{17\pi}{6}^R$

■ *Find an approximation for each trigonometric ratio to four decim*e
places.

Example $\sec 128° \; 12'$

Solution   Set Degree/Radian setting on DEG.

$$\sec 128° \; 12' = \sec 128.2°$$

$$= \frac{1}{\cos 128.2°} \approx -1.6171.$$

55. $\cos 310°$ **56.** $\sin 230°$ **57.** $\tan 263.4°$ **58.** $\cos 138.7°$

59. $\cot 342.5°$ **60.** $\csc 273.6°$ **61.** $\csc 322° \; 24'$ **62.** $\sec 269° \; 4$

Example $\cot 5.36^R$

Solution   Set Degree/Radian setting on RAD.

$$\cot 5.36^R = \frac{1}{\tan 5.36^R}$$

$$\approx -0.7564.$$

63. $\tan 2.10^R$ **64.** $\sin 4.22^R$ **65.** $\cos 1.62^R$ **66.** $\tan 3.48^R$

67. $\sec 4.87^R$ **68.** $\csc 1.84^R$ **69.** $\cot 2.93^R$ **70.** $\sec 6.20^R$

2.3 Periodic Functions; Applications

In Chapter 1 we used trigonometric ratios to solve a variety of applications that involved right triangles. In this section we consider some trigonometric properties that will enable us to solve problems that involve periodic phenomena such as alternating current, pendulum oscillations, and wave motion.

First, however, we will interpret as functions the trigonometric relationships we have previously studied. (See Appendix A.3 for a review of algebraic functions.)

Trigonometric Functions

Recall that a **function** is

*An association between the members of two sets in which each member of one set, called the **domain** of the function, is paired with only one member in the second set, called the **range** of the function.*

Since each of the trigonometric ratios pairs each angle for which the ratio exists with exactly one real number, each trigonometric ratio determines a function with a domain that is a set of angles A with the measures shown in Table 2.4.

TABLE 2.4

Function	Pairings	Domain	
sine	$(\alpha, \sin \alpha)$	$\alpha \in A$	
cosine	$(\alpha, \cos \alpha)$	$\alpha \in A$	
tangent	$(\alpha, \tan \alpha)$	$\alpha \in A,$	$\alpha \neq \dfrac{\pi^{R}}{2} + k\pi^{R}$
cosecant	$(\alpha, \csc \alpha)$	$\alpha \in A,$	$\alpha \neq k\pi^{R}$
secant	$(\alpha, \sec \alpha)$	$\alpha \in A,$	$\alpha \neq \dfrac{\pi}{2} + k\pi^{R}$
cotangent	$(\alpha, \cot \alpha)$	$\alpha \in A,$	$\alpha \neq k\pi^{R}$

The trigonometric ratios $\sin \alpha$, $\cos \alpha$, $\tan \alpha$, $\csc \alpha$, $\sec \alpha$, and $\cot \alpha$, are *real numbers* and are called **function values.**

Periodic Property

Note that the sine function has an infinite number of elements in the domain that are paired with a given element in the range. For example,

$$\sin \frac{\pi^{R}}{6} = \frac{1}{2}, \qquad \sin \frac{13\pi^{R}}{6} = \frac{1}{2}, \qquad \sin\left(-\frac{11\pi}{6}\right)^{R} = \frac{1}{2}.$$

In fact, for all $k \in J$,

$$\sin(\alpha + k \cdot 2\pi)^R = \sin \alpha^R. \qquad (1$$

The property of being *periodic,* illustrated by the example above, i defined as follows.

If there exists a real number $a \neq 0$ such that

$$f(x + a) = f(x)$$

for all x in the domain of f, then f is a **periodic function** and a is a **period.**

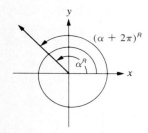

Figure 2.11

Viewing x as the measure of an angle α, we see that the sine and cosin functions are periodic with period $2\pi^R$ as suggested in Figure 2.11.

If $k \in J$, then

I. $\sin(\alpha + k \cdot 2\pi)^R = \sin \alpha^R$

II. $\cos(\alpha + k \cdot 2\pi)^R = \cos \alpha^R$

The smallest positive period of a periodic function is known as th **fundamental period.** Since the terminal side of an angle in standar position must rotate through an angle of $2\pi^R$, before the coordinates of point on the side start to repeat, it follows that the fundamental period c both the sine and cosine function is $2\pi^R$.

Examples

a. $\sin \dfrac{5\pi}{2}^R = \sin\left(\dfrac{\pi}{2} + 2\pi\right)^R$

$= \sin \dfrac{\pi}{2}^R = 1$

b. $\cos 5\pi^R = \cos(\pi + 4\pi)^R$

$= \cos \pi^R = -1$

It can be shown that the points with coordinates (x, y) and $(-x, -$ lie on the terminal sides of angles in standard position whose measure

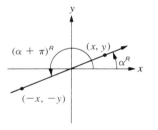

Figure 2.12

differ by π^R (see Figure 2.12). Since for each of these points

$$\tan \alpha = \frac{y}{x} = \frac{-y}{-x},$$

it follows that the fundamental period of the tangent function is π^R.

If $k \in J$, then

$$\tan(\alpha + k\pi)^R = \tan \alpha^R.$$

Examples

a. $\tan \dfrac{7\pi^R}{6} = \tan\left(\dfrac{\pi}{6} + \pi\right)^R$ **b.** $\tan \dfrac{13\pi^R}{4} = \tan\left(\dfrac{\pi}{4} + 3\pi\right)^R$

$\qquad = \tan \dfrac{\pi^R}{6} = \dfrac{1}{\sqrt{3}}$ $\qquad = \tan \dfrac{\pi^R}{4} = 1$

Note that because $\sec x = 1/\cos x$ and $\csc x = 1/\sin x$, the secant and cosecant functions have period $2\pi^R$; because $\cot x = 1/\tan x$, the cotangent function has period π^R.

It follows from the above discussion (where we used radian measure) that in degree measure, the fundamental period of the sine, cosine, secant, and cosecant functions is $360°$, and the fundamental period of the tangent and cotangent functions is $180°$.

Circular Functions   The trigonometric functions have been defined so that the *domains are sets of angles*. Let us now consider six new functions in which the *domains are sets of real numbers*. Each of these new functions pairs *real numbers x* with real numbers trig *x* where the values of *x* in the domain are *radian measures* of angles and the values of trig *x* in the range are the same as the corresponding values of trig x^R. These new functions are given the same names as the corresponding trigonometric functions. For example,

$$\sin \frac{\pi}{6} \text{ is paired with the real number } \frac{\pi}{6}$$

and

$$\cos 1.73 \text{ is paired with the real number } 1.73,$$

where

$$\sin \frac{\pi}{6} = \sin \frac{\pi}{6}^{R} \qquad \text{and} \qquad \cos 1.73 = \cos 1.73^{R}.$$

In general

trig x is paired with x,

where

$$\text{trig } x = \text{trig } x^{R}.$$

Because the elements in the domains of these new function $\{(x, \text{trig } x)\}$ are the measures of angles in radians and because the radia measures of angles are determined by the lengths of arcs on a circle (se Section 2.1), these functions are called **circular functions.** Because thes functions pair real numbers with real numbers, they are sometimes calle **real number functions.**

Since trig $x = \text{trig } x^{R}$ for every element x in their respective domain we can obtain values trig x in the range of the circular functions by usir the same methods that we have been using to find trig α^{R}.

Examples   Find numerical values for each circular function value.

a. $\sin \dfrac{\pi}{3}$ b. $\cos 4$

Solutions
   a. Using the exact values in Table 2.1, we have that

$$\sin \frac{\pi}{3} = \frac{\sqrt{3}}{2}.$$

b. Using a calculator set on the RAD setting, we have that

$$\cos 4 \approx -0.6536.$$

Applications   The usefulness of the circular functions arises from the fact that the are periodic and that *the real numbers that are elements in the domain c. be viewed as measures of things other than angles.* For the latter reaso the variables frequently used to represent unspecified elements in the s R of real numbers (r, s, t, x, y, etc.) are used to represent elements in th domain of the circular functions. For example, in the field of physics, ar oscillatory (periodic) motion satisfying an equation of the form

$$d = A \sin Bt \qquad \text{or} \qquad d = A \cos Bt,$$

where d is a measure of displacement, A and B are constants for the particular motion, and t is a measure of time, is called **simple harmonic motion.** We can now solve some simple problems pertaining to such motion.

If a point is moving so that the motion repeats itself in equal intervals of time, the motion is said to be **periodic,** or **harmonic.** Such motions occur in the vibration of a guitar string, a swinging pendulum, and sound waves, to name a few examples. Using laws of physics, we can show that the mathematical models determining the position of the point at any instant involve the sine and cosine functions. For *simple harmonic motion,* the models take the form

$$d = A \sin Bt \qquad \text{or} \qquad d = A \cos Bt,$$

where d is a measure of displacement, A and B are constants for the particular motion, and t is a measure of time.

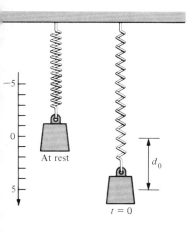

Example   A weighted spring is displaced from its resting position as shown in the figure. The displacement d is given by the equation $d = d_0 \cos \omega t$, where d_0, the initial displacement, and d are expressed in centimeters, t is expressed in seconds, and $\omega = 6$. If the initial displacement $d_0 = 4.2$ centimeters, find the displacement to the nearest tenth when $t = 2.5$ seconds.

Solution   When $t = 2.5$,

$$d = 4.2 \cos 6(2.5) \approx -3.19$$

Hence, because the displacements below the resting positions are positive, displacements above the resting position are negative, and at 2.5 seconds, the end of the spring is approximately 3.2 centimeters *above* the "at rest" position.

Example   The horizontal displacement of the bob on an oscillating pendulum is described by the equation $d = k \sin 2\pi t$ where k is the maximum horizontal displacement from a vertical position, k and d are in centimeters, and t is in seconds. If $k = 10$, find the horizontal displacement to the nearest hundredth for $t = 1.1$ and $t = 1.7$.

Solution   For $t = 1.1$,

$$d = 10 \sin 2\pi(1.1)$$

$$\approx 5.8779.*$$

(continued)

* Unless otherwise specified, a calculator value for $\pi \approx 3.1415927$ is used in computations involving π.

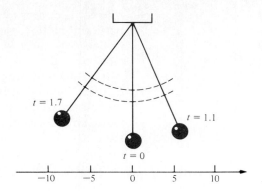

Hence, to the nearest hundredth, the horizontal displacement is 5.88 cen timeters to the *right*.

For $t = 1.7$,

$$d = 10 \sin 2\pi(1.7) \approx -9.5106.$$

Hence, to the nearest hundredth, the horizontal displacement is 9.51 cen timeters to the *left*.

Besides mechanical systems such as the examples above, certai electromagnetic systems can be analyzed using similar mathematic models.

Example The instantaneous current i passing through a certain resi tance at time t is given by

$$i = I \sin 4\pi t,$$

where I is the maximum current; the current is measured in amperes an time in seconds. Find i to the nearest hundredth when $t = 0.02$ secon if $I = 20$ amperes.

Solution For $t = 0.02$,

$$i = 20 \sin 4\pi(0.02)$$

$$\approx 4.9738.$$

Hence, to the nearest hundredth, $i = 4.97$ amperes when $t = 0.0$ second.

Example The instantaneous voltage e in an electrical circuit can b expressed as

$$e = E \sin t,$$

where E is the maximum voltage and t is the time in seconds. If $E = 100$, find e to three significant digits when $t = 2$.

Solution   Since $E = 100$ and $t = 2$,

$$e = E \sin t = 100 \sin 2 \approx 90.9297.$$

Hence, e is approximately 90.9 volts when $t = 2$ seconds.

EXERCISE SET 2.3

A

■ *Write each of the following expressions in the form* $\cos(\alpha + k \cdot 2\pi)$, $\sin(\alpha + k \cdot 2\pi)$, *or* $\tan(\alpha + k\pi)$, *where* $0^R \leq \alpha \leq 2\pi^R$, $k \in J$.

Examples

a. $\cos \dfrac{9\pi}{4} = \cos\left(\dfrac{\pi}{4} + \dfrac{8\pi}{4}\right)$ b. $\tan \dfrac{10\pi}{3} = \tan\left(\dfrac{\pi}{3} + \dfrac{9\pi}{3}\right)$

$\qquad\qquad = \cos\left(\dfrac{\pi}{4} + 2\pi\right)$ $\qquad\qquad = \tan\left(\dfrac{\pi}{3} + 3\pi\right)$

1. $\cos \dfrac{11\pi}{4}$ **2.** $\sin \dfrac{8\pi}{3}$ **3.** $\tan \dfrac{5\pi}{4}$ **4.** $\tan \dfrac{7\pi}{3}$

5. $\sin \dfrac{13\pi}{4}$ **6.** $\cos \dfrac{17\pi}{6}$ **7.** $\tan \dfrac{15\pi}{4}$ **8.** $\cos \dfrac{11\pi}{3}$

9. $\sin \dfrac{19\pi}{6}$ **10.** $\cos \dfrac{17\pi}{4}$ **11.** $\sin \dfrac{13\pi}{3}$ **12.** $\tan \dfrac{13\pi}{6}$

■ *Solve Exercises 13–16 for the conditions for the spring in the example on page 59. Express results to the nearest hundredth.*

13. If the initial displacement is 3, find d when $t = 1.2$ and when $t = 1.8$.

14. If the initial displacement is -3, find d when $t = 1.2$ and when $t = 2.2$.

15. If the initial displacement is 1.5, find d when $t = \frac{1}{2}$ and when $t = \frac{3}{4}$.

16. If the initial displacement is -1.5, find d when $t = \frac{1}{2}$ and when $t = \frac{1}{4}$.

■ *Solve Exercises 17–20 for the conditions of the pendulum in the example on page 60. Express results to the nearest hundredth.*

17. Find d when $k = 20$ and $t = 2.1$.

18. Find d when $k = 16$ and $t = 3.4$.

19. Find d when $k = -12$ and $t = 2\frac{1}{4}$.

20. Find d when $k = -8$ and $t = 3\frac{3}{4}$.

■ *The instantaneous voltage e in a circuit at time t is given by the equation $e = 200 \sin 3700t$ where e is expressed in volts and t in seconds. Find the voltage to three significant digits at each of the following times.*

Example $t = 0.004$

Solution   Evaluate the formula for $t = 0.004$.

$$e = 200 \sin 3700t = 200 \sin 3700(0.004)$$
$$= 200 \sin 14.8 \approx 158.$$

Hence, the voltage is approximately 158 volts when $t = 0.004$.

21. $t = 0.001$ **22.** $t = 0.002$ **23.** $t = 0.005$ **24.** $t = 0.008$

25. $t = 0.012$ **26.** $t = 0.020$

■ *The instantaneous current i in an alternating system is given by the equation $i = I \sin \omega t$, where i is expressed in amperes and t is expressed in seconds. Find the current to three significant digits at each of the following times if $I = 0.05$ and $\omega = 377$.*

27. $t = 0.01$ **28.** $t = 0.03$ **29.** $t = 0.2$ **30.** $t = 0.4$

31. $t = 1.0$ **32.** $t = 3.0$

■ *The pressure P in a traveling sound wave is described by the equation $P = 10 \sin 200\pi[t - (x/1000)]$, where x is the distance from the source of the sound in centimeters, t is the time expressed in seconds, and P is expressed in dynes per square centimeter. Find P to the nearest thousandth for each of the following conditions.*

33. $x = 200$ and $t = 0.06$ **34.** $x = 330$ and $t = 0.01$

35. $x = 660$ and $t = 0.02$ **36.** $x = 1650$ and $t = 0.05$

B

37. The displacement d of a point on a vibrating spring under certain conditions is given by $d = (1/50) \sin 50t - t \cos 50t$, where d is in feet and t is in seconds. Find the displacement when $t = 0.3$ second.

38. It is shown in calculus that an object having simple harmonic motion (described on page 58) has a velocity v given by $v = AB \cos Bt$ and an acceleration given by $a = -AB^2 \sin Bt$. The motion of a certain spring is expressed in $d = 3 \sin(\pi t/2)$, where d is a displacement in centimeters and t is in seconds.
   a. Express the velocity and the acceleration of a point on the free end of the spring in terms of t.
   b. Find the velocity and acceleration when t equals 2, 3, and 4 seconds.

Chapter Summary

[2.1] An angle that has its initial side along the positive x-axis and its vertex at the origin is said to be in **standard position.** Such an angle is said to be in that quadrant in which its terminal side lies.

   The measure of an angle in radians is related to the measure of the angle in degrees by the equations

$$\alpha^\circ = \frac{180}{\pi} \alpha^R \quad \text{and} \quad \alpha^R = \frac{\pi}{180} \alpha^\circ.$$

In particular, $1^R \approx 57.30^\circ$ and $1^\circ \approx 0.01745^R$.

   Angles that have the same initial side and the same terminal side are called **coterminal angles.**

   The length s of the arc of a circle of radius of length r intercepted by an angle α whose vertex is at the center of the circle is given by

$$s = r \cdot \alpha^R.$$

[2.2] For all angles α in standard position, if x and y $[(x, y) \neq (0, 0)]$ are the coordinates of any point on the terminal side of α and $r = \sqrt{x^2 + y^2}$,

$$\text{sine } \alpha = \frac{y}{r}, \qquad\qquad \text{cosecant } \alpha = \frac{r}{y} \quad (y \neq 0),$$

$$\text{cosine } \alpha = \frac{x}{r}, \qquad\qquad \text{secant } \alpha = \frac{r}{x} \quad (x \neq 0),$$

$$\text{tangent } \alpha = \frac{y}{x} \quad (x \neq 0), \qquad \text{cotangent } \alpha = \frac{x}{y} \quad (y \neq 0).$$

The **reference angle** of any nonquadrantal angle α in standard position
the positive acute angle $\tilde{\alpha}$ whose sides are the terminal side of α and a ra
of the x-axis. The absolute values of the trigonometric ratios of an
angle α are equal, respectively, to the analogous trigonometric ratio
for $\tilde{\alpha}$. That is, $|\text{trig } \alpha| = \text{trig } \tilde{\alpha}$.

The use of a reference angle facilitates finding ratios of nonquadranta
angles with measures greater than $90°$ or less than $0°$.

The sign (plus or minus) of a trigonometric ratio depends on th
quadrant in which the terminal side of the angle lies.

The trigonometric ratios for an angle α are equal to the analogou
ratios for *any* angle coterminal with α, including angles with negativ
measures.

[2.3]   Functions in which the elements in the domains are numbers that ar
measures (in radian units) of angles and whose ranges are the same as th
respective ranges of the analogous trigonometric functions are calle
circular functions. A calculator set on RAD or tables in radian measur
that are used to obtain elements in the ranges of the trigonometri
functions can be used to obtain elements in the ranges of the circula
functions.

A function f is **periodic** if there exists a real number a $(a \neq 0$
such that for each x in the domain $x + a$ is in the domain an
$f(x + a) = f(x)$. The smallest positive value of a for which this is tru
is the **period** of the function.

Review Exercises

[2.1] ▪ *Find the degree measure of α to the nearest tenth of a degree.*

1. $\alpha = \dfrac{2\pi}{5}^{\text{R}}$

2. $\alpha = 2.41^{\text{R}}$

▪ *Find the radian measure of α to the nearest hundredth of a radian.*

3. $\alpha = 75°$

4. $\alpha = 257°$

▪ *Find the least positive angle coterminal with each angle α.*

5. $\alpha = 938°$

6. $\alpha = 7.23^{\text{R}}$

▪ *Write a general expression for all angles coterminal with α.*

7. $\alpha = 131°$

8. $\alpha = \dfrac{7\pi}{4}^{\text{R}}$

9. Find the length (to the nearest tenth) of the arc intercepted on a circle by a central angle α with measure 2.2^R if the radius of the circle is 1.3 meters in length.

10. Find the measure (to the nearest tenth) of the central angle that intercepts an arc with length 60.2 centimeters on a circle whose radius is 3.1 meters in length.

▪ *Evaluate each expression* (Exact value).

11. $\tan \dfrac{\pi}{6}^{R} - \sin^2 \dfrac{\pi}{4}^{R}$

12. $\left(\cos^2 \dfrac{\pi}{3}^{R} \right)\left(\tan^2 \dfrac{\pi}{3}^{R} \right)$

▪ *Find an approximation* (*to four decimal places*) *for each trigonometric ratio.*

13. $\cos 1.24^R$

14. $\csc 0.94^R$

▪ *Find an approximation in radians* (*to the nearest tenth*) *for the measure of each angle* α.

15. $\cos \alpha = 0.6421$

16. $\csc \alpha = 2.1409$

[2.2] ▪ *Find the six trigonometric ratios of the angle* α (*positive or negative*) *whose terminal side contains the given point.*

17. $(-4, 3)$ **18.** $(-1, 2)$ **19.** $(\sqrt{3}, -1)$ **20.** $(-8, -15)$

▪ *Determine the quadrant in which the terminal side of each angle* α *lies if* α *is in standard position.*

21. $\sin \alpha < 0$ and $\sec \alpha < 0$ **22.** $\cos \alpha < 0$ and $\cot \alpha > 0$

▪ *Find the measure of* $\tilde{\alpha}$ *in the same units in which each* α *is given.*

23. $210°$ **24.** $330°$ **25.** $\dfrac{7\pi}{4}^{R}$ **26.** $\dfrac{3\pi}{4}^{R}$

▪ *Find the exact value for each of the following.*

27. $\sin 150°$

28. $\tan 330°$

29. $\cos \dfrac{11\pi}{4}^{R}$

30. $\sec\left(-\dfrac{3\pi}{4} \right)^{R}$

■ *Find an approximation for each trigonometric ratio to four decime*
places.

31. $\sin 163°$ **32.** $\cot 331° \ 48'$ **33.** $\tan 3.27^R$ **34.** $\sec 2^R$

[2.3] ■ *Write each expression in the form* $\cos(\alpha + k \cdot 2\pi)$ *or* $\sin(\alpha + k \cdot 2\pi$
where $0^R \le \alpha < 2\pi^R$, $k \in J$. *Specify the value of the expression.*

35. $\cos \dfrac{8\pi}{3}$ **36.** $\sin \dfrac{17\pi}{4}$

37. The displacement of a spring is given by $d = 4 \cos 8\pi t$, wher
d is expressed in centimeters and t is expressed in seconds. Fin
d when $t = 2\frac{1}{2}$ seconds.

38. The horizontal displacement d of the bob on an oscillatin
pendulum is given by the equation $d = 18 \sin 2\pi t$, where d
expressed in centimeters and t is expressed in seconds. Find
when $t = 2\frac{1}{2}$ seconds.

39. The instantaneous voltage e in a circuit at time t is given by th
equation $e = 1.3 \sin 980\pi t$, where e is expressed in volts and t
expressed in seconds. Find the voltage when $t = 3$ seconds.

40. The instantaneous current i in an alternating current system
given by the equation $i = I \sin \omega t$, where i and I are expressed i
amperes, ω in radians per second, and t in seconds. Find the curre
to four decimal places at $t = 0.002$ if $I = 0.04$ and $\omega = 352$.

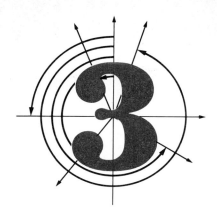

GRAPHS

The periodic characteristics of the trigonometric and circular functions can be exhibited very effectively by graphing these functions in a Cartesian coordinate system. You may want to review the material on graphing in Appendix A.4 before starting this chapter.

3.1 Sine and Cosine Functions

In graphing the trigonometric and circular functions, it is convenient to view line segments on the x-axis of a Cartesian coordinate system as obtained by unwinding the arc on a unit circle with the same length, as shown in Figure 3.1. A similar association could be made for the lengths of arcs in a clockwise direction and line segments on the negative axis.

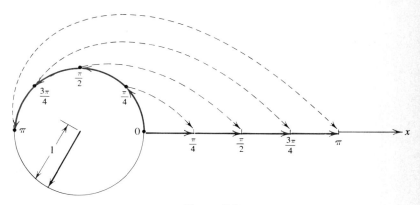

Figure 3.1

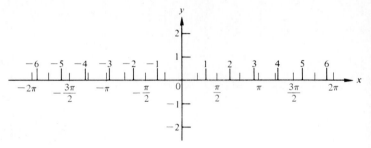

Figure 3.2

Sometimes it is convenient to use integers instead of rational multiple of π as scale numbers on the x-axis. Both of these scale numbers are shown in a Cartesian system in Figure 3.2. The y-axis is scaled in integers.

The graphs of the sine and cosine functions, defined by the two equations $y = \sin x$ and $y = \cos x$, consist of the set of points corresponding to all ordered pairs (x, y) in the respective functions. The replacement set of x is considered to be *either* the set of all angles *or* the set of real numbers.

Although a calculator can be used to find good approximations for $\sin x$ and $\cos x$ function values, the entries in Table 1.1 (page 9) and the notion of a reference angle are usually sufficient to obtain good approximations to the graphs of these functions for $0 \le x \le 2\pi$.

The Sine Graph   We first consider the graph of

$$y = \sin x. \tag{1}$$

In order not to distort the form of the graph of (1) and the following graphs, we will use the same units of measurement on the x-axis and y-axis. To obtain selected ordered pairs $(x, \sin x)$ in the function shown in Figure 3.3a, we first note from Table 1.1 (or by construct

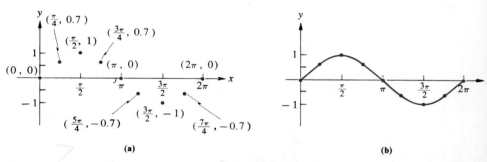

Figure 3.3

tion of an appropriate right triangle, or from memory) that

$$\sin \frac{\pi}{4} = \frac{1}{\sqrt{2}} \approx 0.7;$$

then using a reference angle, we obtain

$$\sin \frac{3\pi}{4} = \sin \frac{\pi}{4} \approx 0.7,$$

$$\sin \frac{5\pi}{4} = -\sin \frac{\pi}{4} \approx -0.7,$$

$$\sin \frac{7\pi}{4} = -\sin \frac{\pi}{4} \approx -0.7.$$

In calculus it is shown that the trigonometric functions are continuous over the intervals for which they are defined. For us this means that since the sine function is defined for all angles (or real numbers), its graph contains no "breaks." Hence, we connect the points in Figure 3.3a with a smooth curve to produce Figure 3.3b. Of course, a calculator can be used to obtain additional points to assist us in graphing the function in the interval 0 to 2π.

Now, because the sine function is periodic with period 2π,

$$\sin(x + k \cdot 2\pi) = \sin x, \quad k \in J.$$

Hence, we repeat the graph in Figure 3.3b in both directions and obtain Figure 3.4 which is part of the graph of $y = \sin x$ for all angles x or real numbers x. The graph is called a **sine wave** or **sinusoidal wave**. The graph over one period is called a **cycle** of the wave.

The amplitude of a sine wave is defined as follows.

The **amplitude** of a sine wave is the absolute value of half the difference of the maximum and minimum ordinates of the wave. This number is also called the amplitude of the function.

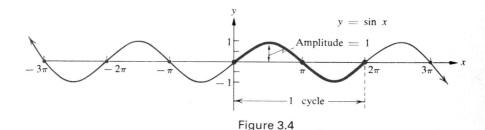

Figure 3.4

Thus, the amplitude of the wave in Figure 3.4 is

$$\left| \frac{1}{2}[1 - (-1)] \right| = 1.$$

Several important notions of the sine function are evident from the graph. Observe that the range of the function (see Appendix A.3), the values of y associated with the values of x in the domain, is given by

$$-1 \le y \le 1.$$

Also observe that the zeros of the function (see Appendix A.5), values of x for which sin x is 0, are $k\pi$, $k \in J$. Note too that these are also the x-coordinates of the points where the curve intersects the x-axis.

The Cosine Graph   We obtain the graph of the cosine function

$$y = \cos x \qquad (2$$

in the same way that we obtained the graph of $y = \sin x$. Some ordered pairs in the function for $0 \le x \le 2\pi$ are obtained for special angles. First noting that

$$\cos \frac{\pi}{4} \approx 0.7,$$

we then obtain

$$\cos \frac{3\pi}{4} = -\cos \frac{\pi}{4} \approx -0.7,$$

$$\cos \frac{5\pi}{4} = -\cos \frac{\pi}{4} \approx -0.7,$$

$$\cos \frac{7\pi}{4} = \cos \frac{\pi}{4} \approx 0.7.$$

The graphs of the ordered pairs are shown in Figure 3.5a. In this case function values decrease as x increases from zero to π, and they increase as x increases from π to 2π. We connect the points in Figure 3.5a with a smooth curve to produce Figure 3.5b. Again, of course, a calculator can be used to obtain additional points for graphing the curve.

Since the cosine function is also periodic with period 2π,

$$\cos(x + k \cdot 2\pi) = \cos x, \quad k \in J.$$

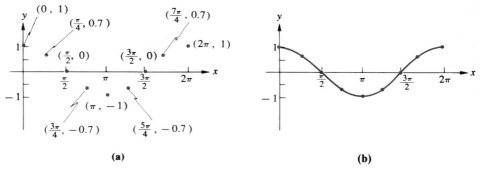

(a)   **(b)**

Figure 3.5

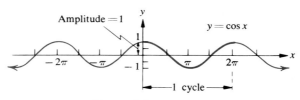

Figure 3.6

Hence, we repeat the pattern in Figure 3.5b and obtain Figure 3.6, which is part of the graph of $y = \cos x$ for all angles x or real numbers x. We observe that, like the sine function, the cosine function has the range $-1 \le y \le 1$.

Also like the sine function, the amplitude of the wave is 1. Furthermore, we note that the zeros of the function, values of x for which $\cos x$ is 0 and which are the x-coordinates of the points where the curve crosses the x-axis, are

$$\frac{\pi}{2} + k\pi, \quad k \in J.$$

Observe that the graphs of the sine and cosine functions have the same "form." For this reason *both* curves are called **sine waves** or **sinusoidal waves.**

In the above discussion we only considered the graphs of

$$y = \sin(x + k \cdot 2\pi) \quad \text{and} \quad y = \cos(x + k \cdot 2\pi).$$

You will graph variations of these basic graphs in the following exercises by plotting points. These graphs will differ from the basic graphs in their amplitude, or zeros, or horizontal position on the x-axis, depending on the constants in the defining equation.

EXERCISE SET 3.1

A

1. Graph the elements of

$$\left\{-2\pi, \frac{-3\pi}{2}, -\pi, \frac{-\pi}{2}, 0, \frac{\pi}{2}, \pi, \frac{3\pi}{2}, 2\pi\right\}$$

on the number line below that is scaled in integral units. Us
$\pi \approx 3.14$.

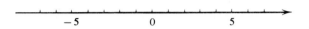

2. Graph the elements of

$$\left\{0, \frac{\pi}{4}, \frac{\pi}{2}, \frac{3\pi}{4}, \pi, \frac{5\pi}{4}, \frac{3\pi}{2}, \frac{7\pi}{4}, 2\pi\right\}$$

on a number line scaled in integral units as shown in Exercise
Use $\pi \approx 3.14$.

3. Graph the elements of

$$\left\{0, \frac{\pi}{4}, \frac{\pi}{2}, \frac{3\pi}{4}, \pi, \frac{5\pi}{4}, \frac{3\pi}{2}, \frac{7\pi}{4}, 2\pi\right\}$$

on the number line below that is scaled in multiples of π.

4. Graph the elements of

$$\left\{-2\pi, \frac{-7\pi}{4}, \frac{-3\pi}{2}, \frac{-5\pi}{4}, -\pi, \frac{-3\pi}{4}, \frac{-\pi}{2}, \frac{-\pi}{4}, 0\right\}$$

on a number line scaled in multiples of π as shown in Exercise 3.

5. Graph the elements of

$$\left\{-2\pi, \frac{-5\pi}{3}, \frac{-4\pi}{3}, -\pi, \frac{-2\pi}{3}, \frac{-\pi}{3}, 0, \frac{\pi}{3}, \frac{2\pi}{3}, \pi, \frac{4\pi}{3}, \frac{5\pi}{3}, 2\pi\right\}$$

on a number line scaled in multiples of π.

6. Graph the elements of

$$\left\{0, \frac{\pi}{6}, \frac{\pi}{3}, \frac{\pi}{2}, \frac{2\pi}{3}, \frac{5\pi}{6}, \pi, \frac{7\pi}{6}, \frac{4\pi}{3}, \frac{3\pi}{2}, \frac{5\pi}{3}, \frac{11\pi}{6}, 2\pi\right\}$$

on a number line scaled in multiples of π.

■ *In Exercises 7–18, (a) graph the function defined by each equation by graphing one cycle and then repeating the pattern and (b) specify the zeros and the amplitude of the function.*

Example $y = 2 \sin x$

Solution

a. Some ordered pairs in the function defined by the equation are obtained in order to graph one cycle. The systematic tabulation of the data shown is helpful. The ordered pairs (x, y) are then graphed as shown, connected with a smooth curve, and the pattern of the wave duplicated. A calculator can be used to obtain additional points to construct the graph.

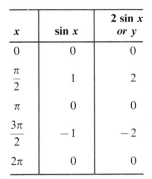

x	$\sin x$	$2 \sin x$ or y
0	0	0
$\dfrac{\pi}{2}$	1	2
π	0	0
$\dfrac{3\pi}{2}$	-1	-2
2π	0	0

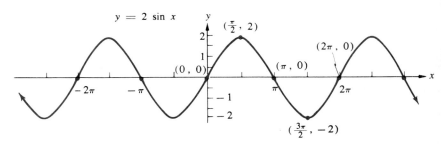

b. Zeros: $k\pi$, $k \in J$;   amplitude: $\left|\dfrac{1}{2}[2 - (-2)]\right| = 2$.

7. $y = 3 \sin x$ **8.** $y = 2 \cos x$ **9.** $y = \dfrac{1}{2}\cos x$

10. $y = \dfrac{1}{3}\sin x$ **11.** $y = -\sin x$ **12.** $y = -\cos x$

Example $y = \cos 2x$

Solution

a. Some ordered pairs in the function defined by the equation are obtained in order to graph one cycle. The ordered pairs (x, y) obtained

(continued)

x	2x	cos 2x or y
0	0	1
$\dfrac{\pi}{4}$	$\dfrac{\pi}{2}$	0
$\dfrac{\pi}{2}$	π	−1
$\dfrac{3\pi}{4}$	$\dfrac{3\pi}{2}$	0
π	2π	1

from the table are then graphed as shown, connected with a smoo curve, and the pattern of the wave duplicated.

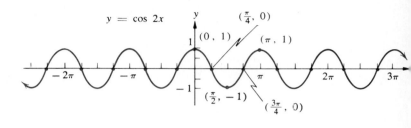

b. Zeros: $x = \dfrac{\pi}{4} + k \cdot \dfrac{\pi}{2}$, $k \in J$;   amplitude: $\left| \dfrac{1}{2}[1 - (-1)] \right| = 1$.

13. $y = \sin 2x$ **14.** $y = \sin \dfrac{1}{2} x$ **15.** $y = \cos \dfrac{1}{2} x$

16. $y = \cos \dfrac{1}{3} x$ **17.** $y = \sin\left(x - \dfrac{\pi}{2}\right)$ **18.** $y = \cos\left(x + \dfrac{\pi}{4}\right)$

■ *Graph two cycles of each of the following.*

Example $y = \cos \dfrac{\pi}{3} x$

Solution   In such cases where the coefficient of x is a multiple of π, it helpful to label the x-axis in rational units and to select rational numbe for x, as shown in the table. Some ordered pairs in the function defined b the equation are graphed as shown and connected with a smooth curv Then the pattern of the wave is duplicated.

x	$\dfrac{\pi}{3} x$	$\cos \dfrac{\pi}{3} x$ or y
0	0	1
1	$\dfrac{\pi}{3}$	$\dfrac{1}{2}$
2	$\dfrac{2\pi}{3}$	$-\dfrac{1}{2}$
3	π	−1
4	$\dfrac{4\pi}{3}$	$-\dfrac{1}{2}$
5	$\dfrac{5\pi}{3}$	$\dfrac{1}{2}$
6	2π	1

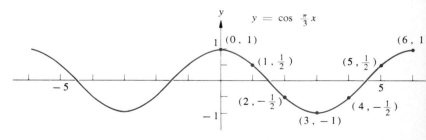

19. $y = \cos \dfrac{\pi}{2} x$ **20.** $y = \sin \dfrac{\pi}{2} x$

21. $y = \sin \pi x$ **22.** $y = \cos \pi x$

B

23. Graph $y = \dfrac{\sin x}{x}$, $-\pi < x < \pi$, and describe the graph as x approaches zero. Is the graph periodic? Why or why not?

24. Graph $y = x \sin x$, $-4\pi \le x \le 4\pi$.

25. Graph $y = \dfrac{\cos x}{x}$, $-\pi < x < \pi$, and describe the graph as x approaches zero. Is the graph periodic? Why or why not?

26. Graph $y = x \cos x$, $-4\pi \le x \le 4\pi$.

■ *The following exercises are examples of noncircular periodic functions. The symbol $[x]$ denotes the number that is the greatest integer not greater than x for each $x \in R$.*
   Graph each function and specify its period.

Examples

a. $[5.91] = [5 + 0.91]$
   $= 5$

b. $[-1.21] = [-2 + 0.79]$
   $= -2$

27. $y = x - [x]$, $-4 \le x < 4$

28. $y = x + [x]$, $-4 \le x < 4$

29. $y = 2x - [2x]$, $-3 \le x < 3$

30. $y = -2x + [2x]$, $-3 \le x < 3$

3.2 Sine Waves

Functions defined by equations of the form

$$y = A \sin B(x + C) + D \qquad \text{or} \qquad y = A \cos B(x + C) + D$$

where $A, B, C, D, x, y \in R$ and $A, B \ne 0$ always have sine waves for their graphs. Depending on the values for A, B, C, and D, these waves have different amplitudes, different periods, and different horizontal and vertical positions with respect to the origin. Analogous graphical features for algebraic functions are reviewed in Appendix A.6.

In this section we consider how values of A affect the amplitude and how values of B affect the period of a sine wave. We will consider the effects of C and D in Section 3.3.

Amplitude   First consider how the value of A in the defining equation affects the graph of the sine function or the cosine function. For each value of x, each ordinate of the graph of

$$y = A \sin x$$

is A times the ordinate of the graph of $y = \sin x$. The effect of A, then, is to multiply the amplitude of the graph of $y = \sin x$ by $|A|$. The graph of $y = A \cos x$ is related to the graph of $y = \cos x$ in a similar way.

Example   Graph $y = 2 \cos x$ over the interval $-2\pi \leq x \leq 4\pi$.

Solution
1. Sketch the graph of $y = \cos x$, over the interval $0 \leq x \leq 2\pi$, as a reference.

2. The amplitude of the graph of $y = 2 \cos x$ is 2. Therefore, sketch the desired graph on the same coordinate system over the same interval by making each ordinate 2 times the ordinate of the graph of $y = \cos x$.

3. Repeat the cycle over the interval $-2\pi \leq x \leq 4\pi$.

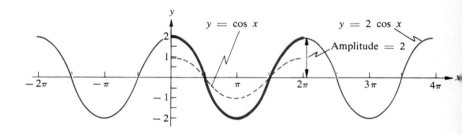

The first cycle is shown sketched with a heavier line for emphasis.

In the above example, the coefficient of $\cos x$ is a positive number. If the coefficient were a negative number, say -2, each ordinate of the graph of $y = -2 \cos x$ would be the negative of the corresponding ordinate of $y = 2 \cos x$. The graph of $y = -2 \cos x$, over the interval $-2\pi \leq x \leq 4\pi$, is shown in Figure 3.7.

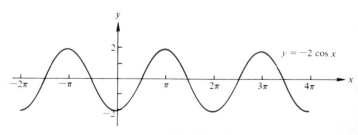

Figure 3.7

Period   Now consider how different values for B affect the graph of

$$y = \sin Bx \qquad \text{or} \qquad y = \cos Bx.$$

A complete cycle of the graph of $y = \sin x$ (or $y = \cos x$) occurs over the period 2π. Hence, a cycle of the graph of $y = \sin Bx$ (or $y = \cos Bx$) is obtained as Bx increases from 0 to 2π. Therefore, the period of either function is the solution of the equation

$$|B|x = 2\pi \qquad \text{or} \qquad x = \frac{2\pi}{|B|}.$$

Hence we have the following result.

> The period P of a function defined by an equation of the form $y = \sin Bx$ or $y = \cos Bx$ is
>
> $$P = \frac{2\pi}{|B|}.$$

The absolute value of B is used so that the period is a positive number.

Example   Graph $y = \sin 3x$, over the interval $0 \le x \le 2\pi$.

Solution
1. Sketch the graph of $y = \sin x$, over the interval $0 \le x \le 2\pi$, as a reference.
2. The period

$$P = \frac{2\pi}{|B|} = \frac{2\pi}{3}.$$

Therefore, sketch a complete cycle of the desired graph on the same set of axes over the interval $0 \le x \le 2\pi/3$.

3. Repeat the cycle obtained in Step 2 over the interval $0 \le x \le 2\pi$.

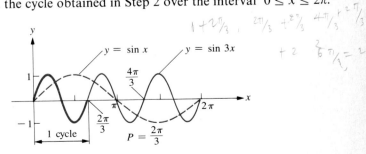

The Period P a Positive Rational Number   If the period P of a graph is a positive rational number, instead of multiple of π such as we encountered in the examples above, it is helpf[ul] to use selected rational values as elements of the domain and to label t[he] x-axis in rational units to facilitate sketching the graph.

Example   Graph $y = \dfrac{1}{2} \cos \pi x$ over the interval $-5 \le x \le 5$.

Solution

1. Compare the equation with

$$y = A \cos Bx.$$

It is evident that the graph of $y = \frac{1}{2} \cos \pi x$ has

$$\text{amplitude: } |A| = \frac{1}{2}; \quad \text{period: } P = \frac{2\pi}{|B|} = \frac{2\pi}{\pi} = 2.$$

2. Sketch the graph, marking the x-axis in rational units.

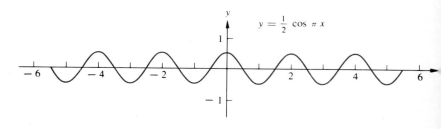

Frequency   The *frequency f* of an object moving in simple harmonic motion is th[e] number of oscillations it makes per unit of time. Because the perio[d] $2\pi/|B|$ is the time of one oscillation (cycle), the frequency is the reciproca[l] of the period.

The frequency of an object moving in simple harmonic motion is

$$f = \frac{1}{2\pi/|B|} = \frac{|B|}{2\pi}.$$

Example The frequency of an object moving in accordance with th[e] function

$$d = d_0 \cos 4\pi t$$

is $4\pi/2\pi$ or 2 oscillations per unit of time.

Variable Amplitude   In our work so far the amplitudes of the functions

$$y = A \sin Bx \quad \text{and} \quad y = A \cos Bx$$

were constants equal to $|A|$. On the other hand, functions such as

$$y = A(x) \sin Bx \quad \text{and} \quad y = A(x) \cos Bx, \tag{1}$$

where $A(x)$ is a function of x, have *variable* amplitudes. As an example we first graph the function

$$y = x \sin \frac{\pi}{2} x, \quad 0 \le x \le 8, \tag{2}$$

by plotting points in the interval specified to obtain the graph in Figure 3.8.

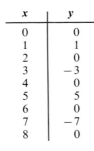

x	y
0	0
1	1
2	0
3	-3
4	0
5	5
6	0
7	-7
8	0

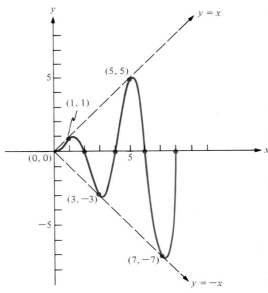

Figure 3.8

Note that as the amplitude of the curve in Figure 3.8 varies, the high and low points are limited by the graph of $y = x$ and the graph of $y = -x$, which is symmetric to the graph of $y = x$ about the x-axis. The graph of $y = -x$ is called the *reflection* of the graph of $y = x$. Furthermore, the period of the curve is

$$\frac{2\pi}{\pi/2} = 4,$$

the same as the period of the function

$$y = \sin \frac{\pi}{2} x.$$

The above example suggests the following procedure to graph functions with variable amplitudes $A(x)$.

1. Graph $y = A(x)$ and its reflection on the x-axis $y = -A(x)$. These graphs are boundary curves between which the required graph lies.
2. Graph $y = \sin Bx$ or $y = \cos Bx$ over the specified domain, but modify the "high" and "low" points of each cycle, which are bounded by the graphs of $y = A(x)$ and $y = -A(x)$.

Example   Graph $y = (x + 1) \cos \pi x$ over the interval $0 \le x \le 8$.

Solution   First graph $y = x + 1$ and its reflection $y = -(x + 1)$ for $x \ge 0$. Then graph $y = \cos \pi x$ with period $2\pi/\pi = 2$, but position the high and low points so that they are bounded by the graphs of $y = x + 1$ and $y = -(x + 1)$.

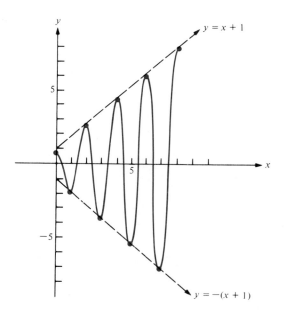

In certain applications the coefficients $A(x)$ of the functions $y = A(x) \sin B(x)$ or $y = A(x) \cos B(x)$ are exponential functions involving the power e^{-x}, where e is an irrational number approximately equal to 2.71828183. Using a scientific calculator, we can readily graph such functions as boundary curves.

Example   The instantaneous current i in a certain circuit is given by

$$i = 10e^{-t} \cos 2\pi t,$$

where t is a unit of time. Graph the function over the interval $0 \le t \le 4$.

Solution   First graph $i = 10e^{-t}$ and its reflection about the t-axis by preparing a table of values as shown. Then graph $y = \cos 2\pi t$ with period $2\pi/2\pi = 1$, but position the high and low points so that they are bounded by the boundary curves.

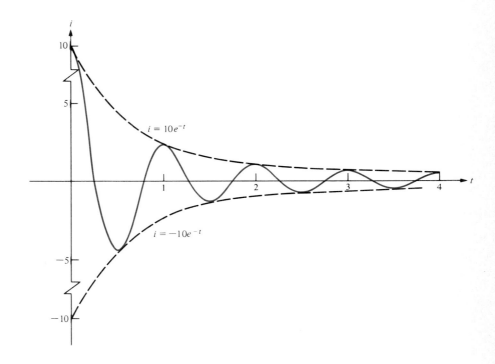

$i = 10e^{-t}$

t	i
0	10.0
1	3.7
2	1.4
3	0.5
4	0.2

Observe how the graph in the above example suggests a rapid drop in the current. The factor e^{-t} that produced this drop is often called a *damping* factor.

The following example illustrates the effect of a damping factor in a mechanical system.

Example   Eighteen inches of a yardstick is clamped to the top of a workbench. The free end of the yardstick is depressed as shown in the figure and then released.

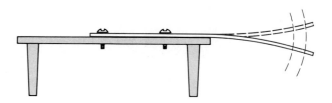

Graph the displacement for $0 \le t \le 0.6$ if the displacement is given by

$$d = -6e^{-4t} \cos 10\pi t,$$

where d is in centimeters and t is in tenths of a second.

Solution   First graph $d = 6e^{-4t}$ and its reflection about the t-axis by preparing a table of values as shown. Then graph $d = -\cos 10\pi t$ with period $2\pi/10\pi = 0.2$, but position the high and low points so that they are bounded by the boundary curves.

$d = 6e^{-4t}$

t	d
0	6.0
0.1	4.0
0.2	2.7
0.3	1.8
0.4	1.2
0.5	0.8
0.6	0.5

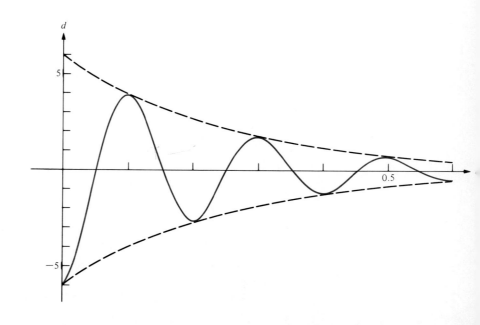

EXERCISE SET 3.2

A

■ *Sketch the graph of each equation over the interval* $-2\pi \le x \le 2\pi$. *Specify the zeros and the amplitude.*

Example $y = \dfrac{1}{2}\cos 2x$

Solution

1. Compare the equation with $y = A \cos Bx$. It is evident that the graph of $y = \frac{1}{2}\cos 2x$ has

$$\text{amplitude: } |A| = \frac{1}{2}; \quad \text{period: } P = \frac{2\pi}{|B|} = \frac{2\pi}{2} = \pi.$$

2. Sketch the graph.

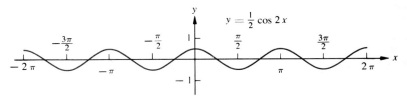

1. $y = 3 \sin x$

2. $y = 3 \cos x$

3. $y = -2 \cos x$

4. $y = -5 \sin x$

5. $y = \dfrac{1}{2}\sin x$

6. $y = \dfrac{1}{3}\cos x$

7. $y = \cos 2x$

8. $y = \sin 3x$

9. $y = \sin \dfrac{1}{2}x$

10. $y = \cos \dfrac{1}{3}x$

11. $y = 2 \sin 2x$

12. $y = \dfrac{1}{3}\cos \dfrac{1}{2}x$

13. $y = -4 \cos 2x$

14. $y = -4 \sin \dfrac{1}{2}x$

■ *Graph two cycles of each of the following with positive elements in the domain.*   Hint: *Scale the x-axis in rational units 1, 2, 3,....*

Example $y = \sin \pi x$

(continued)

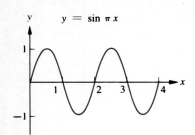

y $y = \sin \pi x$

Solution

1. Compare the equation with $y = A \sin Bx$. It is evident that th graph of $y = \sin \pi x$ has amplitude $|A| = 1$ and period

$$P = \frac{2\pi}{|B|} = \frac{2\pi}{\pi} = 2.$$

2. Sketch the graph on a coordinate system in which the x-axis scaled in rational units.

15. $y = \cos \pi x$ **16.** $y = \sin \dfrac{\pi}{2} x$

17. $y = \dfrac{1}{2} \sin \dfrac{\pi}{3} x$ **18.** $y = 2 \cos \dfrac{\pi}{4} x$

■ *The following equations describe the simple harmonic motion of an objec Find the frequency of the motion to the nearest tenth.*

19. $d = d_0 \sin 6\pi t$ **20.** $d = d_0 \cos 8\pi t$

21. $A = A_0 \cos 12t$ **22.** $A = A_0 \sin 16t$

■ *Graph each equation over the specified interval.*

Example Graph $y = \dfrac{x + 4}{2} \cos \pi x, \ 0 \le x \le 8.$

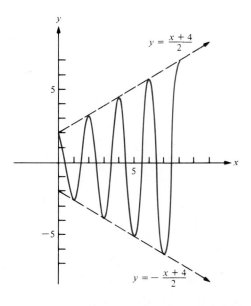

Solution   First graph $y = \dfrac{x + 4}{2}$ $(x \geq 0)$ and its reflection as boundary curves. Then graph $y = \cos \pi x$ with period $2\pi/\pi = 2$, but position the high and low points so that they are bounded by the boundary curves.

23. $y = \dfrac{x + 2}{2} \cos \pi x, 0 \leq x \leq 6$ **24.** $y = \dfrac{6 - x}{2} \sin \dfrac{\pi}{2} x, 0 \leq x \leq 8$

25. $i = 4e^{-t} \cos \pi t, 0 \leq t \leq 4$ **26.** $i = 2e^{-0.2t} \cos \pi t, 0 \leq t \leq 4$

27. $i = 5e^{-0.25t} \cos \pi t, 0 \leq t \leq 6$ **28.** $i = 10e^{-t/8} \cos \pi t, 0 \leq t \leq 6$

■ *The following equations describe the simple harmonic motion of an object, where d is in centimeters and t is in seconds. For each exercise, graph the function over two cycles and give each of the following to the nearest tenth:* (**a**) *the starting position at* $t = 0$; (**b**) *the maximum displacement;* (**c**) *the period and frequency;* (**d**) *the displacement of the specified time.*

Example $d = 3 \sin 4t, \quad t = 4.2$

Solution
   a. At $t = 0$,

$$d = 3 \sin 4(0) = 0.0 \text{ centimeters.}$$

   b. The maximum displacement is the amplitude, 3.0 centimeters.
   c. The period

$$P = \frac{2\pi}{4} \approx 1.6 \text{ seconds;}$$

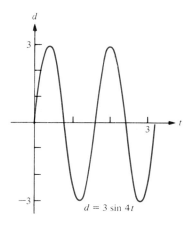

$d = 3 \sin 4t$

the frequency

$$f = \frac{1}{P} = \frac{4}{2\pi} \approx 0.6 \text{ oscillations per second.}$$

d. At $t = 1.5$,

$$d = 3 \sin 4(4.2) \approx -2.7 \text{ centimeters.}$$

29. $d = 3 \cos 2t, \ t = 1.2$ **30.** $d = \frac{1}{2} \cos 3t, \ t = 2.1$

31. $d = \frac{1}{2} \sin 0.5t, \ t = 2.8$ **32.** $d = 4 \sin 0.5t, \ t = 4.2$

33. $d = 2 \cos \pi t, \ t = 0.8$ **34.** $d = 3 \sin \pi t, \ t = 0.6$

B

■ *Approximate an appropriate mathematical model for the given conditions assuming that the "object" moves with simple harmonic motion.*

35. A pendulum moves so that its maximum horizontal displacement from a vertical position is 2.4 meters at time $t = 0$ and makes 3.2 oscillations in one minute. Hint: *A model of the form* $d = A \cos Bt$ *is easier to use than* $d = A \sin Bt$ *because when* $t = 0$, $\cos Bt = 1$ *and the maximum displacement is 2.4 meters.*

36. The pendulum in Exercise 35 starts from a vertical position at $t = 0$

37. A spring *stretched* 4.5 centimeters from its resting position by a weight is released. It makes 1.8 oscillations per second. Find the approximate position of the weight in 3 seconds.

38. The spring in Exercise 37 is compressed 4.5 centimeters and then released. Find the approximate position of the weight in 3 seconds.

39. The amperage of an electric current is dampened by the exponential factor $e^{-0.4t}$, where t is in seconds. If the frequency of the current is 60 cycles per second, find the approximate instantaneous current in 2 seconds.

40. The current in Exercise 39 is dampened by the factor $e^{-1.2t}$. Find the approximate instantaneous current in 2 seconds.

3.3 More About Sine Waves

In Section 3.2 we observed how the graphs of

$$y = A \sin B(x) \qquad \text{and} \qquad y = A \cos B(x) \tag{1}$$

are affected by changes in A and B. In this section we look at equations of the form

$$y = A \sin B(x + C) + D \qquad \text{and} \qquad y = A \cos B(x + C) + D \quad (2)$$

and see how values of C $(C \neq 0)$ shift the graphs of (1) horizontally and how values of D $(D \neq 0)$ shift the graphs vertically. These shifts are called **translations.** See Appendix A.6 for a review of translations for algebraic functions.

We will also graph equations of the form

$$y = A_1 \sin B_1 x + A_2 \cos B_2 x$$

by using a procedure called "addition of ordinates."

Phase Shift; Horizontal Translation   We first consider how different values for C affect the graph of

$$y = \sin(x + C).$$

The graphs of $y = \sin(x + c)$ for $C > 0$ and for $C < 0$ are shown in Figure 3.9 in relation to the graph of $y = \sin x$.

Each of these graphs, which can be verified by graphing several ordered pairs, is the graph of an equation of the form

$$y = \sin(x + C).$$

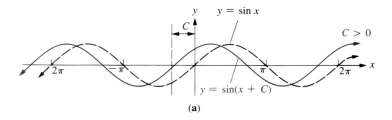

(a)

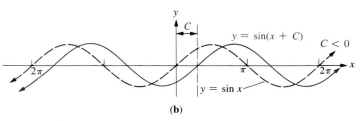

(b)

Figure 3.9

The graph is shifted horizontally C units to the left of the graph $y = \sin x$ if $C > 0$, and C units to the right if $C < 0$.

The graphs of $y = \cos(x + C)$ and $y = \cos x$ are similarly relate

---

The number $|C|$ is the **phase shift** of the graph of periodic functions defined by equations of the form

$$y = \sin(x + C) \qquad or \qquad y = \cos(x + C).$$

The number $|C|$ is also the phase shift of the corresponding function.

A phase shift is an example of a horizontal translation.

Example   Graph $y = \sin\left(x - \dfrac{\pi}{3}\right)$ over the interval $-2\pi \le x \le 4\pi$

Solution
1. Sketch the graph of $y = \sin x$, over the interval $0 \le x \le 2\pi$, as reference.
2. Since $C = -\pi/3$, the graph of $y = \sin x$ is shifted $\pi/3$ units the right to obtain the desired graph.
3. Sketch the graph over the interval $-2\pi \le x \le 4\pi$.

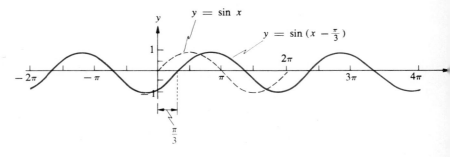

Note that in the above example the graph of $y = \sin(x - \pi/3)$ w shifted $\pi/3$ *units to the right* of the graph of $y = \sin x$ because $-\pi$ the value of C, is *negative*. The graph of $y = \sin(x + \pi/3)$, would shifted *to the left* of the graph of $y = \sin x$ because $\pi/3$, the value of is *positive*. An example is shown in the exercise set on page 93.

We can now use all information about the effects of A, B, and C on t graphs of functions defined by the equation $y = A \sin B(x + C)$ or the equation $y = A \cos B(x + C)$ to facilitate making such graphs.

Example   Graph $y = 3 \cos 2\left(x - \dfrac{\pi}{4}\right)$ over the interval $0 \le x \le 2\pi$.

Solution
1. Compare the given equation with $y = A \cos B(x + C)$ and note that the amplitude is 3, the period is $2\pi/2 = \pi$, and the phase shift is $\pi/4$ units to the right of the graph of $y = \cos x$.
2. Sketch the graph of $y = \cos 2x$, over the interval $0 \le x \le 2\pi$ as a reference.
3. Shift this graph $\pi/4$ units to the right with amplitude 3.

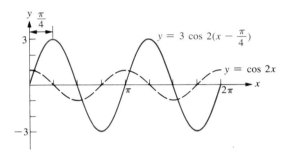

Key Points on the *x*-axis   If a sine or cosine function involves a phase shift, it may be helpful to begin graphing one period of the function by finding some key points on the *x*-axis. If we first graph the phase shift $|C|$ and the sum of $|C|$ and the period on the *x*-axis, these points will locate zeros of the sine function or horizontal positions of high or low points of the cosine function. Other zeros or the horizontal position of other high or low points are spaced equally between these points and can be readily found.

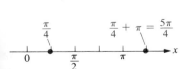

Example   Consider the function $y = 3 \cos 2\left(x - \dfrac{\pi}{4}\right)$ in the above example, where $A = 3$, $P = 2\pi/2 = \pi$, and $C = -\pi/4$ (to the right). If we first graph the phase shift and the sum of the phase shift and period on the *x*-axis, these points will locate the horizontal position of the high points.

The midpoint between $\pi/4$ and $5\pi/4$

$$\dfrac{\dfrac{\pi}{4} + \dfrac{5\pi}{4}}{2} = \dfrac{3\pi}{4}$$

(continued)

is the horizontal position of a low point of the graph. The midpoi
between $\pi/4$ and $3\pi/4$, and $3\pi/4$ and $5\pi/4$,

$$\frac{\dfrac{\pi}{4} + \dfrac{3\pi}{4}}{2} = \frac{\pi}{2} \quad \text{and} \quad \frac{\dfrac{3\pi}{4} + \dfrac{5\pi}{4}}{2} = \pi,$$

then provide zeros of the graph. The graph can then be completed
shown in the previous example.

If an equation were written in the form $y = A \cos(Bx + C)$, it wou
be necessary first to factor B from $Bx + C$ in order to make a judgme
about the period and phase shift of the graph. For example, we wou
write

$$y = 2 \cos\left(3x + \frac{\pi}{2}\right) = 2 \cos 3\left(x + \frac{\pi}{6}\right)$$

from which we would observe that the period is $2\pi/3$ and the phase sh
is $\pi/6$ unit to the left of the graph of $y = 2 \cos 3x$. The graph is shown
the example in the exercise set on page 94.

Vertical Translations

Consider the function

$$y = \sin x + 2.$$

Some ordered pairs in the function are shown in the brief tabulatic
and the graph of $y = \sin x$ and its vertical translation $y = \sin x +$
is shown in Figure 3.10.

Note that the value of y for each x is the sum of the ordinates
$y = \sin x$ and $y = 2$. In general, the value of y for each x in t

x	$\sin x$	$\sin x + 2$
0	0	$0 + 2$
$\dfrac{\pi}{6}$	$\dfrac{1}{2}$	$\dfrac{1}{2} + 2$
$\dfrac{\pi}{4}$	$\dfrac{1}{\sqrt{2}}$	$\dfrac{1}{\sqrt{2}} + 2$
$\vdots$	$\vdots$	$\vdots$

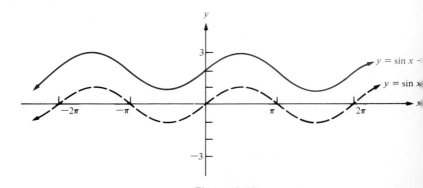

Figure 3.10

function

$$y = A \sin B(x + C) + D \quad \text{or} \quad y = A \cos B(x + C) + D$$

is the sum of the ordinates of

$$y = A \sin B(x + C)$$

or

$$y = A \cos B(x + C) \quad \text{and} \quad y = D.$$

Hence, we can graph such functions by *adding the ordinates graphically* as shown in the following example.

Example   Graph $y = 2 \cos 2x + 3$ for $0 \le x \le 2\pi$.

Solution   First graph $y_1 = 2 \cos 2x$ with amplitude 2 and period $2\pi/2 = \pi$, and $y_2 = 3$ over the given interval.

The ordinate of $y = 2 \cos 2x + 3$ for each x is the algebraic sum of the ordinates of the graph of $y_1 = 2 \cos 2x$ and the graph of $y_2 = 3$. The ordinates can be approximated *graphically* by "adding" vertical directed segments from the x-axis to the respective curves. For example, $\overline{AB} + \overline{AC} = \overline{AD}$. Other ordinates can be obtained in the same way and the points connected in a smooth curve to obtain the graph shown.

You may want to prepare a table of values until you become proficient with this procedure.

x	$2x$	$\cos 2x$	$2 \cos 2x$	$2 \cos 2x + 3$
0	0	1	2	5
$\dfrac{\pi}{2}$	π	-1	-2	1
π	2π	1	2	5
$\vdots$				

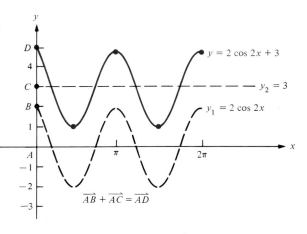

$$\overrightarrow{AB} + \overrightarrow{AC} = \overrightarrow{AD}$$

Sums and Differences   The graphing procedure of geometrically "adding" ordinates illustrated in the example above can be used when functions involve sums and differences of functions.

Example   Graph $y = 2 \sin x + \cos x$ over the interval $0 \le x \le 2\pi$.

Solution   Sketch the graphs of $y_1 = 2 \sin x$ and $y_2 = \cos x$ on the same coordinate system over the given interval, as shown. Note that the ordinate of the graph of $y = 2 \sin x + \cos x$ for each x is the algebraic sum of the ordinates of the graph of $y_1 = 2 \sin x$ and the graph of $y_2 = \cos x$. For example, for $x = \pi/3$,

$$2 \sin x = 2\left(\frac{\sqrt{3}}{2}\right) \approx 1.7 \quad \text{and} \quad \cos x = 0.5,$$

from which

$$y = 2 \sin x + \cos x$$
$$\approx 1.7 + 0.5 = 2.2.$$

The ordinate 2.2 can be approximated graphically by "adding" vertical directed segments, $\overline{AC}$ and $\overline{AB}$, from the x-axis to the respective curves at $x = \pi/3$. Other ordinates can be found graphically in a similar way. The points in the graph of

$$y = 2 \sin x + \cos x$$

for which $2 \sin x$ or $\cos x$ equals zero are particularly easy to locate. Connect the points with a smooth curve to obtain the graph shown.

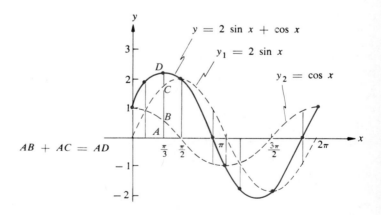

Example   Graph $y = \cos x + x$ over the interval $-\pi \le x \le 2\pi$.

Solution   Sketch the graphs of $y_1 = x$ and $y_2 = \cos x$ on the same set of axes over the given interval. Then "add" directed segments for arbitrary values of x. The points in the graph of $y = x + \cos x$ for which $\cos x = 0$ are easy to locate. Connect the points with a smooth curve to obtain the graph shown.

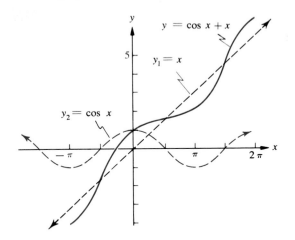

EXERCISE SET 3.3

A

■ *Sketch the graph of each equation.*

Example $y = 3 \sin\left(x + \dfrac{\pi}{2}\right)$

Solution

1. Compare this equation with $y = A \sin B(x + C)$. It is evident that the graph has

    amplitude: $|A| = 3$;   period: $P = \dfrac{2\pi}{|B|} = \dfrac{2\pi}{1} = 2\pi$;

    phase shift: $\pi/2$ units to the left of the graph of $y = 3 \sin x$.

2. Sketch the graph.

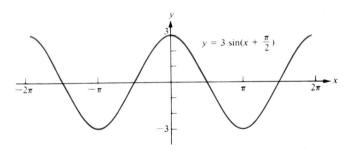

1. $y = \cos(x + \pi)$ **2.** $y = \sin\left(x - \dfrac{\pi}{2}\right)$ **3.** $y = 2\sin\left(x - \dfrac{?}{?}\right)$

4. $y = \dfrac{1}{2}\cos\left(x + \dfrac{\pi}{6}\right)$ **5.** $y = \dfrac{1}{2}\sin\left(x + \dfrac{\pi}{3}\right)$ **6.** $y = \cos\left(x - \dfrac{\pi}{4}\right)$

Example $y = 2\cos\left(3x + \dfrac{\pi}{2}\right)$

Solution

1. Rewrite the equation by factoring 3 from the expression in parenmtheses and then as

$$y = 2\cos 3\left(x + \dfrac{\pi}{6}\right).$$

2. Compare this equation with $y = A\cos B(x + C)$. It is evident that the graph has

amplitude: $|A| = 2$;   period: $P = \dfrac{2\pi}{|B|} = \dfrac{2\pi}{3}$;

phase shift: $\pi/6$ unit to the left of the graph of $y = 2\cos 3x$.

3. Sketch the graph.

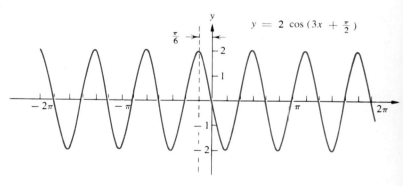

$y = 2\cos\left(3x + \frac{\pi}{2}\right)$

7. $y = 2\cos\left(2x + \dfrac{\pi}{2}\right)$ **8.** $y = \dfrac{1}{2}\sin\left(\dfrac{1}{2}x - \dfrac{\pi}{4}\right)$

9. $y = \dfrac{1}{2}\sin\left(\dfrac{1}{2}x + \dfrac{\pi}{2}\right)$ **10.** $y = 2\cos(2x - \pi)$

11. $y = 3\cos\left(\dfrac{1}{3}x - \dfrac{\pi}{3}\right)$ **12.** $y = 4\sin\left(3x + \dfrac{3\pi}{4}\right)$

■ *Sketch the graph of each equation over the interval* $-\pi \le x \le 2\pi$.

Example $y = 2\sin x + 3$

Solution

1. Graph $y_1 = 2 \sin x$ and $y_2 = 3$ on the same set of axes.
2. Add the ordinates of several pairs of points for several values of x and draw a smooth curve.

If necessary, start a table of values.

x	$\sin x$	$2 \sin x$	$2 \sin x + 3$
$-\pi$	0	0	3
$-\dfrac{\pi}{2}$	-1	-2	1
0	0	0	3
$\vdots$			

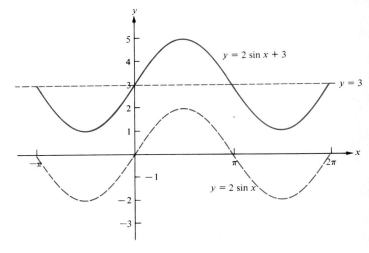

13. $y = 3 \sin x + 1$ **14.** $y = 2 \cos x + 2$

15. $y = 3 \cos x - 2$ **16.** $y = 4 \sin x - 3$

■ *Sketch the graph of each equation over the interval* $-\pi \le x \le 2\pi$.

Example $y = \sin x + \cos x$

Solution

1. Graph $y_1 = \sin x$ and $y_2 = \cos x$ on the same set of axes.
2. Add the ordinates of several pairs of points for several values of x and draw a smooth curve.

If necessary, start a table of values.

x	$\sin x$	$\cos x$	$\sin x + \cos x$
$-\pi$	0	-1	-1
$-\dfrac{3\pi}{4}$	≈ -0.7	≈ -0.7	≈ -1.4
$-\dfrac{\pi}{2}$	-1	0	-1
$\vdots$			

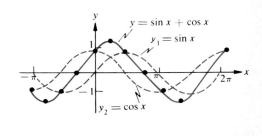

17. $y = 2 \cos x + \sin x$

18. $y = \cos 2x + \dfrac{1}{2} \sin x$

19. $y = \dfrac{1}{2} \cos x + \sin \dfrac{1}{2} x$

20. $y = \cos \dfrac{1}{2} x + \dfrac{1}{3} \sin x$

21. $y = \sin 2x + \dfrac{1}{2} \cos x$

22. $y = \sin \dfrac{1}{2} x + \cos x$

23. $y = 2 \sin x + \cos \dfrac{1}{2} x$

24. $y = \sin \dfrac{1}{2} x + 2 \cos \dfrac{1}{2} x$

25. $y = 2 \sin x - \cos x$

26. $y = \cos x - \dfrac{1}{2} \sin x$

27. $y = \cos \dfrac{1}{2} x - 2 \sin x$

28. $y = 2 \cos x - \dfrac{1}{3} \sin x$

29. $y = \sin x + x$

30. $y = -\cos x + \dfrac{x}{2}$

▪ *For each function in Exercises 31–34, (**a**) specify the amplitude, period and frequency; (**b**) graph one cycle starting with $t = 0$; and (**c**) find the specified values to the nearest tenth.*

31. A horizontal circular disk, suspended by a wire from its center (see figure), is rotated and then. released. The angular displacement θ (in radians) from the "zero" position at time t (in seconds) is given by

$$\theta = 0.4 \cos \pi(t - 0.5).$$

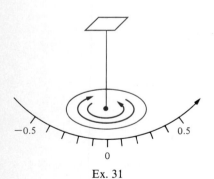

Ex. 31

Find the angular displacement for $t = 1$ and $t = 1.5$.

32. The pressure P in a traveling sound wave is given by

$$P = 10 \sin 200 \, \pi(t - 0.2),$$

where P is expressed in dynes per square centimeter and t is in seconds. Find the pressure for $t = 2.5$ and $t = 4.5$.

33. The instantaneous voltage e in an electric circuit is given by

$$e = 100 \sin(4t - 1.2),$$

where t is in seconds. Find the approximate voltage for $t = 1$ and $t = 3.2$.

34. An alternating current generator produces a current given by

$$i = 10 \sin(60\pi t - 30\pi),$$

where t is in seconds. Find the approximate voltage for $t = 2.5$ and $t = 4.5$.

B

■ *Sketch the graph of each equation over the specified interval.*

35. $y = \cos x + x - 2,\quad 0 \le x \le 2\pi.$
 Hint: First graph $y_1 = x - 2$ and $y_2 = \cos x.$

36. $y = \sin x + 3 - x,\quad 0 \le x \le 2\pi.$
 Hint: First graph $y_1 = 3 - x$ and $y_2 = \sin x.$

37. $y = 2 \cos x + \dfrac{x^2}{12},\quad 0 \le x \le 2\pi$

38. $y = 2 \sin x + \dfrac{1}{x},\quad 0 \le x \le 2\pi$

39. $y = \sin \pi x - x + 4,\quad 0 \le x \le 4$

40. $y = \cos \dfrac{\pi}{2} x - \dfrac{x^2}{4},\quad 0 \le x \le 4$

■ *Sketch one cycle of the graphs of each pair of equations and verify that the graphs are identical.*

41. $y = \cos x$

$$y = \sin\left(x + \frac{\pi}{2}\right)$$

42. $y = \sin x$

$$y = \cos\left(x - \frac{\pi}{2}\right)$$

43. $y = -\sin x$

$$y = \cos\left(x + \frac{\pi}{2}\right)$$

44. $y = -\cos x$

$$y = \sin\left(x - \frac{\pi}{2}\right)$$

3.4 Graphs of Other Circular Functions

We can graph the tangent, cotangent, secant, and cosecant functions on a Cartesian coordinate system in the same way that we graphed the sine and cosine functions. Although the graphs of these functions are not sine waves, the graphs do exhibit the periodic nature and other important features of the functions.

Tangent Function

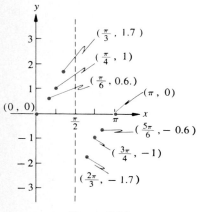

Figure 3.11

The period of the tangent function is π. Thus, to graph $y = \tan x$ we need some values of x only in the interval 0 to π. Using Tables 2.1 and 2.2, we obtain some ordered pairs in the function and graph these ordered pairs as shown in Figure 3.11. Observe that

$$\tan \frac{\pi}{6} = \frac{1}{\sqrt{3}} \approx 0.6,$$

and

$$\tan \frac{\pi}{3} = \sqrt{3} \approx 1.7.$$

It can be shown that the function defined by $y = \tan x$ is continuous in the domain

$$0 \le x < \frac{\pi}{2} \quad \text{and} \quad \frac{\pi}{2} < x \le \pi$$

($\cos x \ne 0$ in these intervals). Thus, we connect the points in Figure 3.11 with a smooth curve to produce one cycle (heavy line), and then we repeat this pattern in both directions to obtain Figure 3.12. Of course, a calculator can be used to obtain additional points to assist in graphing the entire function.

Because $\tan x$ is undefined for $x = \pi/2$, there is no point in the graph of the function when $x = \pi/2$. Notice, however, that $|\tan x|$ increases indefinitely as x is taken closer and closer to $\pi/2$. The vertical dashed lines in Figure 3.11 and Figure 3.12 are called **asymptotes.** The

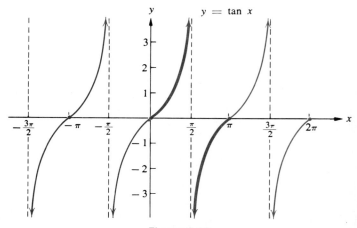

Figure 3.12

entire set of asymptotes to the curve is the set of lines that are the graphs of

$$x = \frac{\pi}{2} + k\pi, \quad k \in J.$$

Furthermore, observe that the zeros of the function that are associated with the points where the curve intersects the x-axis are $k\pi$, $k \in J$.

Cotangent Function   The graph of the function defined by $y = \cot x$ can also be obtained by plotting points. However, for $x \neq k(\pi/2)$, $k \in J$,

$$\cot x = \frac{1}{\tan x}.$$

Thus, each ordinate of the graph of $y = \cot x$, where $x \neq k(\pi/2)$ is the reciprocal of the corresponding ordinate of the graph of $y = \tan x$. With this information we can quickly sketch the graph of $y = \cot x$ if we first sketch the graph of $y = \tan x$ on the same coordinate system. The graphs appear in Figure 3.13. Notice that the zeros of the function defined by $y = \cot x$ are

$$\frac{\pi}{2} + k\pi, \quad k \in J,$$

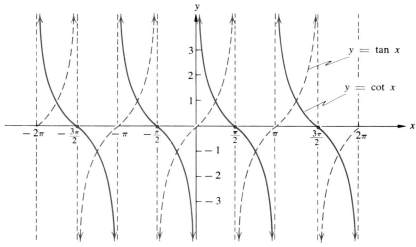

Figure 3.13

and that the asymptotes to the curve are given by

$$x = k\pi, \quad k \in J.$$

Also, note that the range of either the tangent or cotangent function the set of real numbers.

Secant and Cosecant Functions   The graphs of the functions defined by $y = \sec x$ and $y = \csc x$ c also be obtained by using the facts that

$$\sec x = \frac{1}{\cos x}, \quad x \neq \frac{\pi}{2} + k\pi, \quad k \in J,$$

and

$$\csc x = \frac{1}{\sin x}, \quad x \neq k\pi, \quad k \in J.$$

Hence, each ordinate of the graph of $y = \sec x$ is the reciprocal of t corresponding ordinate of the graph of $y = \cos x$, and each ordinate the graph of $y = \csc x$ is the reciprocal of the corresponding ordina of the graph of $y = \sin x$, except for the noted restrictions. Thus, t graphs of

$$y = \sec x \qquad \text{and} \qquad y = \csc x$$

can be obtained directly from the graphs of

$$y = \cos x \qquad \text{and} \qquad y = \sin x,$$

respectively, as shown in Figures 3.14 and 3.15 (see Exercises 7 and 8

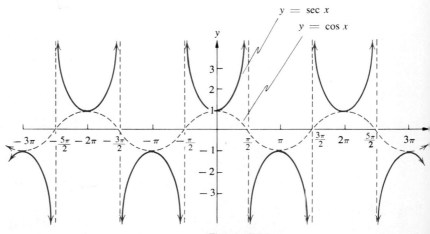

Figure 3.14

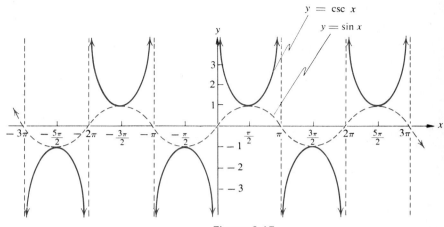

Figure 3.15

Notice that the asymptotes of the graph of $y = \sec x$ are the vertical lines that are the graphs of

$$x = \frac{\pi}{2} + k\pi, \quad k \in J;$$

the x-intercepts of the asymptotes are the zeros of the function defined by $y = \cos x$.

The asymptotes of the graph of $y = \csc x$ are the vertical lines that are the graphs of

$$x = k\pi, \quad k \in J;$$

the x-intercepts of the asymptotes are the zeros of the function defined by $y = \sin x$. The range of both the secant and the cosecant function is $y \le -1$ or $y \ge 1$. Thus, there are no zeros of these functions.

Variations of Constants   Variations in the constants A, B, and C in equations such as

$$y = A \tan B(x + C) \tag{1}$$

affect their graphs in much the same way that these constants affect the graphs of $y = A \sin B(x + C)$ and $y = A \cos B(x + C)$. However, in the graphs of equations of the form of Equation (1), the coefficient A is not referred to as the amplitude.

Domains and Ranges of the Circular Functions   The domains and ranges of the six circular functions have been briefly considered in Chapter 1 and in this chapter. For convenience we summarize this information in Table 3.1 on page 102.

TABLE 3.1

Function	Pairings	Domain	Range
Sine	$(x, \sin x)$	$x \in R$	$-1 \leq \sin x \leq$
Cosine	$(x, \cos x)$	$x \in R$	$-1 \leq \cos x \leq$
Tangent	$(x, \tan x)$	$x \in R \quad \left(x \neq \dfrac{\pi}{2} + k\pi\right)$	$\tan x \in R$
Cotangent	$(x, \cot x)$	$x \in R \quad (x \neq k\pi)$	$\cot x \in R$
Secant	$(x, \sec x)$	$x \in R \quad \left(x \neq \dfrac{\pi}{2} + k\pi\right)$	$|\sec x| \geq 1$
Cosecant	$(x, \csc x)$	$x \in R \quad (x \neq k\pi)$	$|\csc x| \geq 1$

EXERCISE SET 3.4

A

■ *Sketch the graph of each equation over the interval $0 \leq x \leq 2\pi$ and specify (a) the range, (b) the equations of the asymptotes, and (c) the zeros of the function (if they exist).*

Example $y = \tan 2x$

Solution   Since the period of $y = \tan x$ is π, the period of $y = \tan 2$ is $\pi/2$. The graph is sketched as shown.

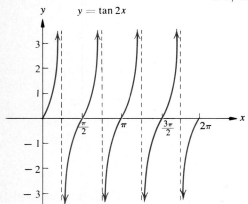

a. Range: $y \in R$

b. Since the equations of the asymptotes of $y = \tan x$ are

$$x = \frac{\pi}{2} + k\pi, \quad k \in J,$$

the equations of the asymptotes of $y = \tan 2x$ are

$$x = \frac{1}{2}\left(\frac{\pi}{2} + k\pi\right) = \frac{\pi}{4} + k \cdot \frac{\pi}{2}, \quad k \in J.$$

c. Zeros: $x = k \cdot \dfrac{\pi}{2}, \ k \in J.$

1. $y = \tan 3x$

2. $y = \tan \dfrac{1}{2} x$

3. $y = \tan\left[x - \left(-\dfrac{\pi}{6}\right)\right]$

4. $y = \tan\left(x - \dfrac{\pi}{4}\right)$

5. $y = \cot \dfrac{1}{2} x$ **6.** $y = \cot 2x$

7. $y = \cot\left(x - \dfrac{\pi}{2}\right)$ **8.** $y = \cot\left(x + \dfrac{\pi}{2}\right)$

9. $y = -\tan 2x$ **10.** $y = -\cot \dfrac{1}{2} x$

11. Without referring to Figure 3.14, graph $y = \sec x$, by using the fact that $\sec x = \dfrac{1}{\cos x}$, $x \neq \dfrac{\pi}{2} + k\pi$, $k \in J$.

12. Without referring to Figure 3.15, graph $y = \csc x$ by using the fact that $\csc x = \dfrac{1}{\sin x}$, $x \neq k\pi$, $k \in J$.

13. Use the graph obtained in Exercise 11 to write the equations of the asymptotes of the graph of $y = \sec x$.

14. Use the graph obtained in Exercise 12 to write the equations of the asymptotes of the graph of $y = \csc x$.

■ *Solve each of the following over the interval* $-\pi \leq x \leq 2\pi$.

15. $y = \dfrac{1}{2} \sec x$ **16.** $y = 2 \csc x$ **17.** $y = \csc 2x$

18. $y = \sec \dfrac{1}{2} x$ **19.** $y = \sec\left(x + \dfrac{\pi}{4}\right)$ **20.** $y = \csc\left(x - \dfrac{\pi}{6}\right)$

B

■ *Sketch each graph over one cycle.*

21. $y = \tan \dfrac{1}{2}\left(x - \dfrac{\pi}{6}\right)$ **22.** $y = \tan 2\left(x + \dfrac{\pi}{6}\right)$

23. $y = \cot\left(2x + \dfrac{2\pi}{3}\right)$ **24.** $y = \sec\left(2x - \dfrac{\pi}{2}\right)$

25. $y = \sec \dfrac{1}{2}(x - \pi) + 2$ **26.** $y = \csc \dfrac{1}{2}(x + \pi) + 2$

Chapter Summary

[3.1]   Graphs of sine or cosine functions are called **sine waves** or **sinusoidal waves**. The part of the graph over one period is called a **cycle** of the wave.

The absolute value of half the difference of the maximum and t[
minimum ordinates is called the **amplitude** of the wave.

Zeros of the sine function, that is, values of x for which $\sin x = 0$, a[
$k\pi$, $k \in J$. Zeros of the cosine function are $(\pi/2) + k\pi$, $k \in J$. The zer[
of a function are the x-coordinates of the points where the grap[
intersects the x-axis (see Figures 3.4 and 3.6).

[3.2–3.3]   Functions defined by equations of the form

$$y = A \sin B(x + C) + D \qquad \text{or} \qquad y = A \cos B(x + C) + D$$

where $A, B, C, D, x, y \in R$ and $A, B \neq 0$ always have sine waves f[
their graphs. The **amplitude** of the wave or function is $|A|$, the **period**[
$2\pi/|B|$, the **phase shift (horizontal translation)** is $|C|$, and the **vertic**[
translation is $|D|$.

If the factor A is a function of x, the sine waves have variab[
amplitudes.

Functions defined by equations of the form

$$y = A_1 \sin B_1 x + A_2 \cos B_2 x$$

can be graphed by graphing $y_1 = A_1 \sin B_1 x$ and $y_2 = A_2 \cos B_2$[
and then "adding" their ordinates graphically for arbitrary values of[

[3.4]   The graphs of the circular functions defined by $y = \tan x$, $y = \cot$[
$y = \sec x$, and $y = \csc x$ exhibit the periodic nature of the functio[
(see Figures 3.12–3.15). The following table shows the zeros and/or t[
equations for the asymptotes of these circular functions.

Function	Zeros	Equations for Asymptotes
Tangent	$k\pi$, $k \in J$	$x = \dfrac{\pi}{2} + k\pi$, $k \in J$
Cotangent	$\dfrac{\pi}{2} + k\pi$, $k \in J$	$x = k\pi$, $k \in J$
Secant	none	$x = \dfrac{\pi}{2} + k\pi$, $k \in J$
Cosecant	none	$x = k\pi$, $k \in J$

Review Exercises

[3.1–3.2] ▪ *Graph each equation over the interval* $-2\pi \leq x \leq 2\pi$. *Specify* **(a)** *[
amplitude and* **(b)** *the period.*

1. $y = 4 \sin x$ 　　　　　　　　　　　**2.** $y = \dfrac{1}{2} \cos 2x$

3. $y = -3 \cos \dfrac{1}{2} x$ **4.** $y = -2 \sin \dfrac{1}{2} x$

■ *Graph two cycles of each of the following.*

5. $y = \cos \dfrac{\pi}{4} x$ **6.** $y = 2 \sin \dfrac{\pi}{3} x$

■ *Graph each of the following over the specified interval.*

7. $y = \dfrac{x + 4}{2} \sin \pi x, \quad 0 \le x \le 4$

8. $y = 4e^{-0.3x} \cos \pi x, \quad 0 \le x \le 4$

[3.3] ■ *Graph each equation over the interval* $-2\pi \le x \le 2\pi$. *Specify* **(a)** *the amplitude,* **(b)** *the period, and* **(c)** *the phase shift.*

9. $y = \sin\left(x + \dfrac{\pi}{2}\right)$ **10.** $y = 2 \cos\left(x - \dfrac{\pi}{4}\right)$

11. $y = 2 \cos\left(\dfrac{1}{2} x - \dfrac{\pi}{6}\right)$ **12.** $y = 3 \sin\left(\dfrac{1}{2} x + \dfrac{\pi}{4}\right)$

■ *Using the method of graphical addition of ordinates, graph the equations over the interval* $0 \le x \le 2\pi$.

13. $y = \sin \dfrac{1}{2} x + 3$ **14.** $y = 3 \cos x - 1$

15. $y = \sin 2x - \dfrac{1}{2} \cos x$ **16.** $y = -\sin x + \dfrac{1}{3} x$

■ *Sketch the graph of the function defined by each equation over the interval* $0 \le x \le 2\pi$. *Specify* **(a)** *the range,* **(b)** *the equations of the asymptotes, and* **(c)** *the set of all zeros of the function (if applicable).*

17. $y = \tan\left(x - \dfrac{\pi}{3}\right)$ **18.** $y = \cot 3x$

19. $y = \csc \dfrac{1}{2} x$ **20.** $y = 2 \sec x$

IDENTITIES

Equations such as

$$a \cdot 1 = a, \qquad a + b = b + a, \qquad \text{and} \qquad a(b + c) = ab + ac,$$

which are true for all real numbers a, b, and c, are called **identities;** *the left-hand member and the right-hand member of each equation are equi*~~valent~~ *alent expressions.** In algebra, we have used such identities to simpli~~fy~~ algebraic expressions whenever we substituted one expression for a~~n~~ equivalent expression.

In this chapter we will review some basic identities that involve tri~~g~~onometric (or circular) function values that have been introduced in pr~~e~~vious chapters, and we will introduce additional ones that will be usef~~ul~~ in our work. We will then use these identities to rewrite expressions tha~~t~~ involve trigonometric function values as equivalent expressions.

In our work with identities we will not explicitly specify the replac~~e~~ment set of the variable in each case. The variables α, β, γ, and θ w~~ill~~ represent elements of the set of all angles, and the variables x and ~~y~~ will be considered as representing elements either of the set of all angl~~es~~ or of the set of real numbers. Furthermore, it should be understo~~od~~ that *an equation that is specified to be an identity without noting the r*~~e~~ *strictions is in fact an identity only for all replacements of the variables f*~~or~~ *which both members of the equation are defined.*

* Also see Appendix A.3.

4.1 Basic Trigonometric Identities

The reciprocal identities

$$\csc \alpha = \frac{1}{\sin \alpha}, \qquad \sec \alpha = \frac{1}{\cos \alpha}, \qquad \text{and} \qquad \cot \alpha = \frac{1}{\tan \alpha},$$

introduced in Chapter 1, have been shown to be consequences of the definition of the trigonometric ratios.

Several other important identities follow from the trigonometric ratios. Since

$$\cos \alpha = \frac{x}{r} \qquad \text{and} \qquad \sin \alpha = \frac{y}{r}, \quad r \neq 0,$$

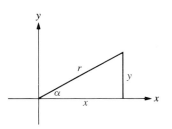

(see Figure 4.1),

$$\sin^2 \alpha + \cos^2 \alpha = \left(\frac{y}{r}\right)^2 + \left(\frac{x}{r}\right)^2$$

$$= \frac{y^2 + x^2}{r^2}.$$

Figure 4.1

However, by the Pythagorean theorem,

$$y^2 + x^2 = r^2.$$

Hence,

$$\sin^2 \alpha + \cos^2 \alpha = 1.$$

Also, since $\sin \alpha = y/r$ and $\cos \alpha = x/r$ $(r \neq 0)$,

$$\frac{\sin \alpha}{\cos \alpha} = \frac{\dfrac{y}{r}}{\dfrac{x}{r}} = \frac{y}{x}, \quad x \neq 0.$$

Since $\tan \alpha = y/x$,

$$\tan \alpha = \frac{\sin \alpha}{\cos \alpha}$$

for those values of α for which $\cos \alpha \neq 0$. Because $\cot \alpha = x/y$ and

$$\frac{\cos \alpha}{\sin \alpha} = \frac{\dfrac{x}{r}}{\dfrac{y}{r}} = \frac{x}{y}, \quad y \neq 0,$$

$$\cot \alpha = \frac{\cos \alpha}{\sin \alpha},$$

for those values of α for which $\sin \alpha \neq 0$.

Two other identities follow from the identity $\sin^2 \alpha + \cos^2 \alpha = 1$. For any α for which $\cos \alpha \neq 0$, we multiply each member by $1/\cos^2 \alpha$ to obtain the equivalent equations

$$\frac{\sin^2 \alpha}{\cos^2 \alpha} + \frac{\cos^2 \alpha}{\cos^2 \alpha} = \frac{1}{\cos^2 \alpha},$$

$$\tan^2 \alpha + 1 = \sec^2 \alpha.$$

Multiplying each member of $\sin^2 \alpha + \cos^2 \alpha = 1$ by $1/\sin^2 \alpha$ for $\sin \alpha \neq 0$, we obtain

$$\frac{\sin^2 \alpha}{\sin^2 \alpha} + \frac{\cos^2 \alpha}{\sin^2 \alpha} = \frac{1}{\sin^2 \alpha},$$

$$1 + \cot^2 \alpha = \csc^2 \alpha.$$

The basic identities that we have stated above should be recognized in equivalent forms. For example, using the symbol "$\leftrightarrow$" to indicate equivalency, we have

$$\sec \alpha = \frac{1}{\cos \alpha} \quad \leftrightarrow \quad (\cos \alpha)(\sec \alpha) = 1 \quad \leftrightarrow \quad \cos \alpha = \frac{1}{\sec \alpha};$$

$$\csc \alpha = \frac{1}{\sin \alpha} \quad \leftrightarrow \quad (\sin \alpha)(\csc \alpha) = 1 \quad \leftrightarrow \quad \sin \alpha = \frac{1}{\csc \alpha};$$

$$\cot \alpha = \frac{1}{\tan \alpha} \quad \leftrightarrow \quad (\tan \alpha)(\cot \alpha) = 1 \quad \leftrightarrow \quad \tan \alpha = \frac{1}{\cot \alpha};$$

$$\sin^2 \alpha + \cos^2 \alpha = 1 \quad \leftrightarrow \quad \sin^2 \alpha = 1 - \cos^2 \alpha \quad \leftrightarrow \quad \cos^2 \alpha = 1 - \sin^2 \alpha;$$

$$\tan^2 \alpha + 1 = \sec^2 \alpha \quad \leftrightarrow \quad \tan^2 \alpha = \sec^2 \alpha - 1;$$

$$\cot^2 \alpha + 1 = \csc^2 \alpha \quad \leftrightarrow \quad \cot^2 \alpha = \csc^2 \alpha - 1.$$

Identities for $(-\alpha)$

Several other useful trigonometric relationships can be obtained from geometric considerations. In Figure 4.2a,

$$\angle AOB = \angle COB, \qquad OB = OB, \qquad \text{and} \qquad \angle OBA = \angle OBC.$$

Hence,

$$\triangle AOB \cong \triangle COB,$$

from which we have that each pair of corresponding sides is congruent and the absolute values of the respective components of the ordered pairs corresponding to points A and C are equal. Because

$$\sin(-\alpha) = \frac{-y}{r} \qquad \text{and} \qquad \sin \alpha = \frac{y}{r},$$

$$\sin(-\alpha) = -\sin \alpha. \tag{1}$$

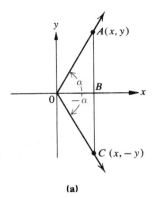

Also, $\cos(-\alpha) = x/r$ and $\cos \alpha = x/r,$ from which

$$\cos(-\alpha) = \cos \alpha, \tag{2}$$

and $\tan(-\alpha) = -y/x$ and $\tan \alpha = y/x,$ from which

$$\tan(-\alpha) = -\tan \alpha. \tag{3}$$

A similar argument for angles α and $-\alpha$ that are in Quadrants II and III, as shown in Figure 4.2b, also leads to the results stated by Equations (1), (2), and (3).

Because identities we have introduced are simply equations in which the left member and right member are equivalent expressions and the function values in these expressions are real numbers, we can sometimes simplify expressions that involve trigonometric function values by using these identities and applying the familiar algebraic properties for real numbers.

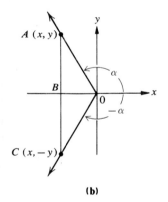

(b)

Figure 4.2

Examples Write each expression in terms of sine and cosine and simplify.

a. $(\tan \alpha)(\cos^2 \alpha)$ 　　b. $\dfrac{\cos \beta}{\cot^2 \beta}$ 　　c. $\tan(-\gamma) + \sec(-\gamma)$

Solutions

a. Substitute $\sin \alpha/\cos \alpha$ for $\tan \alpha$ and simplify.

$$(\tan \alpha)(\cos^2 \alpha) = \frac{\sin \alpha}{\cos \alpha} \cdot \cos^2 \alpha = \sin \alpha \cdot \cos \alpha$$

(continued)

b. Substitute $\cos^2 \beta / \sin^2 \beta$ for $\cot^2 \beta$ and simplify.

$$\frac{\cos \beta}{\cot^2 \beta} = \frac{\cos \beta}{\dfrac{\cos^2 \beta}{\sin^2 \beta}}$$

$$= \cos \beta \cdot \frac{\sin^2 \beta}{\cos^2 \beta} = \frac{\sin^2 \beta}{\cos \beta}$$

c. Substitute $\sin(-\gamma)/\cos(-\gamma)$ for $\tan(-\gamma)$ and $1/\cos(-\gamma)$ for $\sec(-\gamma)$ and simplify.

$$\tan(-\gamma) + \sec(-\gamma) = \frac{\sin(-\gamma)}{\cos(-\gamma)} + \frac{1}{\cos(-\gamma)}$$

$$= \frac{-\sin \gamma}{\cos \gamma} + \frac{1}{\cos \gamma}$$

$$= \frac{1 - \sin \gamma}{\cos \gamma}$$

EXERCISE SET 4.1

A

■ *Use the identities of this section to express each of the following in terms of a single function.*

Examples   a. $1 - \sin^2 \alpha$ b. $\dfrac{\cos^2 \alpha}{\sin^2 \alpha}$

Solutions
 a. Since $\sin^2 \alpha + \cos^2 \alpha = 1$,

$$1 - \sin^2 \alpha = \cos^2 \alpha.$$

 b. Since $\dfrac{\cos \alpha}{\sin \alpha} = \cot \alpha,$

$$\frac{\cos^2 \alpha}{\sin^2 \alpha} = \cot^2 \alpha.$$

1. $1 + \tan^2 \alpha$ **2.** $1 + \cot^2 \alpha$ **3.** $1 - \cos^2 \alpha$

4. $\sec^2 \alpha - 1$ **5.** $\cos^2 \alpha - 1$ **6.** $1 - \sec^2 \alpha$

7. $\dfrac{\sin^3 \alpha}{\cos^3 \alpha}$ **8.** $\dfrac{\cos^4 \alpha}{\sin^4 \alpha}$

▪ *Use the identities of this section to write each expression in terms of sine and cosine and simplify.*

Examples

a. $(\sin \alpha)(\cot^2 \alpha)(\sec^2 \alpha)$

b. $\dfrac{\cot^2 x}{\cos x}$

Solutions

a. Substituting $\cos \alpha / \sin \alpha$ for $\cot \alpha$ and $1/\cos \alpha$ for $\sec \alpha$ yields

$$(\sin \alpha)(\cot^2 \alpha)(\sec^2 \alpha) = (\sin \alpha)\left(\frac{\cos^2 \alpha}{\sin^2 \alpha}\right)\left(\frac{1}{\cos^2 \alpha}\right)$$

$$= \frac{1}{\sin \alpha}.$$

b. Since $\cot x = \cos x / \sin x,$

$$\frac{\cot^2 x}{\cos x} = \frac{\cos^2 x / \sin^2 x}{\cos x}$$

$$= \frac{\cos^2 x}{\sin^2 x} \cdot \frac{1}{\cos x} = \frac{\cos x}{\sin^2 x}.$$

9. $(\sec \alpha)(\cot \alpha)$

10. $(\csc \alpha)(\tan \alpha)$

11. $(\cot \alpha)(\csc \alpha)$

12. $(\tan \alpha)(\sec \alpha)$

13. $1 - \tan^2 \alpha$

14. $1 - \cot^2 \alpha$

15. $1 - \sec^2 \alpha$

16. $1 - \csc^2 \alpha$

17. $(\csc^2 \beta)(\tan \beta)$

18. $(\sec^2 \beta)(\cot \beta)$

19. $(\csc^2 \beta)(\cot \beta)$

20. $(\sec^2 \beta)(\tan \beta)$

21. $\dfrac{\sec \gamma}{\tan \gamma}$

22. $\dfrac{\csc \gamma}{\cot \gamma}$

23. $\dfrac{\cot \gamma}{\sec \gamma}$

24. $\dfrac{\tan \gamma}{\csc \gamma}$

25. $\dfrac{\cot^2 x}{\csc^2 x}$

26. $\dfrac{\tan^2 x}{\sec^2 x}$

27. $\dfrac{\sec^2 x}{\cot^2 x}$

28. $\dfrac{\csc^2 x}{\tan^2 x}$

29. $\tan y + \cot y$

30. $\sec y + \csc y$

31. $\tan^2 \theta + \cot^2 \theta$

32. $\sec^2 \theta + \csc^2 \theta$

33. $\sec \theta + \dfrac{\csc^2 \theta}{\sec \theta}$

34. $\csc \theta = \dfrac{\sec^2 \theta}{\csc \theta}$

35. $(\sin \alpha)(\cot^2 \alpha)(\csc^2 \alpha)$

36. $(\cos \alpha)(\tan^2 \alpha)(\sec^2 \alpha)$

37. $\tan \alpha - \dfrac{\cot^2 \alpha}{\tan \alpha}$

38. $\sec \alpha - \dfrac{\csc^2 \alpha}{\sec \alpha}$

Examples　　a. $\cos(-\alpha) \cdot \tan(-\alpha)$　　b. $\cot(-\alpha) \cdot \sin \alpha$

Solutions

　a. Since $\cos(-\alpha) = \cos \alpha$ and $\tan(-\alpha) = -\tan \alpha$,

$$\cos(-\alpha) \cdot \tan(-\alpha) = \cos \alpha \cdot (-\tan \alpha)$$

$$= \cos \alpha \cdot \left(\frac{-\sin \alpha}{\cos \alpha}\right)$$

$$= -\sin \alpha.$$

　b. Since $\cot(-\alpha) = \dfrac{\cos(-\alpha)}{\sin(-\alpha)}$, $\cos(-\alpha) = \cos \alpha$, and $\sin(-\alpha) = -\sin \alpha$,

$$\cot(-\alpha) \cdot \sin \alpha = \frac{\cos \alpha}{-\sin \alpha} \cdot \sin \alpha$$

$$= -\cos \alpha.$$

39. $\csc(-\alpha) \cdot \tan \alpha$

40. $\sec(-\alpha) \cdot \cot(-\alpha)$

41. $\sec^2(-\beta) \cdot \cot(-\beta)$

42. $\csc^2 \beta \cdot \tan(-\beta)$

43. $\dfrac{\csc x}{\cot(-x)}$

44. $\dfrac{\sec (-x)}{\tan(-x)}$

45. $\cot(-y) + \tan(-y)$

46. $\csc(-y) + \sec y$

B

■ *Simplify each expression.*

47. $\dfrac{(\cos \alpha + \sin \alpha)^2}{\cos \alpha - \sin \alpha} - \dfrac{1}{\dfrac{\cos \alpha - \sin \alpha}{2 \cos \alpha \sin \alpha}}$

48. $\dfrac{\sin \alpha}{\dfrac{\sin \alpha}{\sin \alpha + \cos \alpha}} - \dfrac{2 \sin \alpha \cos \alpha}{\sin \alpha + \cos \alpha}$

49. $\dfrac{\dfrac{1}{\sin \alpha - \cos \alpha}}{\dfrac{\sin \alpha}{\sin^2 \alpha - \cos^2 \alpha}}$

50. $\dfrac{\dfrac{1}{\sin^2 \alpha} + \dfrac{1}{\cos^2 \alpha}}{\dfrac{1}{\sin \alpha \cos \alpha}} - \dfrac{\cos \alpha}{\sin \alpha}$

4.2 Verifying Identities

In Section 4.1 we introduced some basic identities that we used to rewrite and simplify expressions that involved trigonometric function values. In Section 4.1 we always expressed our results in terms of sine and cosine function values. Sometimes we may also want to rewrite trigonometric expressions in terms of other function values or constants.

Example Show that $\csc \alpha \cos \alpha$ is equivalent to $\cot \alpha$.

Solution Since $\csc \alpha = 1/\sin \alpha$,

$$\csc \alpha \cos \alpha = \frac{1}{\sin \alpha} \cdot \cos \alpha$$

$$= \frac{\cos \alpha}{\sin \alpha}.$$

Since $\cos \alpha / \sin \alpha = \cot \alpha$, we have that

$$\csc \alpha \cos \alpha = \cot \alpha.$$

It is customary to express the statement in the above example as follows.

"Show that $\csc \alpha \cos \alpha = \cot \alpha$ is an identity."

Of course, the verification for this identity is exactly as given in the above solution.

Let us consider another example.

Example Show that

$$\frac{\cos \alpha - \sin \alpha}{\cos \alpha} = 1 - \tan \alpha \tag{1}$$

is an identity.

(*continued*)

Solution   We will rewrite the right-hand member in terms of sines and cosines. Substituting $\sin \alpha / \cos \alpha$ for $\tan \alpha$ in the right-hand member

$$\frac{\cos \alpha - \sin \alpha}{\cos \alpha} = 1 - \frac{\sin \alpha}{\cos \alpha}.$$

Writing the right-hand member as a single term yields

$$\frac{\cos \alpha - \sin \alpha}{\cos \alpha} = \frac{\cos \alpha - \sin \alpha}{\cos \alpha}, \tag{1}$$

which is clearly an identity.

Example   Show that

$$1 + \tan^2 \alpha = \frac{1}{\cos^2 \alpha} \tag{}$$

is an identity.

Solution   We will rewrite the left-hand member in terms of sines and cosines. Substituting $\sin^2 \alpha / \cos^2 \alpha$ for $\tan^2 \alpha$ in the left-hand member we obtain

$$1 + \frac{\sin^2 \alpha}{\cos^2 \alpha} = \frac{1}{\cos^2 \alpha}.$$

Writing the left-hand member as a single term yields

$$\frac{\cos^2 \alpha + \sin^2 \alpha}{\cos^2 \alpha} = \frac{1}{\cos^2 \alpha}.$$

Since $\cos^2 \alpha + \sin^2 \alpha$ equals 1, we obtain

$$\frac{1}{\cos^2 \alpha} = \frac{1}{\cos^2 \alpha}$$

which is clearly an identity.

Sometimes it is helpful to express *both* members of an identity in terms of sine and cosines.

Example   Show that

$$\cot \alpha + \tan \alpha = \csc \alpha \sec \alpha \tag{}$$

is an identity.

Solution   Substituting $\cos \alpha/\sin \alpha$ for $\cot \alpha$ and $\sin \alpha/\cos \alpha$ for $\tan \alpha$ in the left-hand member,

$$\frac{\cos \alpha}{\sin \alpha} + \frac{\sin \alpha}{\cos \alpha} = \csc \alpha \sec \alpha,$$

$$\frac{\cos^2 \alpha + \sin^2 \alpha}{\sin \alpha \cos \alpha} = \csc \alpha \sec \alpha,$$

$$\frac{1}{\sin \alpha \cos \alpha} = \csc \alpha \sec \alpha.$$

Substituting $1/\sin \alpha$ for $\csc \alpha$ and $1/\cos \alpha$ for $\sec \alpha$ in the right-hand member,

$$\frac{1}{\sin \alpha \cos \alpha} = \frac{1}{\sin \alpha} \cdot \frac{1}{\cos \alpha},$$

$$\frac{1}{\sin \alpha \cos \alpha} = \frac{1}{\sin \alpha \cos \alpha}.$$

The two members are identical, and the original equation is an identity.

In the above examples members of the equations were written in terms of sine and cosine function values to establish that the equations were indeed identities. Although there is no general method to verify that an equation is an identity, there are three possible ways to proceed using algebraic properties and the basic identities that have been established:

1. Rewrite the right-hand member so that it is identical to the left-hand member. [See Equation (1) above.]
2. Rewrite the left-hand member so that it is identical to the right-hand member. [See Equation (2) above.]
3. Rewrite both members *independently* so that they are identical. [See Equation (3) above.]

Working with identities will give you the opportunity to use the basic identities and hence become familiar with these important relationships, and also to use the trigonometric ratios (which simply represent real numbers) in conjunction with properties of algebra.

The following suggestions may help you determine the best method to use to verify an identity.

1. If one member of an equation appears to be the more complicated, it should be transformed so that it reduces to the simpler member of the equation.

2. In each step, look for any possible application of algebraic procedures such as combining fractions, writing a single fraction a
   two or more fractions, using the distributive law, factoring, o
   multiplying the numerator and denominator of a fraction by th
   conjugate of one or the other.

3. If an elegant approach in which an identity can be established i
   one or two steps is not obvious, as a last resort change all the term
   to expressions involving only sines and/or cosines, and then sim
   plify both members.

In the following examples, we illustrate several of these approaches.

Example   Show that $\dfrac{\sin^2 x}{1 + \cos x} = \dfrac{\tan x - \sin x}{\tan x}$ is an identity.

Solution   In this case we will simplify both members by appropria
substitutions. Expressing the right-hand member in terms of sines an
cosines,

$$\frac{\sin^2 x}{1 + \cos x} = \frac{\tan x - \sin x}{\tan x}$$

$$= \frac{\tan x}{\tan x} - \frac{\sin x}{\tan x}$$

$$= 1 - \frac{\sin x}{\dfrac{\sin x}{\cos x}} = 1 - \cos x.$$

Rewriting the left-hand member by substituting $1 - \cos^2 x$ for $\sin^2$

$$\frac{1 - \cos^2 x}{1 + \cos x} = 1 - \cos x.$$

Factoring $1 - \cos^2 x$ as $(1 - \cos x)(1 + \cos x)$ yields

$$\frac{(1 - \cos x)(1 + \cos x)}{1 + \cos x} = 1 - \cos x$$

$$1 - \cos x = 1 - \cos x.$$

Hence, the original equation is an identity.

Example   Show that $\dfrac{\cos x}{1 + \sin x} = \dfrac{1 - \sin x}{\cos x}$ is an identity.

Solution   We first note that the numerator of the right-hand member is $1 - \sin x$. Hence, we try to obtain this expression in the numerator of the left-hand member by using an algebraic property. In this case, we can multiply the left-hand member by $\dfrac{1 - \sin x}{1 - \sin x}$ to obtain

$$\frac{\cos x(1 - \sin x)}{(1 + \sin x)(1 - \sin x)} = \frac{1 - \sin x}{\cos x},$$

from which

$$\frac{\cos x(1 - \sin x)}{1 - \sin^2 x} = \frac{1 - \sin x}{\cos x}.$$

Substituting $\cos^2 x$ for $1 - \sin^2 x$ in the left-hand member yields

$$\frac{\cos x(1 - \sin x)}{\cos^2 x} = \frac{1 - \sin x}{\cos x}.$$

Then, dividing the numerator and denominator of the left-hand member by $\cos x$,

$$\frac{1 - \sin x}{\cos x} = \frac{1 - \sin x}{\cos x}.$$

Hence, the original equation is an identity.

Example   Show that $\dfrac{\sec^2 \theta - \tan^2 \theta}{\sec \theta + \tan \theta} = \dfrac{1 - \sin \theta}{\cos \theta}$ is an identity.

Solution   If we note that $\sec^2 \theta - \tan^2 \theta$ (of the form $a^2 - b^2$) can be factored as $(\sec \theta - \tan \theta)(\sec \theta + \tan \theta)$, we can make a substitution to obtain

$$\frac{(\sec \theta - \tan \theta)(\sec \theta + \tan \theta)}{\sec \theta + \tan \theta} = \frac{1 - \sin \theta}{\cos \theta}$$

from which

$$\sec \theta - \tan \theta = \frac{1 - \sin \theta}{\cos \theta}.$$

Now substituting $1/\cos \theta$ for $\sec \theta$ and $\sin \theta/\cos \theta$ for $\tan \theta$ and writing the left-hand member as a single fraction, we have that

$$\frac{1}{\cos \theta} - \frac{\sin \theta}{\cos \theta} = \frac{1 - \sin \theta}{\cos \theta}$$

$$\frac{1 - \sin \theta}{\cos \theta} = \frac{1 - \sin \theta}{\cos \theta}.$$

Hence, the original equation is an identity.

Counterexamples   In general, it is a relatively simple matter to show that an equation
not an identity, if such is the case, by finding a *counterexample.* Fo
example, substituting 0 for α in the equation

$$\sin \alpha = \cos \alpha.$$

yields

$$\sin 0 = \cos 0,$$

$$0 = 1,$$

which is a false statement. Thus, the equation $\sin \alpha = \cos \alpha$ is not a
identity.

Of course, in demonstrating a counterexample, we should select
replacement for the variable so that each member of the equation
defined.

EXERCISE SET 4.2

A

▪ *Simplify each expression on the left to show that it is equivalent to th
expression on the right for all values of the variable for which both e.
pressions are defined.*

Example $(1 - \cos^2 \beta)(1 + \cot^2 \beta);$ 1

Solution   Substituting $\sin^2 \beta$ for $1 - \cos^2 \beta$ and $\csc^2 \beta$ for $1 + \cot^2$
yields

$$(1 - \cos^2 \beta)(1 + \cot^2 \beta) = (\sin^2 \beta)(\csc^2 \beta).$$

Since $\csc^2 \beta = 1/\sin^2 \beta,$

$$(1 - \cos^2 \beta)(1 + \cot^2 \beta) = (\sin^2 \beta)\left(\frac{1}{\sin^2 \beta}\right) = 1.$$

1. $\sec \alpha \sin \alpha$; $\tan \alpha$

2. $\tan \alpha \cos \alpha$; $\sin \alpha$

3. $\sin \beta \cot \beta$; $\cos \beta$

4. $\csc \beta \tan \beta$; $\sec \beta$

5. $\dfrac{\sin \gamma}{\tan \gamma}$; $\cos(-\gamma)$

6. $\dfrac{\sin(-\gamma) \sec \gamma}{\tan(-\gamma)}$; 1

7. $\cot^2 \gamma \sin^2 \gamma$; $\cos^2 \gamma$

8. $\tan^2 \alpha \cos^2 \alpha$; $\sin^2 \alpha$

9. $\cos^2 \alpha(1 + \tan^2 \alpha)$; 1

10. $(\csc^2 \beta - 1)(\sin^2 \beta)$; $\cos^2 \beta$

11. $\dfrac{(1 + \sin \beta)(1 - \sin \beta)}{\cos \beta}$; $\cos \beta$

12. $\dfrac{(1 + \cos \gamma)(1 - \cos \gamma)}{\sin \gamma}$; $\sin \gamma$

13. $\dfrac{\cos^2 \gamma}{\cot^2 \gamma}$; $\sin^2 \gamma$

14. $\dfrac{1 + \tan^2 \gamma}{\csc^2 \gamma}$; $\tan^2 \gamma$

15. $\dfrac{\sin^2 \alpha}{\tan^2 \alpha - \sin^2 \alpha}$; $\cot^2 \alpha$

16. $\dfrac{\cos^2 \alpha}{\cot^2 \alpha - \cos^2 \alpha}$; $\tan^2 \alpha$

17. $\dfrac{1}{\tan \alpha + \cot \alpha}$; $\dfrac{\cos \alpha}{\csc \alpha}$

18. $\dfrac{1}{\csc \alpha - \sin \alpha}$; $\dfrac{\tan \alpha}{\cos \alpha}$

▪ *Verify that each equation is an identity.*

19. $\cos x \tan x \csc x = 1$

20. $\cot x \sec x = \csc x$

21. $\sec y - \cos y = \tan y \sin y$

22. $\sec y - \tan y \sin y = \cos y$

23. $\cos^2 x - \sin^2 x = 2 \cos^2 x - 1$

24. $(\sec^2 y - 1)(\csc^2 y - 1) = 1$

25. $\tan^2 y + \sec^2 y = \dfrac{2}{\cos^2 y} - 1$

26. $(\tan x + \cot x)\dfrac{\sin x}{\sec x} = 1$

27. $\dfrac{\cos x + \sin x}{\sin x} = 1 + \cot x$

28. $\dfrac{1 - \tan^2 y}{\tan y} = \cot y - \tan y$

29. $\tan^2 y - \sin^2 y = \tan^2 y \sin^2 y$

30. $\dfrac{1}{1 + \sin x} + \dfrac{1}{1 - \sin x} = \dfrac{2}{\cos^2 x}$

31. $\tan y + \cot y = \sec y \csc y$

32. $\dfrac{1 + \tan^2 y}{\tan^2 y} = \csc^2 y$

33. $\cos^4 x - \sin^4 x = \cos^2 x - \sin^2 x$ *Hint:* Factor left-hand member.

34. $\dfrac{1 + \sec x}{\sec x} = \dfrac{\sin^2 x}{1 - \cos x}$ *Hint:* $\sin^2 x = (1 - \cos x)(1 + \cos x)$.

35. $\dfrac{1 - \cos x}{\sin x} = \dfrac{\sin x}{1 + \cos x}$ *Hint:* $\dfrac{\sin x}{1 + \cos x} = \dfrac{\sin x(1 - \cos x)}{(1 + \cos x)(1 - \cos x)}$.

36. $\dfrac{1 + \sin x}{\cos x} = \dfrac{\cos x}{1 - \sin x}$ *Hint:* $\dfrac{1 + \sin x}{\cos x} = \dfrac{(1 + \sin x)(1 - \sin x)}{\cos x(1 - \sin x)}$

37. $\csc y = \dfrac{\sec y + \csc y}{1 + \tan y}$ **38.** $\dfrac{1}{\sec y - \tan y} = \sec y + \tan y$

39. $\dfrac{\sec y - 1}{\sec y + 1} = \dfrac{1 - \cos y}{1 + \cos y}$ **40.** $\dfrac{\cos y}{\sec y - \tan y} = \dfrac{\cos^2 y}{1 - \sin y}$

41. $\sec^2 y + \csc^2 y = \sec^2 y \csc^2 y$

42. $\sec^4 y - \sec^2 y = \tan^4 y + \tan^2 y$

43. $\cos(-\alpha) \tan(-\alpha) \csc \alpha = -1$

44. $\cot(-\alpha) \sec(-\alpha) = \csc(-\alpha)$

45. $\dfrac{\cos(-\alpha) + \sin \alpha}{\sin(-\alpha)} = -(1 + \cot \alpha)$

46. $\dfrac{\cos(-\alpha) - \sin(-\alpha)}{\cos \alpha} = 1 + \tan \alpha$

47. $\sec(-\alpha) \sin(-\alpha) = -\tan \alpha$ **48.** $\dfrac{\sin(-\alpha)}{\tan(-\alpha)} = \cos \alpha$

49. $\dfrac{\sin(-\alpha) \sec \alpha}{\tan(-\alpha)} = 1$ **50.** $\dfrac{1 + \tan^2(-\alpha)}{\csc^2(-\alpha)} = \tan^2 \alpha$

■ *By using counterexamples show that each equation is not an identity.*

51. $\sin x + \cos x \tan x = 2$ **52.** $\sin^2 x + 2 \cos x - \cos^2 x = 1$

53. $2 \sin y \cos y + \sin y = 0$ **54.** $\tan^2 y - 2 \tan y = 0$

B

■ *Verify that each equation is an identity.*

55. $\sec \alpha = \dfrac{\cot \alpha + \tan \alpha}{\csc \alpha}$ **56.** $\dfrac{\cot \alpha + \tan \alpha}{\sec \alpha} = \csc \alpha$

57. $\sin^2 \alpha = \dfrac{\cos^2 \alpha + \tan^2 \alpha - 1}{\tan^2 \alpha}$ **58.** $\dfrac{\sin^2 \alpha + \cot^2 \alpha - 1}{\cot^2 \alpha} = \cos^2 \alpha$

59. $\sec \beta - 1 = \dfrac{\sec \beta + 1}{(\csc \beta + \cot \beta)^2}$ **60.** $\tan \beta + \sec \beta = \dfrac{1}{\sec \beta - \tan \beta}$

61. $\csc \gamma - \cot \gamma = \dfrac{1}{\cot \gamma + \csc \gamma}$ **62.** $\sec^2 \gamma - \csc^2 \gamma = \dfrac{\tan \gamma}{\cot \gamma} - \dfrac{\cot \gamma}{\tan \gamma}$

63. $\sec^2 x - \tan x = \dfrac{1 + \tan^3 x}{1 + \tan x}$ **64.** $\sec x = \dfrac{\tan^2 x - 2 \tan x}{\sin x \tan x - 2 \sin x}$

65. $\dfrac{\cos y + \sin y}{\cos^2 y + \cos y \sin y} = \dfrac{\cos y + 1}{\cos y + \cos^2 y}$

66. $\sin y + \cos y - 1 = \dfrac{2 \sin y \cos y}{1 + \sin y + \cos y}$

4.3 Sum and Difference Formulas for Cosine

In this section we will establish several trigonometric identities involving the sum or difference of two angles or two real numbers. We first derive a formula for $\cos(\alpha + \beta)$. In Figure 4.3, angles α, $\alpha + \beta$, and $-\beta$ are in standard position, with their sides intersecting the circle of radius 1 at points P_1, P_2, P_3, and P_4. If P_1 has coordinates (x_1, y_1), then, since the radius r of the circle equals 1,

$$\cos \alpha = \frac{x_1}{1} \quad \text{and} \quad \sin \alpha = \frac{y_1}{1}.$$

Hence, the coordinates at P_1 are $x_1 = \cos \alpha$ and $y_1 = \sin \alpha$. We obtain the coordinates at P_2 and P_3 in a similar way. Note that from Equations (1) and (2) in Section 4.1, we have $\sin(-\beta) = -\sin \beta$ and

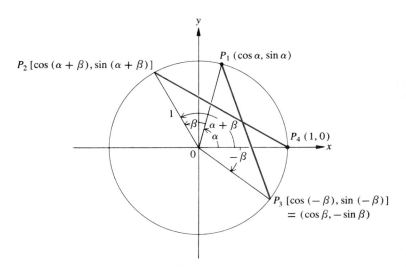

Figure 4.3

$\cos(-\beta) = \cos \beta$. Since all radii are equal and the central angle $\angle P_4OP_2$ and $\angle P_3OP_1$ are equal,

$$\triangle P_4OP_2 \cong \triangle P_3OP_1$$

and

$$P_4P_2 = P_3P_1.$$

From the distance formula (see Appendix A.4), we have

$$
\begin{aligned}
(P_3P_1)^2 &= [\cos \alpha - \cos \beta]^2 + [\sin \alpha - (-\sin \beta)]^2 \\
&= (\cos \alpha - \cos \beta)^2 + (\sin \alpha + \sin \beta)^2 \\
&= \cos^2 \alpha - 2 \cos \alpha \cos \beta + \cos^2 \beta + \sin^2 \alpha \\
&\quad + 2 \sin \alpha \sin \beta + \sin^2 \beta.
\end{aligned}
$$

Since $\cos^2 \alpha + \sin^2 \alpha = 1$ and $\cos^2 \beta + \sin^2 \beta = 1$,

$$(P_3P_1)^2 = 2 - 2 \cos \alpha \cos \beta + 2 \sin \alpha \sin \beta.$$

Again using the distance formula, we have

$$
\begin{aligned}
(P_4P_2)^2 &= [\cos(\alpha + \beta) - 1]^2 + [\sin(\alpha + \beta) - 0]^2 \\
&= \cos^2(\alpha + \beta) - 2 \cos(\alpha + \beta) + 1 + \sin^2(\alpha + \beta).
\end{aligned}
$$

Since $\cos^2(\alpha + \beta) + \sin^2(\alpha + \beta) = 1$,

$$(P_4P_2)^2 = 2 - 2 \cos(\alpha + \beta).$$

Since $P_4P_2 = P_3P_1$, we have

$$2 - 2 \cos(\alpha + \beta) = 2 - 2 \cos \alpha \cos \beta + 2 \sin \alpha \sin \beta,$$

which gives

$$\cos(\alpha + \beta) = \cos \alpha \cos \beta - \sin \alpha \sin \beta. \qquad ($$

Substituting $-\beta$ for β in Equation (1), we have

$$\cos[\alpha + (-\beta)] = \cos \alpha \cos(-\beta) - \sin \alpha \sin(-\beta).$$

Since

$$\cos(-\beta) = \cos \beta \qquad \text{and} \qquad \sin(-\beta) = -\sin \beta,$$

we obtain

$$\cos(\alpha - \beta) = \cos \alpha \cos \beta + \sin \alpha \sin \beta. \qquad ($$

These very important relationships for $\cos(\alpha + \beta)$ and $\cos(\alpha - \beta)$ are called the **sum formula** and the **difference formula,** respectively, for the cosine function.

We can, as illustrated in the following examples, find some function values using the sum and difference formulas. Although using Table II, Table III, or a calculator is a more efficient way of finding these function values, such examples give us practice in applying the formulas.

Example Find the exact value of $\cos 15°$.

Solution Because we know the exact function values of $45°$ and $30°$, we use $\alpha = 45°$ and $\beta = 30°$ in Equation (2) and obtain

$$\cos 15° = \cos(45° - 30°)$$

$$= \cos 45° \cos 30° + \sin 45° \sin 30°$$

$$= \frac{1}{\sqrt{2}} \cdot \frac{\sqrt{3}}{2} + \frac{1}{\sqrt{2}} \cdot \frac{1}{2} = \frac{\sqrt{3} + 1}{2\sqrt{2}}.$$

Example Compute $\cos \dfrac{7\pi}{12}$ exactly.

Solution Since $\dfrac{1}{3} + \dfrac{1}{4} = \dfrac{7}{12}$, we have $\dfrac{7\pi}{12} = \dfrac{\pi}{3} + \dfrac{\pi}{4}$. By Equation (1),

$$\cos \frac{7\pi}{12} = \cos\left(\frac{\pi}{3} + \frac{\pi}{4}\right) = \cos \frac{\pi}{3} \cos \frac{\pi}{4} - \sin \frac{\pi}{3} \sin \frac{\pi}{4}.$$

The appropriate substitutions for the function values yield

$$\cos \frac{7\pi}{12} = \frac{1}{2} \cdot \frac{1}{\sqrt{2}} - \frac{\sqrt{3}}{2} \cdot \frac{1}{\sqrt{2}} = \frac{1 - \sqrt{3}}{2\sqrt{2}}.$$

The sum and difference formulas for the cosine function can be used to establish other identities. Three identities that we will be using are

$$\cos(180° - \alpha) = -\cos \alpha. \tag{3}$$

$$\cos(180° + \alpha) = -\cos \alpha, \tag{4}$$

$$\cos(360° - \alpha) = \cos \alpha. \tag{5}$$

Identities such as these are known as **reduction formulas.** The proof that Equation (3) is an identity is shown here; the proofs of (4) and (5) are left as exercises.

(continued)

Using the difference formula [Equation (2)] and substituting 180° for α and α for β, we obtain

$$\cos(180° - α) = (\cos 180°)(\cos α) + (\sin 180°)(\sin α).$$

Since $\cos 180° = -1$ and $\sin 180° = 0$, we have

$$\cos(180° - α) = (-1) \cos α + (0) \sin α = -\cos α.$$

Reduction formulas can be used, instead of relying on sketches or reference angles, to rewrite a function value of an angle greater than 90° or less than 0° as a function value of an acute angle.

Examples Write each expression as the cosine of an acute angle.

a. cos 150°   b. cos 225°

Solutions
   a. Since $150° = 180° - 30°$, we use Equation (3) to obtain

$$\cos 150° = \cos(180° - 30°) = -\cos 30°.$$

   b. Since $225° = 180° + 45°$, we use Equation (4) to obtain

$$\cos 225° = \cos(180° + 45°) = -\cos 45°.$$

EXERCISE SET 4.3

A

■ *Find the exact value of each expression using Equation (1) or Equation (2) of this section.*

1. cos 75°   (Use: $75° = 120° - 45°$)

2. cos 105°   (Use: $105° = 60° + 45°$)

3. cos 165°   (Use: $165° = 120° + 45°$)

4. cos 195°   (Use: $195° = 135° + 60°$)

5. $\cos \dfrac{\pi}{12}$ $\left(\text{Use: } \dfrac{\pi}{12} = \dfrac{\pi}{3} - \dfrac{\pi}{4}\right)$ **6.** $\cos \dfrac{5\pi}{12}$ $\left(\text{Use: } \dfrac{5\pi}{12} = \dfrac{\pi}{6} + \dfrac{\pi}{4}\right)$

■ *Verify that each equation is correct.*

Example $\cos 110° \cos 70° - \sin 110° \sin 70° = -1$

Solution   From Equation (1), if we replace α with $110°$ and β with $70°$, we obtain

$$\cos 110° \cos 70° - \sin 110° \sin 70° = \cos(110° + 70°)$$
$$= \cos 180° = -1.$$

7. $\cos 40° \cos 140° - \sin 40° \sin 140° = -1$

8. $\cos 20° \cos 160° - \sin 20° \sin 160° = -1$

9. $\cos 35° \cos 55° - \sin 35° \sin 55° = 0.$

10. $\cos 50° \cos 40° - \sin 50° \sin 40° = 0$

11. $\cos 100° \cos 55° + \sin 100° \sin 55° = \dfrac{1}{\sqrt{2}}$

12. $\cos 130° \cos 70° + \sin 130° \sin 70° = \dfrac{1}{2}$

■ *Write each expression as a single function of kx, k an integer.*

Example $\cos 2x \cos 3x - \sin 2x \sin 3x$.

Solution   From Equation (1),

$$\cos 2x \cos 3x - \sin 2x \sin 3x = \cos(2x + 3x) = \cos 5x.$$

13. $\cos 4x \cos 5x - \sin 4x \sin 5x$ **14.** $\cos x \cos 2x - \sin x \sin 2x$

15. $\cos 8x \cos 6x + \sin 8x \sin 6x$ **16.** $\cos 5x \cos x + \sin 5x \sin x$

17. $\cos \dfrac{2}{3} x \cos \dfrac{1}{3} x - \sin \dfrac{2}{3} x \sin \dfrac{1}{3} x$

18. $\cos \dfrac{3}{4} x \cos \dfrac{5}{4} x - \sin \dfrac{3}{4} x \sin \dfrac{5}{4} x$

19. $\cos(x + 70°) \cos(x - 70°) - \sin(x + 70°) \sin(x - 70°)$

20. $\cos(65° + x) \cos(65° - x) + \sin(65° + x) \sin(65° - x)$

■ *If angles α and β have the given function values and are in the indicated quadrant, use Equation (1) to find $\cos(\alpha + \beta)$.*

Example $\cos \alpha = -\dfrac{1}{\sqrt{5}}$ and $\sin \beta = \dfrac{2}{\sqrt{13}}$; α in II and β in I

(*continued*)

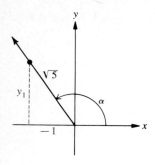

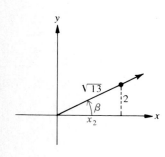

Solution Since α is in Quadrant II, as shown in the sketch, using th
Pythagorean theorem,

$$y_1^2 + (-1)^2 = (\sqrt{5})^2,$$
$$y_1^2 + 1 = 5,$$
$$y_1^2 = 4,$$
$$y_1 = 2 \quad (-2 \text{ does not apply}).$$

Hence, $\sin \alpha = \dfrac{y_1}{\sqrt{5}} = \dfrac{2}{\sqrt{5}}.$

Since β is in Quadrant I, from the sketch,

$$x_2^2 + 2^2 = (\sqrt{13})^2,$$
$$x_2^2 + 4 = 13,$$
$$x_2^2 = 9,$$
$$x_2 = 3 \quad (-3 \text{ does not apply}).$$

Hence, $\cos \beta = \dfrac{x_2}{\sqrt{13}} = \dfrac{3}{\sqrt{13}}.$ Therefore,

$$\cos(\alpha + \beta) = \cos \alpha \cos \beta - \sin \alpha \sin \beta$$

$$= -\frac{1}{\sqrt{5}} \cdot \frac{3}{\sqrt{13}} - \frac{2}{\sqrt{5}} \cdot \frac{2}{\sqrt{13}}$$

$$= -\frac{3}{\sqrt{65}} - \frac{4}{\sqrt{65}} = -\frac{7}{\sqrt{65}}.$$

21. $\cos \alpha = 3/5$ and $\sin \beta = 5/13$; α and β in Quadrant I

22. $\cos \alpha = -4/5$ and $\sin \beta = -7/25$; α in Quadrant III and β in IV

23. $\sin \alpha = 1/\sqrt{10}$ and $\cos \beta = 4/\sqrt{17}$; α in Quadrant II and β in II

24. $\sin \alpha = -5/\sqrt{29}$ and $\cos \beta = 1/\sqrt{26}$; α in Quadrant IV and β in

■ *Use one of the reduction formulas [Equation (3), (4), or (5)] of this section*
to write each expression as a function value of an acute angle.

Example $\cos 135°$

Solution   Using Equation (3), we find that

$$\cos 135° = \cos(180° - 45°) = -\cos 45°.$$

25. cos 120°   **26.** cos 240°   **27.** cos 210°

28. cos 300°   **29.** cos 315°   **30.** cos 330°

▪ *Verify that each equation in Exercises 31–35 is an identity.*

31. $\cos(180° + \alpha) = -\cos \alpha$ **32.** $\cos(360° - \alpha) = \cos \alpha$

33. $\cos(90° - \alpha) = \sin \alpha$ **34.** $\cos(90° + \alpha) = -\sin \alpha$

35. $\cos 2x = \cos^2 x - \sin^2 x.$ *Hint:* $\cos 2x = \cos(x + x).$

36. Use the results of Exercise 35 to show that

$$\cos 2x = 2 \cos^2 x - 1 = 1 - 2 \sin^2 x.$$

37. Show by counterexample that $\cos(\alpha + \beta) = \cos \alpha + \cos \beta$ is not an identity.

38. Show by counterexample that $\cos(\alpha - \beta) = \cos \alpha - \cos \beta$ is not an identity.

39. Show that $\cos\left(\dfrac{\pi}{2} - \alpha\right) = \sin \alpha.$

40. Use the results of Exercise 39 to show that $\sin\left(\dfrac{\pi}{2} - \alpha\right) = \cos \alpha.$

4.4 Sum and Difference Formulas for Sine

Sum and difference formulas for the sine function can be obtained from the sum and difference formulas for the cosine function. First, however, we consider two relationships between function values for the cosine and sine functions, From the difference formula for the cosine function we have

$$\cos\left(\frac{\pi}{2} - \theta\right) = \cos \frac{\pi}{2} \cos \theta + \sin \frac{\pi}{2} \sin \theta.$$

Replacing $\cos(\pi/2)$ with 0 and $\sin(\pi/2)$ with 1, we have

$$\cos\left(\frac{\pi}{2} - \theta\right) = 0 \cdot \cos \theta + 1 \cdot \sin \theta,$$

$$\cos\left(\frac{\pi}{2} - \theta\right) = \sin \theta. \tag{1}$$

Replacing θ with $\pi/2 - \theta$ in Equation (1), we obtain

$$\cos\left[\frac{\pi}{2} - \left(\frac{\pi}{2} - \theta\right)\right] = \sin\left(\frac{\pi}{2} - \theta\right),$$

$$\cos\left[\left(\frac{\pi}{2} - \frac{\pi}{2}\right) + \theta\right] = \sin\left(\frac{\pi}{2} - \theta\right),$$

$$\cos\theta = \sin\left(\frac{\pi}{2} - \theta\right),$$

$$\sin\left(\frac{\pi}{2} - \theta\right) = \cos\theta. \tag{2}$$

Examples

a. $\sin\dfrac{\pi}{3} = \cos\left(\dfrac{\pi}{2} - \dfrac{\pi}{3}\right)$

$\qquad = \cos\dfrac{\pi}{6}$

b. $\cos 46° = \sin(90° - 46°)$

$\qquad\qquad = \sin 44°$

Now we consider the sum and difference formulas for the sine function. If we replace θ with $\alpha + \beta$ in Equation (1), we have

$$\sin(\alpha + \beta) = \cos\left[\frac{\pi}{2} - (\alpha + \beta)\right]$$

$$= \cos\left[\left(\frac{\pi}{2} - \alpha\right) - \beta\right],$$

which by the difference formula for the cosine can be written as

$$\sin(\alpha + \beta) = \cos\left(\frac{\pi}{2} - \alpha\right)\cos\beta + \sin\left(\frac{\pi}{2} - \alpha\right)\sin\beta.$$

By Equations (1) and (2),

$$\cos\left(\frac{\pi}{2} - \alpha\right) = \sin\alpha \qquad \text{and} \qquad \sin\left(\frac{\pi}{2} - \alpha\right) = \cos\alpha.$$

Hence, we have

$$\sin(\alpha + \beta) = \sin\alpha\cos\beta + \cos\alpha\sin\beta. \tag{3}$$

Replacing β with $-\beta$ in Equation (3) yields

$$\sin[\alpha + (-\beta)] = \sin\alpha\cos(-\beta) + \cos\alpha\sin(-\beta).$$

Substituting $\cos \beta$ for $\cos(-\beta)$ and $(-\sin \beta)$ for $\sin(-\beta)$, we have

$$\sin(\alpha - \beta) = \sin \alpha \cos \beta + \cos \alpha(-\sin \beta),$$

from which

$$\sin(\alpha - \beta) = \sin \alpha \cos \beta - \cos \alpha \sin \beta. \tag{4}$$

Example Compute $\sin 15°$.

Solution Since $15° = 45° - 30°$, by Equation (4),

$$\sin 15° = \sin(45° - 30°)$$
$$= \sin 45° \cos 30° - \cos 45° \sin 30°$$
$$= \frac{1}{\sqrt{2}} \cdot \frac{\sqrt{3}}{2} - \frac{1}{\sqrt{2}} \cdot \frac{1}{2} = \frac{\sqrt{3} - 1}{2\sqrt{2}}.$$

Example Compute $\sin \dfrac{\pi}{12}$.

Solution Observe that $\dfrac{\pi}{12} = \dfrac{\pi}{3} - \dfrac{\pi}{4}$. By using Equation (4),

$$\sin \frac{\pi}{12} = \sin\left(\frac{\pi}{3} - \frac{\pi}{4}\right) = \sin \frac{\pi}{3} \cos \frac{\pi}{4} - \cos \frac{\pi}{3} \sin \frac{\pi}{4}.$$

Making the appropriate substitutions for the function values yields

$$\sin \frac{\pi}{12} = \frac{\sqrt{3}}{2} \cdot \frac{1}{\sqrt{2}} - \frac{1}{2} \cdot \frac{1}{\sqrt{2}} = \frac{\sqrt{3} - 1}{2\sqrt{2}}.$$

The same results are obtained if $\pi/12$ is written as $(\pi/4) - (\pi/6)$.

Again we note that in practical applications, approximations for function values such as $\sin(\pi/12)$ can be found readily by using Table II or a calculator.

The sum and difference formulas for the sine function can be used to establish the following reduction formulas.

$$\sin(180° - \alpha) = \sin \alpha, \tag{5}$$
$$\sin(180° + \alpha) = -\sin \alpha, \tag{6}$$
$$\sin(360° - \alpha) = -\sin \alpha. \tag{7}$$

The proof that Equation (5) is an identity is presented here; the proo͟
of (6) and (7) are left as exercises.

Using the difference formula, Equation (4), and substituting 180° for͟
and α for β, we obtain

$$\sin(180° - \alpha) = \sin 180° \cos \alpha - \cos 180° \sin \alpha.$$

Since

$$\sin 180° = 0 \quad \text{and} \quad \cos 180° = -1,$$

we have

$$\sin(180° - \alpha) = 0 \cdot \cos \alpha - (-1)\sin \alpha = \sin \alpha.$$

EXERCISE SET 4.4

A

■ *Find the value of each expression using Equation (3) or (4).*

1. $\sin 105°$ (Use: $105° = 60° + 45°$)

2. $\sin 75°$ (Use: $75° = 120° - 45°$)

3. $\sin 165°$ (Use: $165° = 210° - 45°$)

4. $\sin 195°$ (Use: $195° = 135° + 60°$)

5. $\sin \dfrac{5\pi}{12}$ $\left(\text{Use: } \dfrac{5\pi}{12} = \dfrac{\pi}{6} + \dfrac{\pi}{4}\right)$ 6. $\sin \dfrac{\pi}{12}$ $\left(\text{Use: } \dfrac{\pi}{12} = \dfrac{\pi}{3} - \dfrac{\pi}{4}\right)$

■ *Verify that each equation is correct.*

Example $\sin 100° \cos 80° + \cos 100° \sin 80° = 0$

Solution From Equation (3) of this section, if we replace α with 10͟
and β with 80°,

$$\sin 100° \cos 80° + \cos 100° \sin 80° = \sin(100° + 80°)$$
$$= \sin 180° = 0.$$

7. $\sin 20° \cos 160° + \cos 20° \sin 160° = 0$

8. $\sin 40° \cos 140° + \cos 40° \sin 140° = 0$

9. $\sin 50° \cos 40° + \cos 50° \sin 40° = 1$

10. $\sin 35° \cos 55° + \cos 35° \sin 55° = 1$

11. $\sin 110° \cos 80° - \cos 110° \sin 80° = \dfrac{1}{2}$

12. $\sin 95° \cos 50° - \cos 95° \sin 50° = \dfrac{1}{\sqrt{2}}$

■ *Write each expression as a single function of kx, where k is an integer.*

Example $\sin 2x \cos 3x + \cos 2x \sin 3x$

Solution   Using Equation (3),

$$\sin 2x \cos 3x + \cos 2x \sin 3x = \sin(2x + 3x)$$
$$= \sin 5x.$$

13. $\sin x \cos 2x + \cos x \sin 2x$ **14.** $\sin 4x \cos 5x + \cos 4x \sin 5x$

15. $\sin 5x \cos x - \cos 5x \sin x$ **16.** $\sin 8x \cos 6x - \cos 8x \sin 6x$

17. $\sin \dfrac{3}{4} x \cos \dfrac{5}{4} x + \cos \dfrac{3}{4} x \sin \dfrac{5}{4} x$

18. $\sin \dfrac{4}{3} x \cos \dfrac{1}{3} x - \cos \dfrac{4}{3} x \sin \dfrac{1}{3} x$

19. $\sin(50° + x) \cos(50° - x) - \cos(50° + x) \sin(50° - x)$

20. $\sin(x + 60°) \cos(x - 60°) + \cos(x + 60°) \sin(x - 60°)$

■ *If angles α and β have the given function values and are in the indicated quadrants, use Equation (3) to find sin(α + β).*

Example $\cos \alpha = -\dfrac{1}{\sqrt{5}}, \quad \sin \beta = \dfrac{2}{\sqrt{13}}, \quad \alpha$ in Quadrant II, β in I

Solution   From the example on page 126, we have $\sin \alpha = 2/\sqrt{5}$ and $\cos \beta = 3/\sqrt{13}$. Therefore, from Equation (3),

$$\sin(\alpha + \beta) = \sin \alpha \cos \beta + \cos \alpha \sin \beta$$
$$= \dfrac{2}{\sqrt{5}} \cdot \dfrac{3}{\sqrt{13}} + \left(-\dfrac{1}{\sqrt{5}}\right) \cdot \dfrac{2}{\sqrt{13}}$$
$$= \dfrac{6}{\sqrt{65}} - \dfrac{2}{\sqrt{65}} = \dfrac{4}{\sqrt{65}}.$$

21. $\cos \alpha = -4/5$, $\sin \beta = -7/25$, α in Quadrant III, β in IV

22. $\cos \alpha = 3/5$, $\sin \beta = 5/13$, α and β in Quadrant I

23. $\sin \alpha = -5/\sqrt{29}$, $\cos \beta = 1/\sqrt{26}$, α in Quadrant IV, β in I

24. $\sin \alpha = 1/\sqrt{10}$, $\cos \beta = -4/\sqrt{17}$, α in Quadrant II, β in III

■ *Use one of the reduction formulas [Equation (5), (6), or (7)] of this sectio to write each expression as a function of an acute angle.*

Example $\sin 135°$

Solution   By Equation (5),

$$\sin 135° = \sin(180° - 45°) = \sin 45°.$$

25. $\sin 240°$ 26. $\sin 120°$ 27. $\sin 150°$

28. $\sin 210°$ 29. $\sin 330°$ 30. $\sin 315°$

■ *Verify that each equation is an identity.*

31. $\sin(180° + \alpha) = -\sin \alpha$ 32. $\sin(360° - \alpha) = -\sin \alpha$

33. $\sin(270° - \alpha) = -\cos \alpha$ 34. $\sin(270° + \alpha) = -\cos \alpha$

35. $\sin 2x = 2 \sin x \cos x$.   *Hint:* $\sin 2x = \sin(x + x)$.

36. $\sin(\alpha - 30°) = \dfrac{\sqrt{3} \sin \alpha - \cos \alpha}{2}$

37. Show by a counterexample that $\sin(\alpha - \beta) = \sin \alpha - \sin \beta$ is no an identity.

38. Show by a counterexample that $\sin(\alpha + \beta) = \sin \alpha + \sin \beta$ is no an identity.

4.5 Sum and Difference Formulas for Tangent

Sum and difference formulas for the tangent function follow from th sum and difference formulas for the sine and cosine functions. Since

$$\tan(\alpha + \beta) = \frac{\sin(\alpha + \beta)}{\cos(\alpha + \beta)},$$

by substituting appropriately, we have

$$\tan(\alpha + \beta) = \frac{\sin \alpha \cos \beta + \cos \alpha \sin \beta}{\cos \alpha \cos \beta - \sin \alpha \sin \beta}.$$

Multiplying each term of the numerator and denominator of the right-hand member by $1/\cos \alpha \cos \beta$ yields

$$\tan(\alpha + \beta) = \frac{\dfrac{\sin \alpha \cos \beta}{\cos \alpha \cos \beta} + \dfrac{\cos \alpha \sin \beta}{\cos \alpha \cos \beta}}{\dfrac{\cos \alpha \cos \beta}{\cos \alpha \cos \beta} - \dfrac{\sin \alpha \sin \beta}{\cos \alpha \cos \beta}},$$

from which, by simplifying the right-hand member, we obtain

$$\tan(\alpha + \beta) = \frac{\tan \alpha + \tan \beta}{1 - \tan \alpha \tan \beta}. \tag{1}$$

By substituting $-\beta$ for β in Equation (1), we have

$$\tan[\alpha + (-\beta)] = \frac{\tan \alpha + \tan(-\beta)}{1 - \tan \alpha \tan(-\beta)}.$$

Substituting $-\tan \beta$ for $\tan(-\beta)$ yields

$$\tan[\alpha + (-\beta)] = \frac{\tan \alpha + (-\tan \beta)}{1 - \tan \alpha(-\tan \beta)},$$

$$\tan(\alpha - \beta) = \frac{\tan \alpha - \tan \beta}{1 + \tan \alpha \tan \beta}. \tag{2}$$

Example   Compute $\tan 75°$.

Solution   Observe that $75° = 45° + 30°$. Then, using Equation (1) and replacing α with $45°$ and β with $30°$,

$$\tan 75° = \tan(45° + 30°) = \frac{\tan 45° + \tan 30°}{1 - \tan 45° \tan 30°}$$

$$= \frac{1 + \dfrac{1}{\sqrt{3}}}{1 - 1\left(\dfrac{1}{\sqrt{3}}\right)} = \frac{\left(1 + \dfrac{1}{\sqrt{3}}\right) \cdot \sqrt{3}}{\left(1 - \dfrac{1}{\sqrt{3}}\right) \cdot \sqrt{3}}$$

$$= \frac{\sqrt{3} + 1}{\sqrt{3} - 1}.$$

The sum and difference formulas established in this section can be used to establish the following reduction formulas.

$$\tan(180° - \alpha) = -\tan \alpha, \qquad ($$

$$\tan(180° + \alpha) = \tan \alpha, \qquad (4$$

$$\tan(360° - \alpha) = -\tan \alpha. \qquad (5$$

The proof that Equation (3) is an identity is presented here; the proo
of (4) and (5) are left as exercises.

Using the difference formula [Equation (2)] and substituting 180° fo
α and α for β, we obtain

$$\tan(180° - \alpha) = \frac{\tan 180° - \tan \alpha}{1 + \tan 180° \tan \alpha}$$

$$= \frac{0 - \tan \alpha}{1 + (0) \tan \alpha} = -\tan \alpha.$$

EXERCISE SET 4.5

A

■ *Use Equation (1) or (2) to find the value of each expression.*

1. $\tan 105°$ (Use: $105° = 45° + 60°$)

2. $\tan 15°$ (Use: $15° = 45° - 30°$)

3. $\tan 195°$ (Use: $195° = 240° - 45°$)

4. $\tan 165°$ (Use: $165° = 120° + 45°$)

5. $\tan \dfrac{\pi}{12}$ $\left(\text{Use: } \dfrac{\pi}{12} = \dfrac{\pi}{3} - \dfrac{\pi}{4}\right)$ **6.** $\tan \dfrac{3\pi}{4}$ $\left(\text{Use: } \dfrac{3\pi}{4} = \pi - \dfrac{\pi}{4}\right.$

■ *Verify that each equation is correct.*

Example $\dfrac{\tan 20° + \tan 25°}{1 - \tan 20° \tan 25°} = 1$

Solution   From Equation (1), if we replace α with 20° and β with 25°

$$\frac{\tan 20° + \tan 25°}{1 - \tan 20° \tan 25°} = \tan(20° + 25°) = \tan 45° = 1.$$

7. $\dfrac{\tan 50° + \tan 130°}{1 - \tan 50° \tan 130°} = 0$ **8.** $\dfrac{\tan 100° + \tan 50°}{1 - \tan 100° \tan 50°} = -\dfrac{1}{\sqrt{}}$

9. $\dfrac{\tan 140° + \tan 100°}{1 - \tan 140° \tan 100°} = \sqrt{3}$ **10.** $\dfrac{\tan 95° + \tan 40°}{1 - \tan 95° \tan 40°} = -1$

11. $\dfrac{\tan 110° - \tan 50°}{1 + \tan 110° \tan 50°} = \sqrt{3}$ **12.** $\dfrac{\tan 115° - \tan 70°}{1 + \tan 115° \tan 70°} = 1$

■ *Write each expression as a single function of kx where k is an integer.*

Example $\dfrac{\tan 2x + \tan 3x}{1 - \tan 2x \tan 3x}$

Solution   Using Equation (1), we obtain

$$\frac{\tan 2x + \tan 3x}{1 - \tan 2x \tan 3x} = \tan(2x + 3x) = \tan 5x.$$

13. $\dfrac{\tan 4x + \tan 5x}{1 - \tan 4x \tan 5x}$ **14.** $\dfrac{\tan 2x + \tan x}{1 - \tan 2x \tan x}$

15. $\dfrac{\tan 8x - \tan 6x}{1 + \tan 8x \tan 6x}$ **16.** $\dfrac{\tan 5x - \tan x}{1 + \tan 5x \tan x}$

17. $\dfrac{\tan \frac{4}{3}x - \tan \frac{1}{3}x}{1 + \tan \frac{4}{3}x \tan \frac{1}{3}x}$ **18.** $\dfrac{\tan \frac{5}{4}x + \tan \frac{3}{4}x}{1 - \tan \frac{5}{4}x \tan \frac{3}{4}x}$

■ *Use one of the reduction formulas [Equation (3), (4), or (5)] of this section to write each expression as a function of an acute angle.*

Example $\tan 135°$

Solution   Using Equation (3),

$$\tan 135° = \tan(180° - 45°) = -\tan 45°.$$

19. $\tan 120°$ **20.** $\tan 240°$ **21.** $\tan 210°$

22. $\tan 150°$ **23.** $\tan 315°$ **24.** $\tan 330°$

■ *Verify that each equation is an identity.*

25. $\tan(180° + \alpha) = \tan \alpha$ **26.** $\tan(360° - \alpha) = -\tan \alpha$

27. $\tan(90° - \alpha) = \cot \alpha$ **28.** $\cot(90° - \alpha) = \tan \alpha$

29. $\tan 2x = \dfrac{2 \tan x}{1 - \tan^2 x}$.   *Hint:* $\tan 2x = \tan(x + x)$.

30. $\tan\left(x + \dfrac{\pi}{4}\right) = \dfrac{1 + \tan x}{1 - \tan x}$

B

31. $\dfrac{\tan(45° - \alpha)}{\tan(45° + \alpha)} = \dfrac{(1 - \tan \alpha)^2}{(1 + \tan \alpha)^2}$ 　　**32.** $\tan \beta = \dfrac{\tan(45° + \beta) - 1}{1 + \tan(45° + \beta)}$

33. $\cot(\alpha + \beta) = \dfrac{\cot \alpha \cot \beta - 1}{\cot \beta + \cot \alpha}$ 　　**34.** $\cot(\alpha - \beta) = \dfrac{\cot \alpha \cot \beta + 1}{\cot \beta - \cot \alpha}$

4.6 Double-Angle Formulas

Additional identities follow from the sum and difference formula
developed in Sections 4.2 and 4.3. For example, if α is substituted for β i

$$\cos(\alpha + \beta) = \cos \alpha \cos \beta - \sin \alpha \sin \beta,$$

we obtain

$$\cos(\alpha + \alpha) = \cos \alpha \cos \alpha - \sin \alpha \sin \alpha,$$
$$\mathbf{\cos 2\alpha = \cos^2 \alpha - \sin^2 \alpha.} \qquad ($$

The right-hand member can be written in a variety of convenient form
Thus, if $1 - \sin^2 \alpha$ is substituted for $\cos^2 \alpha$, we have

$$\cos 2\alpha = (1 - \sin^2 \alpha) - \sin^2 \alpha,$$
$$\mathbf{\cos 2\alpha = 1 - 2 \sin^2 \alpha.} \qquad ($$

If $1 - \cos^2 \alpha$ is substituted for $\sin^2 \alpha$ in Equation (1), we have

$$\cos 2\alpha = \cos^2 \alpha - (1 - \cos^2 \alpha),$$
$$\mathbf{\cos 2\alpha = 2 \cos^2 \alpha - 1.} \qquad ($$

Identities (1), (2), and (3), and similar identities that follow for the sir
and tangent functions, called **double-angle formulas,** are often helpful fo
rewriting expressions in more useful forms. You will prove addition
identities in the exercises to become familiar with these importan
relationships.

Example　Show that $\dfrac{1 - \cos 2x}{2 \sin x \cos x} = \tan x$ is a identity.

Solution　From Equation (2), we can substitute $1 - 2 \sin^2 x$ for $\cos 2$
in the left-hand member to obtain

$$\frac{1 - (1 - 2 \sin^2 x)}{2 \sin x \cos x} = \tan x$$

$$\frac{2 \sin^2 x}{2 \sin x \cos x} = \tan x$$

$$\frac{\sin x}{\cos x} = \tan x$$

$$\tan x = \tan x.$$

Hence, the original equation is an identity.

Example   Show that $2 \sec^2 \beta = \dfrac{5 + \cos 2\beta}{1 + \cos 2\beta} - 1$ is an identity.

Solution   From Equation (3), we substitute $2 \cos^2 \beta - 1$ for $\cos 2\beta$ in the right-hand member to obtain

$$2 \sec^2 \beta = \frac{5 + (2 \cos^2 \beta - 1)}{1 + (2 \cos^2 \beta - 1)} - 1 = \frac{4 + 2 \cos^2 \beta}{2 \cos^2 \beta} - 1$$

$$= \frac{4 + 2 \cos^2 \beta - 2 \cos^2 \beta}{2 \cos^2 \beta} = \frac{4}{2 \cos^2 \beta};$$

$$2 \sec^2 \beta = 2 \sec^2 \beta.$$

Hence, the original equation is an identity.

If α is substituted for β in

$$\sin(\alpha + \beta) = \sin \alpha \cos \beta + \cos \alpha \sin \beta,$$

we obtain

$$\sin(\alpha + \alpha) = \sin \alpha \cos \alpha + \cos \alpha \sin \alpha,$$

$$\mathbf{\sin 2\alpha = 2 \sin \alpha \cos \alpha.} \tag{4}$$

Example   Show that $\dfrac{2}{\sin 2\theta} = \tan \theta + \cot \theta$ is an identity.

Solution   Substituting $2 \sin \theta \cos \theta$ for $\sin 2\theta$ in the left-hand member and $\sin \theta/\cos \theta$ for $\tan \theta$ and $\cos \theta/\sin \theta$ for $\cot \theta$ in the right-hand member yields

$$\frac{2}{2 \sin \theta \cos \theta} = \frac{\sin \theta}{\cos \theta} + \frac{\cos \theta}{\sin \theta} = \frac{\sin^2 \theta + \cos^2 \theta}{\sin \theta \cos \theta}$$

$$= \frac{1}{\sin \theta \cos \theta} = \frac{2}{2 \sin \theta \cos \theta}$$

Hence, the original equation is an identity.

If α is substituted for β in

$$\tan(\alpha + \beta) = \frac{\tan \alpha + \tan \beta}{1 - \tan \alpha \tan \beta},$$

we obtain

$$\tan(\alpha + \alpha) = \frac{\tan \alpha + \tan \alpha}{1 - \tan \alpha \tan \alpha},$$

$$\mathbf{\tan 2\alpha = \frac{2 \tan \alpha}{1 - \tan^2 \alpha}.} \qquad ($$

Example   Show that $\tan 2x = \dfrac{2 \tan x}{2 - \sec^2 x}$ is an identity.

Solution   Substituting $\tan^2 x + 1$ for $\sec^2 x$ yields

$$\tan 2x = \frac{2 \tan x}{2 - (\tan^2 x + 1)}$$

$$= \frac{2 \tan x}{1 - \tan^2 x},$$

which is an identity by Equation (5).

   Relationships similar to those developed in this section for sin
cosine, and tangent can be derived for cotangent, secant, and cosecar
However, they are of less importance and will not be considered.

EXERCISE SET 4.6

A

■ *Express each of the following as a single function of kx, or $k\alpha$, where
is a positive integer. Assume that x or α takes on no value for which t
expression is undefined.*

Examples

a. $1 - 2 \sin^2 x$

b. $\dfrac{\tan 3x}{1 - \tan^2 3x}$

Solutions
   a. From Equation (2),

$$1 - 2 \sin^2 x = \cos 2x.$$

b. From Equation (5),

$$\frac{\tan 3x}{1 - \tan^2 3x} = \frac{1}{2} \cdot \frac{2 \tan 3x}{1 - \tan^2 3x}$$

$$= \frac{1}{2} \tan 2(3x) = \frac{1}{2} \tan 6x.$$

1. $2 \sin x \cos x$ **2.** $2 \cos^2 x - 1$ **3.** $\cos^2 x - \sin^2 x$

4. $\dfrac{2 \tan x}{1 - \tan^2 x}$ **5.** $2 \sin 2x \cos 2x$ **6.** $\dfrac{4 \tan 2x}{1 - \tan^2 2x}$

7. $\sin \dfrac{x}{2} \cos \dfrac{x}{2}$ **8.** $4 \sin 3x \cos 3x$ **9.** $\cos^2 5\alpha - \sin^2 5\alpha$

10. $1 - 2 \sin^2 4\alpha$ **11.** $4 \cos^2 6\alpha - 2$ **12.** $2 \sin 5\alpha \cos 5\alpha$

13. $\sin 2\alpha \cos \alpha - \cos 2\alpha \sin \alpha$ **14.** $\sin 3\alpha \cos 2\alpha + \cos 3\alpha \sin 2\alpha$

15. $\cos 3x \cos 5x + \sin 3x \sin 5x$ **16.** $\cos 4x \cos 2x - \sin 4x \sin 2x$

▪ *Verify that each equation is an identity.*

17. $\cos^2 \alpha = \dfrac{1 + \cos 2\alpha}{2}$ **18.** $\sin^2 \alpha = \dfrac{1 - \cos 2\alpha}{2}$

19. $\cos 2x + 2 \sin^2 x = 1$ **20.** $\sin 2x \csc x = 2 \cos x$

21. $\dfrac{\sin 2x}{1 + \cos 2x} = \tan x$ **22.** $\sin 2x = \dfrac{2 \tan x}{1 + \tan^2 x}$

23. $\sec^2 x = \dfrac{2}{1 + \cos 2x}$ **24.** $\cot x = \dfrac{1 + \cos 2x}{\sin 2x}$

25. $\cos 2\beta = \dfrac{1 - \tan^2 \beta}{1 + \tan^2 \beta}$ **26.** $\cot^2 \beta = \dfrac{1 + \cos 2\beta}{1 - \cos 2\beta}$

27. $\cot 4x = \dfrac{1 - \tan^2 2x}{2 \tan 2x}$ **28.** $\cos^4 x - \sin^4 x = \cos 2x$

29. $\sin 3x = 3 \sin x - 4 \sin^3 x$
Hint: First rewrite $\sin 3x$ as $\sin(2x + x)$.

30. $\cos 3x = 4 \cos^3 x - 3 \cos x$

31. $\sin 4\alpha = (\cos \alpha)(4 \sin \alpha - 8 \sin^3 \alpha)$

32. $\cos 4\alpha = 1 - 8 \sin^2 \alpha \cos^2 \alpha$

Example If $\sin x = 3/5$ and $\cos x < 0$, find $\sin 2x$.

Solution To evaluate

$$\sin 2x = 2 \sin x \cos x,$$ $($

the values for $\cos x$ for those values of x for which $\sin x = 3/5$ an
$\cos x < 0$ are needed. Substituting $3/5$ for $\sin x$ in

$$\sin^2 x + \cos^2 x = 1$$

gives

$$\left(\frac{3}{5}\right)^2 + \cos^2 x = 1,$$

$$\cos^2 x = 1 - \frac{9}{25} = \frac{16}{25}.$$

Because $\cos x < 0$, $\cos x = -4/5$. Substituting $3/5$ for $\sin x$ and $-4/$
for $\cos x$ in $\sin 2x = 2 \sin x \cos x$ yields

$$\sin 2x = 2\left(\frac{3}{5}\right)\left(-\frac{4}{5}\right) = -\frac{24}{25}.$$

33. If $\sin x = 4/5$ and $\cot x > 0$, find $\sin 2x$.

34. If $\sin x = 12/13$ and $\sec x < 0$, find $\cos 2x$.

35. If $\cos 2\alpha = -8/17$ and $180° \leq 2\alpha \leq 270°$, find $\sin \alpha$.

36. If $\cos 2\alpha = 5/13$ and $270° \leq 2\alpha \leq 360°$, find $\cos \alpha$.

37. If $\tan \beta = -3/4$, find $\tan 2\beta$.

38. If $\tan \beta = 5/12$, find $\tan 2\beta$.

4.7 Additional Identities

The identities in this section are also useful in rewriting expression
involving trigonometric function values.

Identities for
$\cos^2 \alpha/2$ and $\sin^2 \alpha/2$

In Section 4.6 we observed that

$$\cos 2\alpha = 2 \cos^2 \alpha - 1.$$

Solving this for $\cos^2 \alpha$, we have

$$1 + \cos 2\alpha = 2 \cos^2 \alpha,$$

from which

$$\cos^2 \alpha = \frac{1 + \cos 2\alpha}{2}. \tag{1}$$

We also noted in Section 4.6 that

$$\cos 2\alpha = 1 - 2 \sin^2 \alpha.$$

Solving this for $\sin^2 \alpha$, we have

$$2 \sin^2 \alpha = 1 - \cos 2\alpha,$$

from which

$$\sin^2 \alpha = \frac{1 - \cos 2\alpha}{2}. \tag{2}$$

These identities enable us to write the second-degree terms, $\cos^2 \alpha$ and $\sin^2 \alpha$, in terms of first-degree terms. They also enable us to write additional useful identities.

Half-Angle Formulas   Substituting $\alpha/2$ for α in Equation (1) yields

$$\cos^2 \frac{\alpha}{2} = \frac{1 + \cos 2(\alpha/2)}{2},$$

from which

$$\cos^2 \frac{\alpha}{2} = \frac{1 + \cos \alpha}{2}. \tag{3}$$

Substituting $\alpha/2$ for α in Equation (2) yields

$$\sin^2 \frac{\alpha}{2} = \frac{1 - \cos 2(\alpha/2)}{2},$$

from which

$$\sin^2 \frac{\alpha}{2} = \frac{1 - \cos \alpha}{2}. \tag{4}$$

Values for $\cos(\alpha/2)$ and $\sin(\alpha/2)$ are positive or negative, depending on the value of $\alpha/2$.

Example   Use Equation (4) to find $\sin \dfrac{\pi}{8}$.

Solution   With $\alpha/2 = \pi/8$ and $\alpha = 2(\pi/8) = \pi/4$, we obtain

$$\sin^2 \frac{\pi}{8} = \frac{1 - \cos \dfrac{\pi}{4}}{2}.$$

Because $0 < \pi/8 < \pi/2$, we use the positive square root for $\sin(\pi/8)$
Then, since $\cos(\pi/4) = 1/\sqrt{2}$,

$$\sin \frac{\pi}{8} = \sqrt{\frac{1 - \dfrac{1}{\sqrt{2}}}{2}} = \sqrt{\frac{\sqrt{2} - 1}{2\sqrt{2}}} = \frac{1}{2}\sqrt{2 - \sqrt{2}}.$$

Formulas for $\tan(\alpha/2)$ also follow directly from identities we have
developed. Substituting $\alpha/2$ for α in $\tan \alpha = (\sin \alpha)/(\cos \alpha)$, we have

$$\tan \frac{\alpha}{2} = \frac{\sin \dfrac{\alpha}{2}}{\cos \dfrac{\alpha}{2}}.$$

Multiplying both the numerator and the denominator of the right-hand
member by $2 \sin(\alpha/2)$ yields

$$\tan \frac{\alpha}{2} = \frac{2 \sin^2 \dfrac{\alpha}{2}}{2 \sin \dfrac{\alpha}{2} \cos \dfrac{\alpha}{2}}.$$

Substituting $(1 - \cos \alpha)/2$ for $\sin^2(\alpha/2)$ and substituting $\sin 2(\alpha/2)$ for
$2 \sin(\alpha/2) \cos(\alpha/2)$, we obtain

$$\tan \frac{\alpha}{2} = \frac{2\left(\dfrac{1 - \cos \alpha}{2}\right)}{\sin 2\left(\dfrac{\alpha}{2}\right)},$$

$$\tan \frac{\alpha}{2} = \frac{1 - \cos \alpha}{\sin \alpha}. \tag{5}$$

By multiplying the numerator and the denominator of the right-hand member of Equation (5) by $1 + \cos \alpha$, we obtain

$$\tan \frac{\alpha}{2} = \frac{(1 - \cos \alpha)(1 + \cos \alpha)}{\sin \alpha(1 + \cos \alpha)}$$

$$= \frac{\sin^2 \alpha}{\sin \alpha(1 + \cos \alpha)},$$

from which

$$\tan \frac{\alpha}{2} = \frac{\sin \alpha}{1 + \cos \alpha}. \tag{6}$$

Example   Use Equation (5) to find $\tan \dfrac{\pi}{8}$.

Solution   If $\alpha/2 = \pi/8$, then $\alpha = \pi/4$ and

$$\tan \frac{\pi}{8} = \frac{1 - \cos \dfrac{\pi}{4}}{\sin \dfrac{\pi}{4}} = \frac{1 - \dfrac{1}{\sqrt{2}}}{\dfrac{1}{\sqrt{2}}}$$

$$= \sqrt{2} - 1.$$

Just as identities such as $\sin 2\alpha = 2 \sin \alpha \cos \alpha$ are sometimes called double-angle formulas, the identities developed above are sometimes called **half-angle formulas.**

Relationships similar to those developed above for sine, cosine, and tangent can be derived for cotangent, secant, and cosecant. They are of less importance and will not be considered.

Product, Sum, and Difference Formulas   The following identities are derived from the sum and difference formulas for the cosine and sine that were developed in Sections 4.3 and 4.4. The proofs are left as exercises.

$$\cos \alpha \cos \beta = \tfrac{1}{2}\left[\cos(\alpha + \beta) + \cos(\alpha - \beta)\right], \tag{7}$$

$$\sin \alpha \sin \beta = -\tfrac{1}{2}\left[\cos(\alpha + \beta) - \cos(\alpha - \beta)\right], \tag{8}$$

$$\sin \alpha \cos \beta = \tfrac{1}{2}\left[\sin(\alpha + \beta) + \sin(\alpha - \beta)\right], \tag{9}$$

$$\cos \alpha \sin \beta = \tfrac{1}{2}\left[\sin(\alpha + \beta) - \sin(\alpha - \beta)\right]. \tag{10}$$

Example   Express $\cos 5x \sin 2x$ as a sum or difference.

Solution   From Equation (10),

$$\cos 5x \sin 2x = \tfrac{1}{2}[\sin(5x + 2x) - \sin(5x - 2x)]$$
$$= \tfrac{1}{2}(\sin 7x - \sin 3x).$$

The following four identities enable us to express sums and differences of function values in terms of products.

$$\cos \alpha + \cos \beta = 2 \cos\left(\frac{\alpha + \beta}{2}\right) \cos\left(\frac{\alpha - \beta}{2}\right), \qquad (11)$$

$$\cos \alpha - \cos \beta = -2 \sin\left(\frac{\alpha + \beta}{2}\right) \sin\left(\frac{\alpha - \beta}{2}\right), \qquad (12)$$

$$\sin \alpha + \sin \beta = 2 \sin\left(\frac{\alpha + \beta}{2}\right) \cos\left(\frac{\alpha - \beta}{2}\right), \qquad (13)$$

$$\sin \alpha - \sin \beta = 2 \cos\left(\frac{\alpha + \beta}{2}\right) \sin\left(\frac{\alpha - \beta}{2}\right). \qquad (14)$$

Example   Express $\sin 4\alpha + \sin 2\alpha$ as a product.

Solution   From Equation (13),

$$\sin 4\alpha + \sin 2\alpha = 2 \sin\left(\frac{4\alpha + 2\alpha}{2}\right) \cos\left(\frac{4\alpha - 2\alpha}{2}\right)$$
$$= 2 \sin 3\alpha \cos \alpha.$$

EXERCISE SET 4.7
A

■ *Use the half-angle formulas to find the exact value for each of the following.*

Example $\sin 105°$

Solution   From Equation (4), with $\alpha/2 = 105°$ and $\alpha = 210°$,

$$\sin 105° = \sqrt{\frac{1 - \cos 210°}{2}},$$

where the positive square root is chosen because 105° is in Quadrant II and sin 105° is therefore positive. Since $\cos 210° = -\sqrt{3}/2$,

$$\sin 105° = \sqrt{\frac{1 - (-\sqrt{3}/2)}{2}} = \frac{\sqrt{2 + \sqrt{3}}}{2}.$$

1. $\sin 15°$ **2.** $\cos 15°$ **3.** $\tan 15°$ **4.** $\sin 75°$

5. $\cos 75°$ **6.** $\tan 75°$ **7.** $\cos \dfrac{\pi}{8}$ **8.** $\sin \dfrac{\pi}{12}$

9. $\tan \dfrac{5\pi}{12}$ **10.** $\cos \dfrac{\pi}{12}$ **11.** $\cos \dfrac{7\pi}{12}$ **12.** $\sin \dfrac{7\pi}{12}$

▪ *Verify each identity.*

13. $\tan \dfrac{x}{2} = \csc x - \cot x$ **14.** $\sin \dfrac{\theta}{2} \cos \dfrac{\theta}{2} = \dfrac{\sin \theta}{2}$

15. $\left(\cos \dfrac{\alpha}{2} - \sin \dfrac{\alpha}{2} \right)^2 = 1 - \sin \alpha$ **16.** $\sin \beta - \cos \beta \tan \dfrac{\beta}{2} = \tan \dfrac{\beta}{2}$

17. $\sin^2 \dfrac{\gamma}{2} = \dfrac{\sec \gamma - 1}{2 \sec \gamma}$ **18.** $\tan \dfrac{x}{2} \sin x = \dfrac{\tan x - \sin x}{\sin x \sec x}$

▪ *In Exercises 19–24, assume $0° \le x < 360°$.*

Example If $\tan x = -\dfrac{3}{4}$ and $\cos x$ is negative, find $\sin \dfrac{x}{2}$.

Solution Since $\cos x$ and $\tan x$ are both negative, x is an angle in Quadrant II, and because $\tan x = -3/4$, a point on the terminal ray has the x-coordinate -4 and the y-coordinate 3. From the Pythagorean formula,

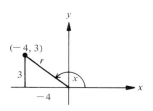

$$r = \sqrt{x^2 + y^2}$$
$$= \sqrt{(-4)^2 + 3^2} = 5.$$

From Equation (4) with $\cos x = -4/5$, we obtain

$$\sin \dfrac{x}{2} = \sqrt{\frac{1 - (-4/5)}{2}} = \frac{3}{\sqrt{10}},$$

where the positive square root is taken because x is in Quadrant II and therefore $x/2$ must be in Quadrant I.

19. If $\sin x = \dfrac{\sqrt{3}}{2}$ and $\cos x > 0$, find $\tan \dfrac{x}{2}$.

20. If $\sin x = \dfrac{1}{2}$ and $\sec x > 0$, find $\sin \dfrac{x}{2}$.

21. If $\cos x = -\dfrac{1}{3}$ and $\tan x > 0$, find $\sin \dfrac{x}{2}$.

22. If $\tan x = 3$ and $\sin x > 0$, find $\cos \dfrac{x}{2}$.

23. If $\tan x = -1$ and $\sin x < 0$, find $\cos \dfrac{x}{2}$.

24. If $\sin x = \dfrac{1}{\sqrt{2}}$ and $\cos x < 0$, find $\tan \dfrac{x}{2}$.

■ *Write each expression as a sum or difference of function values.*

Example $\sin 3x \sin x$

Solution   From Equation (8),

$$\sin 3x \sin x = -\tfrac{1}{2}[\cos(3x + x) - \cos(3x - x)]$$
$$= -\tfrac{1}{2}(\cos 4x - \cos 2x).$$

25. $\sin 5° \cos 12°$ **26.** $\cos 7° \cos 20°$ **27.** $\cos 2x \sin x$

28. $\sin 2x \sin 5x$ **29.** $\cos 2x \cos 5x$ **30.** $\sin 6x \cos x$

■ *Write each expression as a product.*

Example $\cos 3\alpha - \cos \alpha$

Solution   From Equation (12).

$$\cos 3\alpha - \cos \alpha = -2 \sin\left(\frac{3\alpha + \alpha}{2}\right) \sin\left(\frac{3\alpha - \alpha}{2}\right)$$
$$= -2 \sin 2\alpha \sin \alpha.$$

31. $\cos 20° + \cos 4°$ **32.** $\sin 18° - \sin 6°$ **33.** $\sin 3\alpha + \sin 5\alpha$

34. $\cos \alpha + \cos 5\alpha$ **35.** $\sin 5\alpha - \sin \alpha$ **36.** $\cos 6\alpha - \cos 2\alpha$

B

■ *Verify each of the following identities that are listed in this section.*

37. Equation (7) **38.** Equation (8) **39.** Equation (9)

40. Equation (10) **41.** Equation (11) **42.** Equation (12)

43. Equation (13) **44.** Equation (14)

Chapter Summary

[4.1–4.7] Equations are **identities** if they are satisfied for all real number replacements of the variable for which both members of the equation are defined.

An equation can be shown to be an identity by showing that the left-hand member and right-hand member are equivalent expressions.

In addition to the reciprocal identities,

$$\csc \alpha = \frac{1}{\sin \alpha}, \qquad \sec \alpha = \frac{1}{\cos \alpha}, \qquad \text{and} \qquad \cot \alpha = \frac{1}{\tan \alpha}$$

introduced in Chapter 1, the following identities are basic.

$$\cos^2 \alpha + \sin^2 \alpha = 1$$

$$\tan^2 x + 1 = \sec^2 x; \qquad \cot^2 x + 1 = \csc^2 x$$

$$\sin(-x) = -\sin x; \qquad \cos(-x) = \cos x; \qquad \tan(-x) = -\tan x$$

$$\cos(\alpha + \beta) = \cos \alpha \cos \beta - \sin \alpha \sin \beta$$

$$\cos(\alpha - \beta) = \cos \alpha \cos \beta + \sin \alpha \sin \beta$$

$$\cos\left(\frac{\pi}{2} - x\right) = \sin x; \qquad \sin\left(\frac{\pi}{2} - x\right) = \cos x$$

$$\sin(\alpha + \beta) = \sin \alpha \cos \beta + \cos \alpha \sin \beta$$

$$\sin(\alpha - \beta) = \sin \alpha \cos \beta - \cos \alpha \sin \beta$$

$$\tan(\alpha + \beta) = \frac{\tan \alpha + \tan \beta}{1 - \tan \alpha \tan \beta}$$

$$\tan(\alpha - \beta) = \frac{\tan \alpha - \tan \beta}{1 + \tan \alpha \tan \beta}$$

$$\cos 2x = \cos^2 x - \sin^2 x = 1 - 2 \sin^2 x$$

$$= 2 \cos^2 x - 1 \qquad \text{(continued)}$$

$$\cos^2 x = \frac{1 + \cos 2x}{2}; \qquad \cos^2 \frac{x}{2} = \frac{1 + \cos x}{2}$$

$$\sin 2x = 2 \sin x \cos x$$

$$\sin^2 x = \frac{1 - \cos 2x}{2}; \qquad \sin^2 \frac{x}{2} = \frac{1 - \cos x}{2}$$

$$\tan 2x = \frac{2 \tan x}{1 - \tan^2 x}; \qquad \tan \frac{x}{2} = \frac{1 - \cos x}{\sin x}$$

An equation can be shown *not* to be an identity by replacing th
variable with a real number for which the resulting statement is false.

*A list of symbols introduced in this chapter is shown inside the front cove
and important properties are shown inside the back cover.*

Review Exercises

[4.1–4.5] ▪ *Express each of the following in terms of the function values* $\sin \alpha$ *and/
* $\cos \alpha$ *and simplify the expression.*

1. $\csc \alpha \tan \alpha$ **2.** $\cot^2 \alpha \sec^2 \alpha$

3. $\sin^2 \alpha(\csc^2 \alpha - 1)$ **4.** $\cot \alpha \cos \alpha \sec^2 \alpha$

5. $\sin(-\alpha) \cot \alpha$ **6.** $\sec^2(-\alpha) \cos(-\alpha)$

▪ *Verify that each equation is an identity.*

7. $\sin^2 \alpha(1 + \cot^2 \alpha) = 1$ **8.** $\dfrac{\cos \alpha \csc \alpha}{\cot \alpha} = 1$

9. $\tan \alpha = \dfrac{1}{\cos \alpha \csc \alpha}$ **10.** $1 + \tan^2 \alpha = \csc^2 \alpha \tan^2 \alpha$

11. $\sin(-\alpha) = -\dfrac{\tan \alpha}{\sec \alpha}$ **12.** $\tan(-\alpha) = \dfrac{\cos(-\alpha) \sin(-\alpha)}{1 - \sin^2 \alpha}$

13. $\cos(270° + x) = \sin x$ **14.** $\cos(270° - x) = -\sin x$

15. $\sin(90° + x) = \cos x$ **16.** $\sin(\alpha - 45°) = \dfrac{\sin \alpha - \cos \alpha}{\sqrt{2}}$

17. $\cos(\alpha - 45°) = \dfrac{\cos \alpha + \sin \alpha}{\sqrt{2}}$ **18.** $\tan(\alpha - 45°) = \dfrac{\tan \alpha - 1}{1 + \tan \alpha}$

[4.6]   **19.** $\cos 2x + 1 = 2\cos^2 x$ **20.** $\sin 2x = 2\sin^2 x \cos x \csc x$

21. $\csc x = 2\cos x \csc 2x$ **22.** $\cos 2x = \cos^4 x - \sin^4 x$

■ *By counterexamples, show that each equation is not an identity.*

23. $\sin \alpha \sec \alpha = 2 \tan \alpha$ **24.** $\dfrac{1 - \cos \alpha}{\sin \alpha} = \dfrac{1 + \sin \alpha}{\cos \alpha}$

[4.7] ■ *Use half-angle formulas to find the exact value for each of the following.*

25. $\cos 105°$ **26.** $\tan 165°$

■ *Verify the identities in Exercises 27 and 28.*

27. $\cos^2 \dfrac{x}{2} = \dfrac{\sin^2 x}{2(1 - \cos x)}$ **28.** $\sin^2 \dfrac{x}{2} = \dfrac{\csc x - \cot x}{2 \csc x}$

29. Express $\cos x \cos 5x$ in terms of sums of function values.

30. Express $\sin 6x + \sin 4x$ in terms of products of function values.

INVERSE FUNCTIONS; CONDITIONAL EQUATIONS

In Chapter 1 we solved simple trigonometric equations involving any one of the six trigonometric ratios

$$\text{trig } \alpha = y$$

for *acute* angles α and *positive* values y. We also introduced inverse notation in order to express α explicitly in terms of its related trigonometric ratio. Thus, the equation above can be written as

$$\alpha = \text{Trig}^{-1} y \qquad \text{or} \qquad \alpha = \text{Arctrig } y.$$

Since α was an *acute angle* and there was *only one* α for a given trigonometric ratio y, the equations also defined functions. Such functions are called **inverse trigonometric functions.** (See Appendix A.7 for a discussion of inverse algebraic functions.)

Since we now have extended the domains (values of α) for the trigonometric functions and these functions are periodic, there are an *infinite number* of values of α for a specified element in the range (values of the trigonometric ratios). In this chapter we will restrict the domain of each of the trigonometric functions in such a way that each trigonometric ratio (positive or negative) will be associated with *one and only one value* of α and hence each trigonometric function will have an inverse.

5.1 Inverse Functions

Consider the graph of $y = \sin x$ in Figure 5.1 and note that the *least positive x* for which

$$\sin x = \frac{1}{2}$$

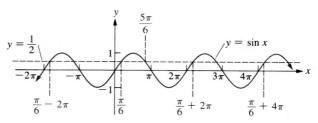

Figure 5.1

is $\pi/6$. However, each value of x such that

$$x = \frac{\pi}{6} + k \cdot 2\pi, \quad k \in J,$$

is also a solution of $\sin x = 1/2$ because the sine function has period 2π. Furthermore, $\sin x = 1/2$ for each value x such that

$$x = \frac{5\pi}{6} + k \cdot 2\pi, \quad k \in J.$$

Inverse of the Function $y = \sin x$, $-\pi/2 \leq x \leq \pi/2$

If we restrict the domain of the sine function

$$y = \sin x \qquad (1)$$

to the interval $-\pi/2 \geq x \geq \pi/2$ (the graph is shown in color in Figure 5.1), we obtain a function that has only one x value, an element in its domain, for any value y in its range. Such a function is called a *one-to-one function* and has an inverse.

As noted in Section A.7, the inverse of a one-to-one function is the function obtained by interchanging the components of every ordered pair in the original function. Hence, the defining equation of the inverse function can be obtained by interchanging the variables in the defining equation of the original function. Furthermore, the graph of the inverse of a function is the reflection of the graph of the one-to-one function with respect to the graph of the equation $y = x$.

If we interchange the variables in Equation (1), we obtain

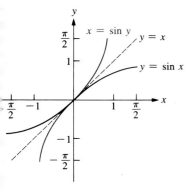

Figure 5.2

$$x = \sin y, \quad -\frac{\pi}{2} \leq y \leq \frac{\pi}{2}, \qquad (2)$$

which defines the inverse of the sine function. Notice that graphs of

$$x = \sin y, \ -\pi/2 \leq y \leq \pi/2, \ \text{and} \ y = \sin x, \ -\pi/2 \leq x \leq \pi/2,$$

are symmetric with respect to the graph of $y = x$ (Figure 5.2).

Inverse Notation   Generally, we wish to write the defining Equation (2) in a form i which y is expressed explicitly in terms of x. This is done by using th inverse notation introduced in Section 1.3. For example,

$$x = \sin y, \quad -\frac{\pi}{2} \le y \le \frac{\pi}{2} \quad \leftrightarrow \quad y = \text{Arcsin } x \quad \leftrightarrow \quad y = \text{Sin}^{-1} x.$$

Similar notation is used for the other inverse circular functions whe the restrictions specified for y limit x to a single value.

The inverse notation corresponding to the six inverse trigono-metric functions is

$$y = \text{Sin}^{-1} x \quad \leftrightarrow \quad x = \sin y, \quad -\frac{\pi}{2} \le y \le \frac{\pi}{2},$$

$$y = \text{Cos}^{-1} x \quad \leftrightarrow \quad x = \cos y, \quad 0 \le y \le \pi,$$

$$y = \text{Tan}^{-1} x \quad \leftrightarrow \quad x = \tan y, \quad -\frac{\pi}{2} < y < \frac{\pi}{2},$$

$$y = \text{Csc}^{-1} x \quad \leftrightarrow \quad x = \csc y, \quad -\frac{\pi}{2} \le y < 0 \quad \text{or} \quad 0 < y \le \frac{\pi}{2},$$

$$y = \text{Sec}^{-1} x \quad \leftrightarrow \quad x = \sec y, \quad 0 \le y < \frac{\pi}{2} \quad \text{or} \quad \frac{\pi}{2} < y \le \pi,$$

$$y = \text{Cot}^{-1} x \quad \leftrightarrow \quad x = \cot y, \quad -\frac{\pi}{2} \le y < 0 \quad \text{or} \quad 0 < y \le \frac{\pi}{2}.$$

Notice that the ranges (values of y) of the inverse functions a specified in the definition rather than following the usual procedure stating the domain (values of x). Furthermore, the choices for the rang are somewhat arbitrary. Those that have been chosen here involve sma values of y and include values between 0 and $\pi/2$, have relatively simp graphs, and yield a one-to-one correspondence between elements in th domain and range. They are also the values that are obtained with mo calculators.

As we noted in Section 1.3, it is sometimes easier to interpret a rel tionship expressed in inverse notation if it is rewritten as an equatic without using such notation.

Example   The equation $y = \text{Arcsin}(-1/2)$ can be rewritten equiv lently without inverse notation as

$$\sin y = -\frac{1}{2}, \quad -\frac{\pi}{2} \le y \le \frac{\pi}{2}.$$

This equation is satisfied for $y = -\pi/6$. Hence,

$$y = \text{Arcsin}\left(-\frac{1}{2}\right) = -\frac{\pi}{6}.$$

The result could also be expressed in degree measure as $-30°$.

With a little practice you will be able to find an inverse function value directly from the expression involving the inverse notation.

Examples   Find the exact value of each of the following.

a. $\text{Arccos}\dfrac{1}{\sqrt{2}}$ b. $\text{Tan}^{-1}\dfrac{1}{\sqrt{3}}$ c. $\text{Csc}^{-1}\dfrac{2}{\sqrt{3}}$

Solutions
a. From the definition on page 152, $\pi/4$ is the number between 0 and $\pi/2$ such that $\cos(\pi/4) = 1/\sqrt{2}$. Hence,

$$\text{Arccos}\,\frac{1}{\sqrt{2}} = \frac{\pi}{4}.$$

b. $\pi/6$ is the number between 0 and $\pi/2$ such that $\tan(\pi/6) = 1/\sqrt{3}$. Hence,

$$\text{Tan}^{-1}\,\frac{1}{\sqrt{3}} = \frac{\pi}{6}.$$

c. Because $\csc x = 1/\sin x$, we want a number x between 0 and $\pi/2$ such that $\csc x = 2/\sqrt{3}$ or $\sin x = \sqrt{3}/2$. Since $\pi/3$ is that number,

$$\text{Csc}^{-1}\,\frac{2}{\sqrt{3}} = \frac{\pi}{3}.$$

Careful attention should be given to the ranges specified in the definition on page 152 when finding values for $\text{Trig}^{-1} x$ for negative values of x.

Examples   Find the exact value of each of the following.

a. $\operatorname{Arccos}\left(-\dfrac{1}{\sqrt{2}}\right)$ b. $\operatorname{Tan}^{-1}\left(-\dfrac{1}{\sqrt{3}}\right)$ c. $\operatorname{Csc}^{-1}\left(-\dfrac{2}{\sqrt{3}}\right)$

Solutions

a. Since $3\pi/4$ is the number between $\pi/2$ and π such that $\cos 3\pi/4 = -1/\sqrt{2}$,

$$\operatorname{Arccos}\left(-\dfrac{1}{\sqrt{2}}\right) = \dfrac{3\pi}{4}.$$

b. Since $-\pi/6$ is the number between 0 and $-\pi/2$ such that $\tan(-\pi/6) = -1/\sqrt{3}$,

$$\operatorname{Tan}^{-1}\left(-\dfrac{1}{\sqrt{3}}\right) = -\dfrac{\pi}{6}.$$

c. Since $-\pi/3$ is the number between 0 and $-\pi/2$ such that $\csc(-\pi/3) = -2/\sqrt{3}$,

$$\operatorname{Csc}^{-1}\left(-\dfrac{2}{\sqrt{3}}\right) = -\dfrac{\pi}{3}.$$

A calculator (or tables) can be used to find approximations for values of the six trigonometric inverse functions for elements in their respective domains that are not exact values. Values for Arcsin x, Arccos x, and Arctan x can be found directly, as we did in Chapter 1.

Examples   Find an approximation for each of the following. Use a calculator in radian mode (or Table III in the Appendix).

a. Arcsin 0.2377   b. $\operatorname{Tan}^{-1}(-0.6989)$

Solutions

a. Arcsin $0.2377 \approx 0.24$.   b. $\operatorname{Tan}^{-1}(-0.6989) \approx -0.61$

Values for Arcsec x, Arccsc x, and Arccot x cannot be obtained directly when using a calculator. However, we can obtain them by using the reciprocal relationships

$$\cos x = \dfrac{1}{\sec x}, \qquad \sin x = \dfrac{1}{\csc x}, \qquad \text{and} \qquad \tan x = \dfrac{1}{\cot x},$$

as we did in Chapter 1. For example, if $y = \text{Arcsec } x$, then $x = \sec y$, so

$$\cos y = \frac{1}{x} \quad \text{or} \quad y = \text{Arccos}\left(\frac{1}{x}\right).$$

Thus,

$$\text{Arcsec } x = \text{Arccos } \frac{1}{x}.$$

Similarly,

$$\text{Arccot } x = \text{Arctan } \frac{1}{x}$$

and

$$\text{Arccsc } x = \text{Arcsin } \frac{1}{x}.$$

Examples Find an approximation for each of the following.

a. Arcsec 1.016   b. $\text{Cot}^{-1}\,1.163$

Solutions

a. $\text{Arcsec } 1.016 = \text{Arccos } \dfrac{1}{1.016}$ b. $\text{Cot}^{-1}\,1.163 = \text{Tan}^{-1}\,\dfrac{1}{1.163}$

$\approx 0.18.$ $\approx 0.71.$

EXERCISE SET 5.1

A

■ *Rewrite each expression without using inverse notation and then specify the exact value of y for which the statement is true.*

Example $y = \text{Arccot } 1$

Solution From the definition of inverse notation (page 152):

$$y = \text{Arccot } 1 \quad \leftrightarrow \quad \cot y = 1, \quad 0 < y \le \frac{\pi}{2}.$$

Hence, $y = \pi/4$.

1. $y = \text{Arccos } \dfrac{\sqrt{3}}{2}$ **2.** $y = \text{Arctan } 1$ **3.** $y = \text{Sin}^{-1}\,1$

4. $y = \mathrm{Csc}^{-1} 2$ **5.** $y = \mathrm{Cot}^{-1} \dfrac{1}{\sqrt{3}}$ **6.** $y = \mathrm{Sec}^{-1} \sqrt{2}$

■ *Find the exact value of each of the following.*

Example $\mathrm{Cos}^{-1}\left(-\dfrac{\sqrt{3}}{2}\right)$

Solution   Let $y = \mathrm{Cos}^{-1}(-\sqrt{3}/2)$. Then, from the definition on pag[e]
152,

$$\cos y = -\frac{\sqrt{3}}{2}, \quad 0 \le y \le \pi,$$

from which

$$y = \mathrm{Cos}^{-1}\left(-\frac{\sqrt{3}}{2}\right) = \frac{5\pi}{6}.$$

7. $\mathrm{Arcsin}\ \dfrac{1}{\sqrt{2}}$ **8.** $\mathrm{Arccos}\ \dfrac{\sqrt{3}}{2}$ **9.** $\mathrm{Arctan}\ 1$

10. $\mathrm{Arcsin}\ 0$ **11.** $\mathrm{Tan}^{-1} \sqrt{3}$ **12.** $\mathrm{Cos}^{-1} 0$

13. $\mathrm{Cos}^{-1} 1$ **14.** $\mathrm{Tan}^{-1} 0$ **15.** $\mathrm{Sec}^{-1} \dfrac{2}{\sqrt{3}}$

16. $\mathrm{Csc}^{-1} 1$ **17.** $\mathrm{Cot}^{-1} 1$ **18.** $\mathrm{Sec}^{-1} 1$

19. $\mathrm{Sin}^{-1}\left(-\dfrac{1}{2}\right)$ **20.** $\mathrm{Cos}^{-1}(-1)$ **21.** $\mathrm{Tan}^{-1}(-\sqrt{3})$

22. $\mathrm{Tan}^{-1}(-1)$ **23.** $\mathrm{Cos}^{-1}\left(-\dfrac{1}{2}\right)$ **24.** $\mathrm{Sin}^{-1}(-1)$

■ *Find an approximation to the nearest hundredth for each of the following[.]*
Use a calculator in radian mode (or Table III in the Appendix).

Examples

a. $\mathrm{Sin}^{-1} 0.8772$ b. $\mathrm{Arcsec}\ 1.238$

Solutions

a. $\mathrm{Sin}^{-1} 0.8772 \approx 1.07$.   b. $\mathrm{Arcsec}\ 1.238 = \mathrm{Arccos}\ \dfrac{1}{1.238}$

$$\approx 0.63.$$

25. Arctan 0.5726 **26.** Arcsin 0.6442 **27.** Arccos 0.8961

28. Arctan 0.1104 **29.** Sin^{-1} 0.3523 **30.** Cos^{-1} 0.6600

31. Tan^{-1} 2.912 **32.** Sin^{-1} 0.8724 **33.** Sec^{-1} 8.299

34. Cot^{-1} 0.6563 **35.** Csc^{-1} 1.077 **36.** Sec^{-1} 1.004

37. $\text{Tan}^{-1}(-0.5463)$ **38.** $\text{Sin}^{-1}(-0.6518)$ **39.** $\text{Cos}^{-1}(-0.9759)$

40. $\text{Cos}^{-1}(-0.3153)$ **41.** $\text{Sin}^{-1}(-0.9927)$ **42.** $\text{Tan}^{-1}(-1.827)$

5.2 More About Inverse Functions

A circular function value is sometimes combined with an inverse function value. Such "composite" function values can be evaluated in the same way that we evaluated each value in earlier sections.

Examples Find the exact value of each of the following.

a. $\text{Cos}^{-1}\left(\sin\dfrac{\pi}{2}\right)$ b. $\text{Tan}^{-1}(\cos 0)$ c. $\sin(\text{Tan}^{-1}\sqrt{3})$

Solutions
a. Since $\sin \pi/2 = 1$,

$$\text{Cos}^{-1}\left(\sin\frac{\pi}{2}\right) = \text{Cos}^{-1}(1) = 0.$$

b. Since $\cos 0 = 1$,

$$\text{Tan}^{-1}(\cos 0) = \text{Tan}^{-1}(1) = \frac{\pi}{4}.$$

c. Since $\text{Tan}^{-1}\sqrt{3} = \pi/3$,

$$\sin(\text{Tan}^{-1}\sqrt{3}) = \sin\frac{\pi}{3} = \frac{\sqrt{3}}{2}.$$

Examples Find approximations for each of the following to the nearest hundredth.

a. $\tan(\text{Cos}^{-1} 0.4321)$ b. $\text{Sin}^{-1}(\cos 0.52)$

Solutions Use a calculator or tables.
a. Since $\text{Cos}^{-1} 0.4321 \approx 1.1240$,

$$\tan(\text{Cos}^{-1} 0.4321) \approx \tan(1.1240) \approx 2.09.$$

(continued)

b. Since cos 0.52 $\approx$ 0.8678,

$$\text{Sin}^{-1}(\cos 0.52) \approx \text{Sin}^{-1} 0.8678 \approx 1.05.$$

Writing Equivalent Equations   Inverse notation for the circular functions enables us to solve equations for one variable in terms of the other variables and constants.

Example   Solve $4 \cos 3x = 1$ for x using inverse notation. Assume restrictions specified in the definition of inverse notation.

Solution   We can multiply each member of

$$4 \cos 3x = 1$$

by 1/4 to obtain

$$\cos 3x = \frac{1}{4}.$$

By the definition on page 152,

$$\cos 3x = \frac{1}{4} \quad \leftrightarrow \quad 3x = \text{Cos}^{-1} \frac{1}{4}.$$

Then, multiplying each member by 1/3, we obtain

$$x = \frac{1}{3} \text{Cos}^{-1} \frac{1}{4}.$$

Of course, if we want to, we can now obtain a decimal approximation for x by first finding an approximation for $\text{Cos}^{-1}(1/4)$.

Example   Solve $y = \frac{1}{3} \tan \frac{x}{2}$ explicitly for x in terms of y using inverse notation. Assume restrictions specified in the definition of inverse notation.

Solution   We can multiply each member of

$$y = \frac{1}{3} \tan \frac{x}{2}$$

by 3 to obtain

$$3y = \tan \frac{x}{2} \quad \leftrightarrow \quad \frac{x}{2} = \text{Tan}^{-1} 3y.$$

Then multiplying each member by 2, we obtain

$$x = 2 \operatorname{Tan}^{-1} 3y.$$

Graphs of the Inverse Trigonometric Functions   The graph of the Arcsine function shown in color in Figure 5.2 on page 151 is shown in Figure 5.3a. The graph of the Arccosine function in (b) is obtained by reflecting the graph of $y = \cos x$, $0 \le y \le \pi$. (See Section 3.1 for the graph of $y = \cos x$.)

The graphs of the other inverse functions in Figure 5.3 are obtained by reflecting the graphs of the respective functions over the appropriate domains. (See Section 3.4 for the graphs of these functions.)

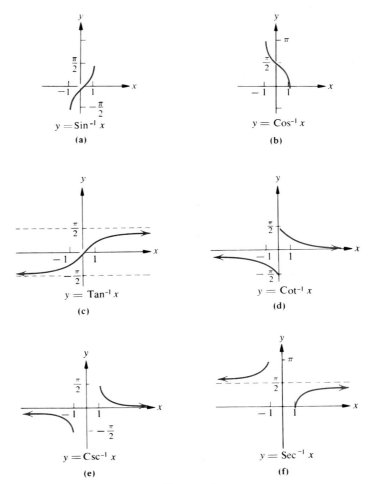

Figure 5.3

EXERCISE SET 5.2

A

■ *Find the exact value of each expression.*

Example $\cos\left(\mathrm{Sin}^{-1}\dfrac{1}{2}\right)$

Solution $\mathrm{Sin}^{-1}\left(\dfrac{1}{2}\right) = \dfrac{\pi}{6};$ therefore

$$\cos\left(\mathrm{Sin}^{-1}\frac{1}{2}\right) = \cos\frac{\pi}{6} = \frac{\sqrt{3}}{2}.$$

1. $\cos\left(\mathrm{Cos}^{-1}\dfrac{\sqrt{3}}{2}\right)$ **2.** $\tan(\mathrm{Tan}^{-1}\,1)$

3. $\tan(\mathrm{Sin}^{-1}\,0)$ **4.** $\cos(\mathrm{Sin}^{-1}\,1)$

5. $\sin[\mathrm{Arctan}(-1)]$ **6.** $\sin\left[\mathrm{Cos}^{-1}\left(-\dfrac{1}{2}\right)\right]$

7. $\mathrm{Sin}^{-1}\left(\sin\dfrac{\pi}{4}\right)$ **8.** $\mathrm{Cos}^{-1}\left(\cos\dfrac{\pi}{3}\right)$

9. $\mathrm{Cos}^{-1}\left(\sin\dfrac{\pi}{3}\right)$ **10.** $\mathrm{Sin}^{-1}\left(\cos\dfrac{\pi}{6}\right)$

11. $\mathrm{Tan}^{-1}\left(\cos\dfrac{\pi}{2}\right)$ **12.** $\mathrm{Csc}^{-1}(\cos\pi)$

13. $\mathrm{Arcsec}\left(\tan\dfrac{3\pi}{4}\right)$ **14.** $\mathrm{Arccot}\left(\sin\dfrac{3\pi}{2}\right)$

■ *Find an approximation to four decimal places for each of the following*

15. $\tan(\mathrm{Sin}^{-1}\,0.8134)$ **16.** $\cos(\mathrm{Tan}^{-1}\,1.432)$

17. $\sin(\mathrm{Cos}^{-1}\,0.4321)$ **18.** $\sin(\mathrm{Tan}^{-1}\,0.632)$

19. $\cot(\mathrm{Cos}^{-1}\,0.5321)$ **20.** $\sec(\mathrm{Sin}^{-1}\,0.6432)$

21. $\csc(\mathrm{Arctan}\,1.632)$ **22.** $\cot(\mathrm{Arccos}\,0.8214)$

23. $\cot[\mathrm{Arccos}(0.3421)]$ **24.** $\csc[\mathrm{Arcsin}(0.6321)]$

■ *Rewrite each equation so that x is expressed explicitly in terms of y and/or the constants using inverse notation. Assume restrictions specified in the definition of inverse notation.*

Example $3 \sin 2x = 1$

Solution   Multiplying each member by 1/3 yields $\sin 2x = 1/3$.

$$\sin 2x = \frac{1}{3} \quad \leftrightarrow \quad 2x = \text{Sin}^{-1} \frac{1}{3}.$$

Then, multiplying each member by 1/2, we obtain

$$x = \frac{1}{2} \text{Sin}^{-1} \frac{1}{3}.$$

25. $4 \cos x = 3$ **26.** $2 \tan x = 3$ **27.** $\sin 3x = \dfrac{1}{4}$

28. $\cos 2x = \dfrac{1}{3}$ **29.** $2 \tan 2x = 5$ **30.** $3 \sin 4x = 2$

31. $2 \cos 3x = 1.24$ **32.** $5 \tan 2x = 2.25$ **33.** $y = 2 \tan 3x$

34. $y = 3 \cos 2x$ **35.** $y = 4 \sin \dfrac{x}{2}$ **36.** $y = 2 \tan \dfrac{x}{3}$

B

■ *Write each of the following as an exact numerical value or an algebraic, expression without using inverse notation. Assume values for the variables for which each function value exists.*

Example $\sin\left(\text{Arccos} \dfrac{3}{5}\right)$

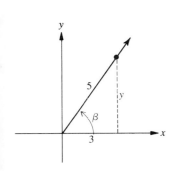

Solution   Because the value of Arccos 3/5 is not evident by inspection, we let $\beta = \text{Arccos } 3/5$, from which $\cos \beta = 3/5$. The sketch shows a right triangle in which $\cos \beta = 3/5$. From the Pythagorean theorem, we have

$$y^2 + 3^2 = 5^2$$
$$y^2 = 25 - 9 = 16$$
$$y = 4.$$

(continued)

Hence,

$$\sin \beta = \sin\left(\text{Arccos } \frac{3}{5} \right)$$

$$= \frac{y}{5} = \frac{4}{5}.$$

37. $\cos\left(\text{Arcsin } \dfrac{3}{5} \right)$

38. $\sin\left(\text{Arccos } \dfrac{5}{13} \right)$

39. $\sin\left(\text{Arcsin } \dfrac{2}{5} \right)$

40. $\cos\left(\text{Arccos } \dfrac{1}{3} \right)$

41. $\tan\left(\text{Arccos } \dfrac{4}{5} \right)$

42. $\tan\left(\text{Arcsin } \dfrac{5}{13} \right)$

43. $\cos\left(\text{Arcsin } \dfrac{y}{3} \right)$

44. $\tan\left(\text{Arccos } \dfrac{x}{4} \right)$

45. $\sin(\text{Arccos } x)$

46. $\sin(\text{Arctan } y)$

■ *Evaluate each of the following.*

Example $\cos\left(\text{Arccot } \dfrac{1}{\sqrt{3}} + \text{Arcsin } \dfrac{1}{\sqrt{2}} \right)$

Solution   Let

$$\alpha = \text{Arccot } \frac{1}{\sqrt{3}} \quad \text{and} \quad \beta = \text{Arcsin } \frac{1}{\sqrt{2}}. \tag{1}$$

Then, using symbols α and β and the sum formula on page 122, we have

$$\cos(\alpha + \beta) = \cos \alpha \cos \beta - \sin \alpha \sin \beta.$$

By inspection of Equations (1), $\alpha = (\pi/3)^{\text{R}}$. Hence,

$$\cos \alpha = \frac{1}{2} \quad \text{and} \quad \sin \alpha = \frac{\sqrt{3}}{2}.$$

Also by inspection of (1), $\beta = (\pi/4)^{\text{R}}$. Thus, we have that

$$\sin \beta = \frac{1}{\sqrt{2}} \quad \text{and} \quad \cos \beta = \frac{1}{\sqrt{2}}.$$

Therefore,

$$\cos\left(\text{Arccot}\,\frac{1}{\sqrt{3}} + \text{Arcsin}\,\frac{1}{\sqrt{2}}\right) = \frac{1}{2}\cdot\frac{1}{\sqrt{2}} - \frac{\sqrt{3}}{2}\cdot\frac{1}{\sqrt{2}}$$

$$= \frac{1-\sqrt{3}}{2\sqrt{2}}.$$

47. $\cos\left(\text{Arcsin}\,\dfrac{1}{2} + \text{Arcsec}\,2\right)$

48. $\cos\left(\text{Arccos}\,\dfrac{1}{2} + \text{Arccsc}\,2\right)$

49. $\cos\left(\text{Arctan}\,1 - \text{Arccos}\,\dfrac{\sqrt{3}}{2}\right)$

50. $\cos\left(\text{Arcsin}\,\dfrac{1}{\sqrt{2}} - \text{Arcsec}\,1\right)$

51. $\sin(\text{Arcsin}\,0 + \text{Arccos}\,1)$

52. $\sin\left(\text{Arcsin}\,\dfrac{1}{2} - \text{Arcsec}\,1\right)$

■ *Graph each function.*   Hint: *Use the graphs shown in Figure 5.3.*

53. $y = \text{Arcsin}\,x + \dfrac{\pi}{2}$

54. $y = \text{Arccos}\,x + \pi$

55. $y = \text{Arctan}\,x + \dfrac{\pi}{2}$

56. $y = \text{Arctan}\,x + \pi$

5.3 Conditional Equations

In Sections 4.1–4.7 we considered identities—equations that are satisfied by all angle or all real number replacements for which both of the members are defined. We now consider equations that are *conditional*—that is, they are satisfied only for particular replacements for the variable.

Finding Values of Angles α, $0° \le \alpha < 360°$

We have already solved many simple conditional equations, such as

$$\sin \alpha = \frac{1}{2}, \tag{1}$$

in Chapter 1 over the interval $0° \le \alpha \le 90°$. In this case

$$\alpha = \text{Arcsin}\,\frac{1}{2} = 30°.$$

However, over the interval $0°$ to $360°$, there are two elements in the domain of the sine function corresponding to $1/2$ in the range; either

$$\alpha_1 = 30° \quad \text{or} \quad \alpha_2 = 150°$$

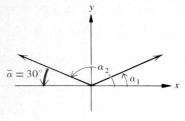

Figure 5.4

satisfies Equation (1), as shown in Figure 5.4, where $\tilde{\alpha} = 30°$. Thus, th
solution set of (1) is $\{30°, 150°\}$.

In general, for most elements in the range of a trigonometric functio
there are two corresponding elements in its domain over the interval (
to 360° or 0^R to $2\pi^R$. A reference angle can be used to find all elements
the domain of a trigonometric function over those intervals that satisfy
given condition. For any positive function value, the angle obtaine
from a table or a calculator is an angle in the first quadrant and is th
reference angle that can be used to find a possible second value tha
satisfies the specified condition.

Example   Given that $\tan \alpha = 1$, find the following for which the stat
ment is true.

    a. The least nonnegative angle α in degrees.

    b. All α such that $0° \le \alpha < 360°$.

Solution   The reference angle $\tilde{\alpha}$, for which

$$\tan \tilde{\alpha} = \tan \alpha = 1,$$

is

$$\tilde{\alpha} = \text{Tan}^{-1} 1 = 45°.$$

Since $\tan \alpha$ (equal to 1) is positive, α is in Quadrant I or Quadrant II
A sketch showing a reference angle is helpful.

    a. The least nonnegative angle α is

$$\alpha = \tilde{\alpha} = 45°.$$

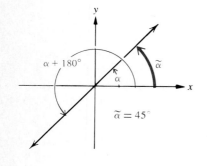

    b. Because the period of the tangent function is 180°,

$$\alpha + 180° = 45° + 180° = 225°.$$

Hence, the solution set is $\{45°, 225°\}$.

A procedure similar to the one used in the above example can also b
used in cases in which the function value is negative. In these cases it
helpful to work with the absolute value of the function value to obtai
the reference angle.

Example   Given that $\sin \alpha = -0.6494$, find the following for whic
the statement is true.

    a. The least nonnegative angle α in degrees.

    b. All α such that $0° \le \alpha < 360°$.

Solution   The reference angle $\tilde{\alpha}$, for which

$$\sin \tilde{\alpha} = |\sin \alpha| = |-0.6494| = 0.6494,$$

is

$$\tilde{\alpha} = \text{Sin}^{-1} \, 0.6494 \approx 40.5°.$$

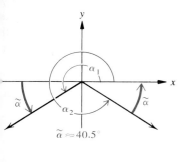

Since $\sin \alpha$ is negative, α is in Quadrant III or IV. A sketch showing the reference angle in each of these quadrants is helpful.

a. The least nonnegative angle is

$$\alpha_1 = 180° + \tilde{\alpha}$$

$$\approx 180° + 40.5° = 220.5°.$$

b. In Quadrant IV,

$$\alpha_2 = 360° - \tilde{\alpha}$$

$$\approx 360° - 40.5° = 319.5°.$$

Hence, the solution set is $\{220.5°, 319.5°\}$.

Example   Given that $\cos \alpha = -0.7518$, find the following for which the statement is true.

a. The least nonnegative angle α in radians.

b. All α such that $0^R \leq \alpha < 2\pi^R$.

Solution   The reference angle $\tilde{\alpha}$, for which

$$\cos \tilde{\alpha} = |\cos \alpha| = |-0.7518| = 0.7518,$$

is

$$\tilde{\alpha} = \text{Cos}^{-1} \, 0.7518 \approx 0.72^R.$$

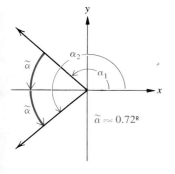

Since $\cos \alpha$ is negative, α is in Quadrant II or Quadrant III. A sketch showing the reference angle in each of these quadrants is helpful.

a. The least nonnegative angle is

$$\alpha_1 = \pi^R - \tilde{\alpha}$$

$$\approx (3.14 - 0.72)^R = 2.42^R.$$

b. In Quadrant III,

$$\alpha_2 = \pi^R + \tilde{\alpha}$$

$$\approx (3.14 + 0.72)^R = 3.86^R.$$

Hence, the solution set is $\{2.42^R, 3.86^R\}$.

Finding All Values of Angles   The trigonometric functions are periodic. Hence, we can obtain th
infinite set of values for angles α for a given trigonometric function valu
by first finding the value of α in the interval $0° \leq \alpha < 360°$ (or in th
interval $0^R \leq x < 2\pi^R$) by the methods used in the above examples.

Example   Find the set of all angles α (in degree measure) for whic
$\cos \alpha = 1/\sqrt{2}$.

Solution   The reference angle $\tilde{\alpha}$, for which

$$\cos \tilde{\alpha} = |\cos \alpha| = \frac{1}{\sqrt{2}}$$

is

$$\tilde{\alpha} = \text{Cos}^{-1} \frac{1}{\sqrt{2}} = 45°.$$

Since $\cos \alpha$ is positive, α is in Quadrants I or IV. The figure show
the reference angles in each of these quadrants. The least nonnegativ
angle is

$$\alpha_1 = 45°.$$

In Quadrant IV,

$$\alpha_2 = 360° - \tilde{\alpha}$$

$$= 360° - 45° = 315°.$$

Hence, the solutions are

$$\alpha = 45° + k \cdot 360° \qquad \text{or} \qquad \alpha = 315° + k \cdot 360°,$$

where $k \in J$.

Finding Values of x, $x \in R$ In the above discussion and examples we viewed the elements in th
domain associated with a given element in the range as angles wit
degree or radian measures. Similar procedures are used if the elements i
the domain are viewed as real numbers.

Example   Find the set of all real numbers for which $\tan x = -\sqrt{}$

Solution   The reference arc $\tilde{x}$, for which

$$\tan \tilde{x} = |\tan x| = \sqrt{3}$$

(figure annotations: y, x, α_2, α_1, $\tilde{\alpha}$, $\tilde{\alpha}$, $\tilde{\alpha} = 45°$)

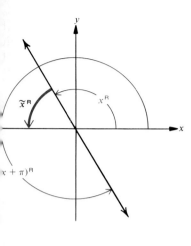

is

$$\tilde{x} = \text{Tan}^{-1} \sqrt{3} = \frac{\pi}{3}.$$

Since tan x is negative, x is a real number that can be associated with an angle in Quadrants II or IV. The figure shows $\tilde{x}$. The least nonnegative x is

$$x = \pi - \frac{\pi}{3} = \frac{2\pi}{3}.$$

Since the tangent function has period π,

$$x = \frac{2\pi}{3} + k \cdot \pi, \quad k \in J.$$

EXERCISE SET 5.3

A

■ *For each function value, find* (**a**) *the exact value of the least nonnegative angle* α *in degrees and* (**b**) *all* α *such that* $0° \le \alpha < 360°$ *for which the statement is true.*

Example $\sin \alpha = -\dfrac{\sqrt{3}}{2}$

Solution   The reference angle $\tilde{\alpha}$, for which

$$\sin \tilde{\alpha} = |\sin \alpha| = \left| -\frac{\sqrt{3}}{2} \right| = \frac{\sqrt{3}}{2},$$

is

$$\tilde{\alpha} = \text{Sin}^{-1} \frac{\sqrt{3}}{2} = 60°.$$

Since $\sin \alpha$ (equal to $-\sqrt{3}/2$) is negative, α is in Quadrant III or Quadrant IV. A sketch showing the reference angle in each of these quadrants is helpful.

a. The least nonnegative angle in Quadrant III is

$$\alpha_1 = 180° + \tilde{\alpha} = 180° + 60° = 240°.$$

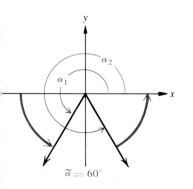

(*continued*)

b. The angle in Quadrant IV is

$$\alpha_2 = 360° - 60° = 300°.$$

Hence, the solution set is $\{240°, 300°\}$.

1. $\cos \alpha = \dfrac{\sqrt{3}}{2}$ **2.** $\sin \alpha = \dfrac{1}{\sqrt{2}}$ **3.** $\tan \alpha = -\dfrac{1}{\sqrt{3}}$

4. $\cos \alpha = \dfrac{1}{2}$ **5.** $\cot \alpha = \sqrt{3}$ **6.** $\tan \alpha = 1$

7. $\sec \alpha = \sqrt{2}$ **8.** $\cot \alpha = -1$ **9.** $\csc \alpha = -2$

10. $\sec \alpha = 2$ **11.** $\sin \alpha = -\dfrac{1}{2}$ **12.** $\csc \alpha = \dfrac{2}{\sqrt{3}}$

13. $\cos \alpha = -\dfrac{1}{\sqrt{2}}$ **14.** $\sin \alpha = -\dfrac{\sqrt{3}}{2}$ **15.** $\csc \alpha = \sqrt{2}$

16. $\cos \alpha = -1$ **17.** $\sin \alpha = 0$ **18.** $\tan \alpha = -\sqrt{3}$

19. $\sec \alpha = -\dfrac{2}{\sqrt{3}}$ **20.** $\cot \alpha = 0$ **21.** $\csc \alpha$ undefined

22. $\sec \alpha$ undefined   **23.** $\cot \alpha$ undefined   **24.** $\tan \alpha$ undefined

■ *For each function value, find an approximation in degrees for* (**a**) *the lea nonnegative angle* α *and* (**b**) *all* α *such that* $0° \le \alpha < 360°$ *for which eac statement is true.*

Example $\tan \alpha = -0.8693$

Solution   The reference angle $\tilde{\alpha}$, for which

$$\tan \tilde{\alpha} = |\tan \alpha| = |-0.8693| = 0.8693,$$

is

$$\tilde{\alpha} = \text{Tan}^{-1}\, 0.8693 \approx 41°.$$

Since $\tan \alpha$ (equal to -0.8693) is negative, α is in Quadrant II or Qua rant IV. A sketch showing a reference angle is helpful.

a. The least nonnegative angle in Quadrant II is

$$\alpha = 180° - \tilde{\alpha}$$

$$\approx 180° - 41° = 139°.$$

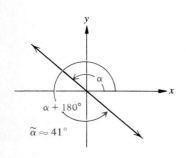

$\tilde{\alpha} \approx 41°$

b. Because the period of the tangent function is 180°,

$$\alpha + 180° \approx 139° + 180° = 319°.$$

Hence, the solution set is $\{139°, 319°\}$.

25. $\sin \alpha = 0.2419$	**26.** $\cos \alpha = 0.5150$	**27.** $\tan \alpha = -2.6051$
28. $\cot \alpha = -2.1445$	**29.** $\sec \alpha = 1.0320$	**30.** $\csc \alpha = 1.1570$
31. $\cos \alpha = -0.9085$	**32.** $\sin \alpha = -0.3535$	**33.** $\cot \alpha = 2.1060$
34. $\tan \alpha = 0.3134$	**35.** $\csc \alpha = -1.2796$	**36.** $\sec \alpha = -3.8140$
37. $\sin \alpha = -0.5476$	**38.** $\cot \alpha = 0.6032$	**39.** $\cos \alpha = 0.3955$
40. $\csc \alpha = -2.3400$	**41.** $\tan \alpha = 1.0212$	**42.** $\sec \alpha = 1.6316$

■ *For each function value, find an approximation in radians for* (**a**) *the least nonnegative angle* α *and* (**b**) *all* α *such that* $0^R \leq \alpha < 2\pi^R$ *for which each statement is true.*

Example $\cos \alpha = -0.9131$

Solution   The reference angle $\tilde{\alpha}$, for which

$$\cos \tilde{\alpha} = |\cos \alpha| = |-0.9131| = 0.9131,$$

is

$$\tilde{\alpha} = \text{Cos}^{-1} \, 0.9131 \approx 0.42^R.$$

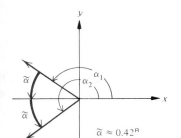

Since $\cos \alpha$ (equal to -0.9131) is negative, α is in Quadrant II or Quadrant III. A sketch showing the reference angle in each of these quadrants is helpful.

a. The least nonnegative angle in Quadrant II is

$$\alpha_1 = \pi^R - \tilde{\alpha} \approx \pi^R - 0.42^R \approx 2.72^R.$$

b. The angle in Quadrant III is

$$\alpha_2 \approx \pi^R + 0.42^R \approx 3.56^R.$$

43. $\sin \alpha = 0.6442$	**44.** $\cos \alpha = 0.5898$	**45.** $\tan \alpha = 0.2341$
46. $\sin \alpha = 0.2280$	**47.** $\sec \alpha = 1.795$	**48.** $\csc \alpha = 1.467$
49. $\cos \alpha = -0.9689$	**50.** $\sin \alpha = -0.5480$	**51.** $\sin \alpha = -0.1197$
52. $\tan \alpha = -1.050$	**53.** $\csc \alpha = -2.125$	**54.** $\sec \alpha = -5.273$

■ *For each function value, find an approximation to the nearest tenth of a degree for* **(a)** *the least nonnegative angle* α *and* **(b)** *all* α *such that* $0° \le \alpha < 360°$ *for which each statement is true.*

Example $\tan \alpha = -0.2162$

Solution   The reference angle $\tilde{\alpha}$, for which

$$\tan \tilde{\alpha} = |\tan \alpha| = |-0.2162| = 0.2162,$$

is

$$\tilde{\alpha} = \text{Tan}^{-1}\, 0.2162 \approx 12.2°.$$

Since $\tan \alpha$ (equal to -0.2162) is negative, α is in Quadrant II or IV. The figure shows a reference angle. The least nonnegative angle is

$$\alpha = 180° - \tilde{\alpha}$$
$$\approx 180° - 12.2° = 167.8°.$$

Because the period of the tangent function is 180°,

$$\alpha \approx 167.8 + k \cdot 180°, \quad k \in J.$$

55. $\sin \alpha = 0.2790$ 　　　　　　**56.** $\cos \alpha = 0.9311$

57. $\cot \alpha = -1.2799$ 　　　　　**58.** $\tan \alpha = -1.4882$

59. $\csc \alpha = -1.2813$ 　　　　　**60.** $\sec \alpha = -1.1164$

61. $\tan \alpha = 0.3019$ 　　　　　　**62.** $\sin \alpha = -0.9426$

5.4 Additional Conditional Equations

In Section 5.3 we obtained the solution set over various domains of simple conditional equations such as

$$\sin x = \frac{1}{2}, \qquad \cos x = -0.3824, \qquad \text{and} \qquad \tan x = 1.752$$

in which the left-hand member contained only one function value and that function value was of first degree. In the case $\sin x = 1/2$, we obtained solutions by inspection.

Perhaps you can determine the solution of

$$\sin x = \cos x, \quad 0 < x < \frac{\pi}{2},$$

by inspection—perhaps not. However, you can generate equivalent equations, and hopefully you will generate one from which the solution(s) will be evident by inspection. Generally, solutions can be determined most easily from equations that involve *values of one function only*. The identities developed in the preceding sections are useful in obtaining such equivalent equations from equations involving values of more than one function. For example, in (1) above, for $0 < x < \pi/2$, $\cos x \neq 0$. Thus, we can multiply each member by $1/\cos x$ to obtain

$$\frac{\sin x}{\cos x} = 1.$$

Substituting $\tan x$ for $\sin x/\cos x$, we obtain

$$\tan x = 1.$$

Since, in this case, values of x are restricted to $0 < x < \pi/2$,

$$x = \text{Tan}^{-1}\, 1 = \frac{\pi}{4}.$$

Example   Solve $2 \sin x \cos x - \sin x = 0$ over $0 \leq x \leq \pi/2$.

Solution   Factoring the left-hand member, we obtain the equivalent equation

$$\sin x(2 \cos x - 1) = 0.$$

The left-hand member equals zero if

$$\sin x = 0 \qquad \text{or} \qquad \cos x = \frac{1}{2},$$

from which

$$x = \text{Sin}^{-1}\, 0 = 0 \qquad \text{or} \qquad x = \text{Cos}^{-1}\, \frac{1}{2} = \frac{\pi}{3},$$

and the solution set over the interval $0 \leq x \leq \pi/2$ is $\left\{0, \dfrac{\pi}{3}\right\}$.

Because function values ($\cos x$, $\sin x$, etc.) are real numbers, equations that are quadratic in a given trigonometric function value, such as

$$\cos^2 x + \cos x - 2 = 0, \quad x \in A, \tag{2}$$

can be solved by first solving the quadratic equation for the functio
value involved. To emphasize that we are using a standard algebrai
method for solving a quadratic equation, we also show each step c
the process, where we let $\cos x = u$ and write Equation (2) as

$$u^2 + u - 2 = 0. \tag{3}$$

Since the left-hand members of Equations (2) and (3) are factorable, th
equations can be written equivalently as

$$(\cos x + 2)(\cos x - 1) = 0 \qquad \text{or} \qquad (u + 2)(u - 1) = 0.$$

Since $\cos x + 2$ cannot equal 0 (there is no angle x such tha
$\cos x = -2$), we seek values of x for which

$$\cos x - 1 = 0 \qquad \text{or} \qquad u - 1 = 0.$$

From either equation, we have

$$\cos x = 1.$$

The least nonnegative value of x is given by

$$x = \text{Cos}^{-1} 1 = 0.$$

Hence, the solutions of Equation (2) are

$$x = k \cdot 360° \qquad \text{or} \qquad x = k \cdot 2\pi^R, \quad k \in J.$$

EXERCISE SET 5.4

A

■ *Solve each equation. Find all solutions in the interval* $0 \le x \le \pi/2$.

Examples

a. $(2 \cos x - \sqrt{3})(\sin x - 1) = 0$ \qquad b. $\sqrt{2} \sin x \cos x - \cos x = 0$

Solutions

a. The left-hand member equals 0 if

$$2 \cos x - \sqrt{3} = 0 \qquad \text{or} \qquad \sin x - 1 = 0.$$

From these equations

$$\cos x = \frac{\sqrt{3}}{2} \quad \text{or} \quad \sin x = 1,$$

and the solution set is $\{\pi/6, \pi/2\}$.

b. Factoring the left-hand member

$$\cos x(\sqrt{2} \sin x - 1) = 0.$$

The left-hand member equals 0 if

$$\cos x = 0 \quad \text{or} \quad \sin x = \frac{1}{\sqrt{2}},$$

and the solution set is $\{\pi/2, \pi/4\}$.

1. $2 \sin x - 1 = 0$

2. $2 \cos x - 1 = 0$

3. $2 \cos^2 x = 1$

4. $4 \sin^2 x = 3$

5. $3 \csc^2 x - 3 = 0$

6. $\sec^2 x = 4$

7. $(2 \cos^2 x - 1)(\sin x - 1) = 0$

8. $(\sin^2 x - 1)(\sec x - 2) = 0$

9. $\cos x \sin x = 0$

10. $3 \tan x \sec x = 0$

11. $2 \cos^3 x - \cos x = 0$

12. $2 \sin^3 x - \sin x = 0$

13. $2 \sin x \cos x + \sin x = 0$

14. $2 \sin x \cos x - \cos x = 0$

■ *Solve each equation. Find all solutions in the interval* $0 \le x < 2\pi$.

Example $2 \cos^2 x + \cos x - 1 = 0$

Solution   Factoring the left-hand member,

$$(2 \cos x - 1)(\cos x + 1) = 0,$$

$$\cos x = \frac{1}{2} \quad \text{or} \quad \cos x = -1,$$

and the solution set is $\{\pi/3, 5\pi/3, \pi\}$.

15. $2 \cos^2 x - \cos x - 1 = 0$

16. $2 \cos^2 x + 3 \cos x - 2 = 0$

17. $9 \cos^2 x - 21 \cos x - 8 = 0$

18. $25 \sin^2 x - 20 \sin x + 4 = 0$

▪ *Solve each equation* (**a**) *in the interval* $0° \leq x < 360°$ *and* (**b**) *for a* $x \in A.$

Example $\sec x - 2 \cos x = 1$

Solution
   a. First write each term using the same function value. Substitut $1/\cos x$ for $\sec x$ to obtain

$$\frac{1}{\cos x} - 2 \cos x = 1.$$

Multiplying each member by $\cos x$ $(\cos x \neq 0),$

$$1 - 2 \cos^2 x = \cos x,$$

from which

$$2 \cos^2 x + \cos x - 1 = 0.$$

This equation is the same as the one in the preceding example. Thus, th solution set over the interval $0° \leq x < 360°$ is $\{60°, 300°, 180°\}.$
   b. For all $x \in A,$ the solutions are

$$x = 60° + k \cdot 360°$$
$$x = 180° + k \cdot 360°$$
$$x = 300° + k \cdot 360°, \quad k \in J.$$

19. $\csc x + \sin x + 2 = 0$

20. $2 \cos x + \sec x - 3 = 0$

21. $2 \sin^2 x - \cos x - 1 = 0$

22. $2 \cos^2 x - 3 \sin x - 3 = 0$

23. $2 \sin x + \csc x - 3 = 0$

24. $\tan x + \cot x + 2 = 0$

B

▪ *Solve each equation* (**a**) *in the interval* $0° \leq x < 360°$ *and* (**b**) *fo all* $x \in A.$ *Approximate answers to the nearest tenth of a degree Hint: Use the quadratic formula.*

25. $\sin^2 x - \sin x - 1 = 0$

26. $\cos^2 x + 5 \cos x + 2 = 0$

27. $3 \cos x - 1 = \sec x$

28. $4 \sin x \cos x - 2 \sin x = \tan x$

■ *Solve each equation in the interval* $0 \leq x \leq 2\pi$.   *Hint: Square each member.*

29. $1 + \cos x = \sqrt{3} \sin x$ **30.** $\cos x + \sin x = 1$

31. $\cot x = 1 + \csc x$ **32.** $1 + \tan x = -\sqrt{2} \sec x$

■ *Solve each equation for all* $x \in R$.

33. $2 \cot x \sec x + \cot x + 2 \sec x = -1$

34. $\cos x \tan x - \cos x + \sin x \tan x - \sin x = 0$

5.5 Conditional Equations for Multiples

Conditional equations such as

$$2 \sin x \cos x = \frac{\sqrt{3}}{2}, \tag{1}$$

$$\tan 3x = 1, \tag{2}$$

and

$$\cos 2x = \cos^2 x - 1 \tag{3}$$

need further discussion.

The identities involving multiples of the variable are often useful to help solve certain kinds of equations. For example, the solution set of

$$2 \sin x \cos x = \frac{\sqrt{3}}{2}, \tag{1}$$

over the interval $0 \leq x \leq \pi/2$, can readily be found if $\sin 2x$ is substituted for the left-hand member, $2 \sin x \cos x$, to obtain

$$\sin 2x = \frac{\sqrt{3}}{2}.$$

By inspection we have that

$$2x = \frac{\pi}{3} + k \cdot 2\pi \quad \text{or} \quad 2x = \frac{2\pi}{3} + k \cdot 2\pi,$$

$$x = \frac{\pi}{6} + k \cdot \pi \qquad\qquad x = \frac{\pi}{3} + k \cdot \pi.$$

If $k = 0$, $x = \pi/6$ or $x = \pi/3$; if $k \geq 1$, then $x > \pi$. Hence, th
solution set of Equation (1) over the interval from 0 to $\pi/2$ is $\{\pi/6, \pi/3$
Now consider the solution set of

$$\tan 3x = 1 \tag{}$$

over the interval $0 \leq x < 2\pi$. By inspection we have that

$$3x = \frac{\pi}{4} + k \cdot \pi,$$

and, dividing each term by 3,

$$x = \frac{\pi}{12} + k \cdot \frac{\pi}{3}, \quad k \in J.$$

To find the solution set over the interval $0 \leq x < 2\pi$, we take 0, 1, 2, 3,
and 5 as replacements for k and obtain

$$\left\{ \frac{\pi}{12}, \frac{5\pi}{12}, \frac{3\pi}{4}, \frac{13\pi}{12}, \frac{17\pi}{12}, \frac{7\pi}{4} \right\}.$$

Note that if $k > 5$, then $x > 2\pi$.
Now consider the equation

$$\cos 2x = \cos^2 x - 1. \tag{}$$

One method of solving this equation for $x \in R$ consists of substitutin;
$2 \cos^2 x - 1$ for $\cos 2x$ to obtain the equivalent equations

$$2 \cos^2 x - 1 = \cos^2 x - 1,$$
$$\cos^2 x = 0.$$

Then,

$$x = \frac{\pi}{2} + k\pi, \quad k \in J.$$

Alternatively, we could have substituted $(1 + \cos 2x)/2$ for $\cos^2 x$ in
the right-hand member of Equation (3) to obtain the equivalen
equations

$$\cos 2x = \frac{1 + \cos 2x}{2} - 1,$$
$$2 \cos 2x = 1 + \cos 2x - 2,$$
$$\cos 2x = -1.$$

By inspection, we have that

$$2x = \pi + k \cdot 2\pi, \quad k \in J,$$

from which the solution is

$$x = \frac{\pi}{2} + k\pi, \quad k \in J,$$

as we found above using the other method.

EXERCISE SET 5.5

A

■ *Solve each equation for all x, such that* $0 \leq x \leq \pi/2$.

Example $\cos 3x = \dfrac{1}{2}$

Solution   Using the fact that the cosine of an angle in Quadrant I or IV is positive, we first consider all possible solutions and obtain

$$3x = \frac{\pi}{3} + k \cdot 2\pi \quad \text{or} \quad 3x = \frac{5\pi}{3} + k \cdot 2\pi.$$

From which, dividing each term by 3,

$$x = \frac{\pi}{9} + k \cdot \frac{2\pi}{3} \quad \text{or} \quad x = \frac{5\pi}{9} + k \cdot \frac{2\pi}{3}.$$

Since the problem restricts x to the interval $0 \leq x \leq \pi/2$, we need only the value 0 for k. If $k = 0$, $x = \pi/9$ or $x = 5\pi/9$. Furthermore, since $5\pi/9$ is not in the required interval, the solution set is $\{\pi/9\}$.

1. $\sin 3x = 0$

2. $\cos 3x = 1$

3. $\tan 4x = -\sqrt{3}$

4. $\cot 4x = -1$

5. $\sec 5x = 2$

6. $\csc 5x = \sqrt{2}$

7. $\cos 2x - \sin x = 0$

8. $\cos 2x + \cos x = 0$

9. $\sin 2x + \sin x = 0$

10. $\sin 2x - \cos x = 0$

11. $\sin 2x = \cos 2x$

12. $\sin \dfrac{x}{2} = \tan \dfrac{x}{2}$

13. $\tan \dfrac{x}{2} - \cos x = -1$

14. $\cos x = \sin \dfrac{x}{2}$

■ *Solve each equation in the interval* $0° \leq x < 360°$.

Example $\sin 4x = \dfrac{\sqrt{3}}{2}$

Solution   By inspection we have $4x = 60°$. However, since $\sin \theta$ is als[o] positive for angles in Quadrant II, we know that $4x = 120°$ will als[o] satisfy the equation. Thus, for $k \in J$.

$$4x = 60° + k \cdot 360° \qquad \text{or} \qquad 4x = 120° + k \cdot 360°.$$

Dividing each member of both equations by 4, we obtain

$$x = 15° + k \cdot 90° \qquad \text{or} \qquad x = 30° + k \cdot 90°.$$

Since the problem restricts x to the interval $0° \leq x < 360°$, we nee[d] only the values of 0, 1, 2, and 3 for k.

If $k = 0$, $x = 15°$ or $x = 30°$,

if $k = 1$, $x = 105°$ or $x = 120°$,

if $k = 2$, $x = 195°$ or $x = 210°$,

if $k = 3$, $x = 285°$ or $x = 300°$,

and the solution set is $\{15°, 30°, 105°, 120°, 195°, 210°, 285°, 300°\}$.

15. $\sin 3x = \dfrac{1}{2}$

16. $\cos 4x = -1$

17. $\sin 3x = \cos 3x$

18. $\tan 2x = \cot 2x$

19. $\sin^2 2x = 1$

20. $\cos^2 2x = 1$

21. $\tan^2 3x = 1$

22. $\cot^2 3x = 3$

23. $\left(1 + \cos \dfrac{x}{2}\right)\cos \dfrac{x}{2} = 0$

24. $\sin^2 \dfrac{x}{4} - \dfrac{1}{2}\sin \dfrac{x}{4} = 0$

25. $5 \sec 3x = 2 \cos 3x - 3$

26. $\tan x \tan 2x = 1$

B

■ *Solve each equation for all* x, $x \in R$.

27. $\sin 2x = \cos x$

28. $2 \cos 2x - 1 = -3 \sin^2 x$

29. $\sin x \cos 3x = 0$

30. $(\sin 2x - 1)(\cos 3x + 1) = 0$

31. $\tan 2x - 3 \sec^2 2x = -5$ **32.** $\cos^2 2x + 3 \sin 2x - 3 = 0$

33. $\sin 2x \cos x + \cos 2x \sin x = \dfrac{1}{\sqrt{2}}$

34. $\sin 3x \cos x - \cos 3x \sin x = \dfrac{\sqrt{3}}{2}$

35. Use graphical methods to approximate a positive solution for

$$\cos 2x = x, \quad x \in R.$$

Hint: Graph $y = \cos 2x$ and $y = x$ on the same axes.

36. Use graphical methods to approximate a positive solution for

$$x^2 - 3 \sin \frac{x}{2} = 0, \quad x \in R.$$

5.6 Systems; Parametric Equations

In Sections 5.3–5.5 we solved trigonometric equations in one variable. The solutions were sets of numbers or angles. In this section we consider equations in two variables and in particular systems of such equations. The solution of a system is the intersection of the solution sets of the equations in the system. In many cases, we can obtain solutions by algebraic methods; we can check the solutions by graphical methods.

Example
  a. Solve the system

$$y = 2 \sin x \tag{1}$$

$$y = \frac{3}{2} \tag{2}$$

algebraically in the interval $0 \le x \le 2\pi$.

  b. Graph the system.

Solution
  a. Substituting 3/2 for y in (1), we have

$$\frac{3}{2} = 2 \sin x,$$

(*continued*)

from which

$$\sin x = \frac{3}{4}$$

and

$$x = \mathrm{Sin}^{-1}\frac{3}{4} \approx 0.85.$$

Since x can also be in the interval $\pi/2 \leq x \leq \pi$,

$$x \approx 3.14 - 0.85 = 2.29.$$

For both values of x, $y = 3/2 = 1.5$. Hence, the solution set is

$$\{(0.8, 1.5), (2.3, 1.5)\},$$

where the values are given to the nearest tenth.

b.

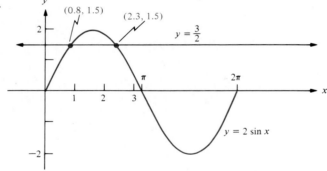

Example
a. Solve the system

$$y = \sin x \qquad (1)$$

$$y = \cos 2x \qquad (2)$$

algebraically in the interval $0 \leq x \leq \pi$.

b. Graph the system.

Solution
a. Substituting $\cos 2x$ for y in (1), we have

$$\cos 2x = \sin x,$$

from which

$$1 - 2 \sin^2 x = \sin x,$$

$$2 \sin^2 x + \sin x - 1 = 0,$$

$$(2 \sin x - 1)(\sin x + 1) = 0.$$

Hence,

$$2 \sin x - 1 = 0 \qquad \text{or} \qquad \sin x + 1 = 0$$

$$x = \text{Sin}^{-1} \frac{1}{2} = \frac{\pi}{6} \qquad\qquad x = \text{Sin}^{-1}(-1) = -\frac{\pi}{2}$$

or, in Quadrant II, $x = 5\pi/6$. $(-\pi/2$ is not in the specified interval.) Substituting these values of x in (1), we obtain

$$y = \sin \frac{\pi}{6} = \sin \frac{5\pi}{6} = \frac{1}{2}.$$

Hence, the solution set is

$$\left\{ \left(\frac{\pi}{6}, \frac{1}{2} \right), \left(\frac{5\pi}{6}, \frac{1}{2} \right) \right\}.$$

b.

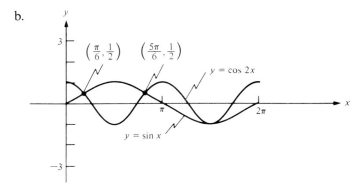

Parametric Equations   Sometimes problems arise that can best be described by mathematical models that involve expressing two or more variables in terms of a different variable, usually t for time or α for an angle. A pair of equations of the form

$$x = f(t) \qquad \text{and} \qquad y = h(t)$$

t	x	y	(x, y)
0	0	0	$(0, 0)$
1	2	1	$(2, 1)$
2	4	4	$(4, 4)$
3	6	9	$(6, 9)$

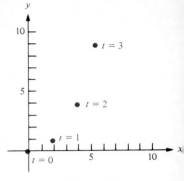

Figure 5.5

is often involved. Such equations are called **parametric equations;** t called the **parameter.**

For example, consider a particle moving on a path that is not ne essarily a straight line. During this motion the particle will, in genera move in horizontal and vertical directions simultaneously. For exam ple, if a particle is tracked for 3 seconds (see Figure 5.5) and its horizon tal travel (in meters) is observed to be always equal to twice the elapse time (in seconds), then at any time t, where $0 \le t \le 3$, we could wri $x = 2t$. Similarly, if the vertical travel (in meters) is observed to l always equal to the square of the elapsed time (in seconds), we cou write $y = t^2$. Thus, we have the system of parametric equations

$$\left. \begin{array}{l} x = 2t \\ y = t^2 \end{array} \right\} \quad 0 \le t \le 3,$$

with the parameter t.

In the above example we can eliminate the parameter t to obtain Cartesian equation in x and y. Solving Equation (1) for t, we obta $t = x/2$. Substituting $x/2$ for t in (2) yields

$$y = \left(\frac{x}{2} \right)^2 = \frac{1}{4} x^2,$$

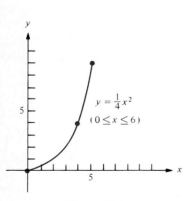

Figure 5.6

an equation of a parabola. Since $0 \le t \le 3$, from Equation (1) it follo that $0 \le x \le 6$. The graph of the motion of the particle can be obtaine from Equation (3) with the restriction $0 \le x \le 6$ as shown in Figure 5.

Some parametric equations that involve trigonometric ratios can al be written equivalently as a Cartesian equation, whose graph can easily obtained.

Example   Find a Cartesian equation equivalent to the parametric equations

$$x = 3 \cos t \quad \text{and} \quad y = 4 \sin t. \tag{1}$$

Solution   Motivated by the fact that $\cos^2 t + \sin^2 t = 1$, we first solve the given equations for $\cos t$ and $\sin t$.

$$x = 3 \cos t \quad \leftrightarrow \quad \cos t = \frac{x}{3};$$

$$y = 4 \sin t \quad \leftrightarrow \quad \sin t = \frac{y}{4}.$$

Hence,

$$\cos^2 t + \sin^2 t = \frac{x^2}{9} + \frac{y^2}{16}.$$

Because $\cos^2 t + \sin^2 t = 1$,

$$\frac{x^2}{9} + \frac{y^2}{16} = 1$$

from which

$$16x^2 + 9y^2 = 144. \tag{2}$$

We can obtain the Cartesian graph of the parametric equations (1) above by graphing Equation (2). The graph is an ellipse shown in Figure 5.7 with x-intercepts 3 and -3, and y-intercepts 4 and -4.

Although it is not always convenient, or even possible, to eliminate a parameter in order to get a Cartesian equation, we can obtain a graphical description in a Cartesian system.

Example   If a projectile is fired with an initial velocity v_0 at an angle α as shown in the figure, then for any time t in seconds, its approximate position is given by

$$x = (v_0 \cos \alpha)t \tag{1}$$

$$y = -16t^2 + (v_0 \sin \alpha)t, \tag{2}$$

where x and y are measured in feet. Graph the motion if $\alpha = 60°$ and $v_0 = 100$ ft/sec.

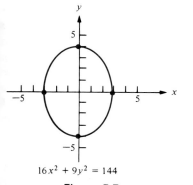

$16x^2 + 9y^2 = 144$

Figure 5.7

(*continued*)

Solution   First, we will use a calculator to tabulate some values for
and y for the system

$$x = (100 \cos 60°)t,$$
$$y = -16t^2 + (100 \sin 60°)t.$$

From the table of values and the graph in the above example, you ca
approximate the position of the projectile at times other than tho
listed, how far from the firing point it strikes the ground, and its hig
point.

t	x	y	(x, y)
0	0	0	$(0, 0)$
1	50	70.6	$(50, 70.6)$
2	100	109.2	$(100, 109.2)$
3	150	115.8	$(150, 115.8)$
4	200	90.4	$(200, 90.4)$
5	250	33.0	$(250, 33.0)$
6	300	-56.4	$(300, -56.4)$

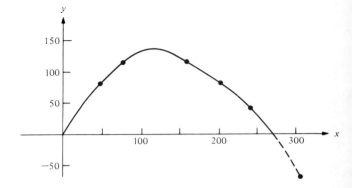

EXERCISE SET 5.6

A

■ **(a)** *Solve each system (to the nearest tenth) over the interval* $0 \le x \le 2$
(b) *Check the results by graphing the system.*

Example $y = \sin x$ (

$y = 2 \cos x$ (

Solution
a. Substitute $\sin x$ for y in (2) to obtain the equation in one variab

$$\sin x = 2 \cos x,$$

from which

$$\frac{\sin x}{\cos x} = 2,$$

$$\tan x = 2.$$

Hence, for $0 \leq x \leq \pi/2$,

$$x = \text{Tan}^{-1} 2 \approx 1.1$$

and from Equation (1)

$$y = \sin 1.1 \approx 0.9.$$

For $\pi \leq x \leq 3\pi/2$,

$$x = \pi + 1.1 \approx 4.2 \qquad \text{and} \qquad y = \sin 4.2 \approx -0.9.$$

Hence, the solution set is

$$\{(1.1, 0.9), (4.2, -0.9)\}$$

where the components are given to the nearest tenth.

b.

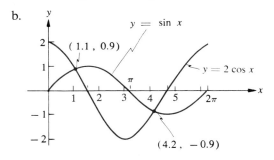

1. $y = \cos x$ 2. $y = \sin 2x$ 3. $y = \cos \frac{1}{2}x$ 4. $y = 2 \sin x$
   $y = 1$ $y = -1$ $y = \frac{1}{2}$ $y = -\frac{1}{2}$

5. $y = \tan x$ 6. $y = \cot x$ 7. $y = \sin 2x$ 8. $y = \cos 2x$
   $y = -1$ $y = \sqrt{3}$ $y = \frac{1}{2} \sin x$ $y = 2 \sin x$

■ *Write each parametric system as a Cartesian equation.*

Example $x = 4 \cos t, \ y = 2 \sin t; \ 0 \leq t \leq 2\pi$

Solution   Bearing in mind that $\cos^2 t + \sin^2 t = 1$, first solve the given equations for $\cos t$ and $\sin t$.

$$x = 4 \cos t \qquad \text{or} \qquad \cos t = \frac{x}{4}$$

$$y = 2 \sin t \qquad \text{or} \qquad \sin t = \frac{y}{2}$$

(continued)

Hence,

$$\cos^2 t + \sin^2 t = \frac{x^2}{16} + \frac{y^2}{4}.$$

Because $\cos^2 t + \sin^2 t = 1$,

$$4x^2 + 16y^2 = 144.$$

9. $x = 3t + 3, \ y = 2t - 5; \ -2 \leq t \leq 3$

10. $x = t - 3, \ y = t^2; \ -3 \leq t \leq 3$

11. $x = 2 \cos t, \ y = 2 \sin t; \ 0 \leq t \leq 2\pi$

12. $x = 2 \cos t, \ y = 5 \sin t; \ 0 \leq t \leq 2\pi$

13. $x = \cos t, \ y = \sin^2 t; \ 0 \leq t \leq \pi$

14. $x = \cos^2 t, \ y = 2 \sin t; \ -\dfrac{\pi}{2} \leq t \leq \dfrac{\pi}{2}$

15. $x = \sin 2t, \ y = \cos 2t; \ 0 \leq t \leq \dfrac{\pi}{2}$

16. $x = \sin \dfrac{1}{2} t, \ y = \cos \dfrac{1}{2} t; \ -\pi \leq t \leq \pi$

■ *Graph the equations in the example on page 183 in an x, y-coordinate system for the conditions specified in each exercise.*

17. A missile is fired with an initial velocity of 80 ft/sec at a 42° angle

18. A missile is fired with an initial velocity of 120 ft/sec at a 36.4° angle

19. A curve called the **cycloid** is determined by the equations

$$x = a(\alpha - \sin \alpha),$$
$$y = a(1 - \cos \alpha).$$

Graph these equations in an x, y-coordinate system for $a = 1$ and $0 \leq \alpha \leq 2\pi$.

20. Graph the equations in Exercise 19 for $a = 2$ and for $a = 3$.

21. A curve called the **hypocycloid** is obtained by the equations

$$x = a \cos^3 \alpha,$$
$$y = a \sin^3 \alpha.$$

Graph these equations in an x, y-coordinate system for $a = 1$ and $0 \leq \alpha \leq 2\pi$.

22. Graph the equations in Exercise 21 for $a = \frac{1}{2}$ and $a = 2$.

B

▪ *Write each parametric system as a Cartesian equation.*

23. $x = 2 \cot t$
    $y = 2 \sin^2 t$

24. $x = \sin t$
    $y = \cos^2 t(2 + \sin t)$

25. $x = \sec t$
    $y = \cos t$

26. $x = \sec t$
    $y = \tan t$

27. $x = a \cos t + h$
    $y = b \sin t + k$

28. $x = a \sec t + h$
    $y = b \csc t + k$

Chapter Summary

[5.1] A function that has only one element in its domain for any value in its range is called a **one-to-one function**. The **inverse** of such a function is obtained by interchanging the components of each ordered pair in the function.

[5.2] A trigonometric equation

$$y = \text{trig } x$$

can be written equivalently in the form

$$x = \text{Trig}^{-1} y \quad \text{or} \quad x = \text{Arctrig } y.$$

The ranges of the inverse trigonometric functions can be observed from the graphs of the functions shown on page 159.

[5.3–5.5] **Conditional equations** involving circular function values can be solved by inspection or by generating equivalent equations from which the solutions are evident by inspection. The identities introduced in this chapter are useful in obtaining such equivalent equations.

[5.6] The solutions of a system of equations are ordered pairs of numbers that can be found by first obtaining a single equation in one variable that includes as solutions one component of the solution of the system.
    A pair of equations of the form

$$x = f(t) \quad \text{and} \quad y = g(t)$$

in which two variables, in this case x and y, are each expressed in term of a third variable, are called **parametric equations.** The parameter, i this case t, can sometimes be eliminated to obtain a single Cartesia equation.

   The relation between x and y can be graphed by arbitrarily choosin replacements for the parameter and then obtaining the correspondin values for x and y.

A list of symbols introduced in this chapter is shown inside the front cover and important properties are shown inside the back cover.

Review Exercises

[5.1] ■ *Rewrite each expression without using inverse notation and then specif the exact value of y for which the statement is true.*

1. $y = \text{Arccos}\,\dfrac{1}{\sqrt{2}}$

2. $y = \text{Arccot}(-1)$

3. $y = \text{Csc}^{-1}(-2)$

4. $y = \text{Sec}^{-1}\,\dfrac{2}{\sqrt{3}}$

■ *Find the exact value of each expression.*

5. Arcsin 1

6. $\text{Arccot}(-\sqrt{3})$

7. $\text{Cos}^{-1}\left(-\dfrac{1}{\sqrt{2}}\right)$

8. $\text{Tan}^{-1}\,\dfrac{1}{\sqrt{3}}$

■ *Find an approximation to the nearest hundredth for each of the following*

9. Arccos 0.1205

10. Arctan 1.398

11. $\text{Sin}^{-1}\,0.4529$

12. $\text{Cot}^{-1}(-0.6142)$

[5.2] ■ *Rewrite each equation using inverse notation so that x is expressed expli citly in terms of y and/or the constants. Assume restrictions specified o page 152.*

13. $5\cos x = 3$

14. $\sin 2x = \dfrac{1}{5}$

15. $y = 3\tan 2x$

16. $y = 2\cos\dfrac{x}{2}$

■ *Find the exact value of each of the following.*

17. $\tan\left(\mathrm{Cos}^{-1}\dfrac{1}{2}\right)$ **18.** $\cos(\mathrm{Arccot}\ \sqrt{3})$

19. $\mathrm{Sin}^{-1}\left(\tan\dfrac{\pi}{4}\right)$ **20.** $\mathrm{Tan}^{-1}(\cos 0)$

[5.3] ■ *For each function value, find (**a**) the exact value of the least nonnegative angle α in degrees and (**b**) all α such that $0° \leq \alpha < 360°$ for which the statement is true.*

21. $\sin\alpha = \dfrac{\sqrt{3}}{2}$ **22.** $\tan\alpha = -1$

■ *For each function value, find an approximation in degrees for (**a**) the least nonnegative angle α and (**b**) all α such that $0° \leq \alpha < 360°$ for which each statement is true.*

23. $\cos\alpha = -0.4695$ **24.** $\csc\alpha = 1.6972$

■ *For each function value, find an approximation in radians for (**a**) the least nonnegative angle α and (**b**) all α such that $0^R \leq \alpha < 2\pi^R$ for which each statement is true.*

25. $\sec\alpha = 1.0720$ **26.** $\cot\alpha = -0.2439$

27. $\sin\alpha = 0.3107$ **28.** $\cos\alpha = 0.8965$

[5.4] ■ *Solve each equation in the interval $0 \leq x < \pi/2$.*

29. $2\sin x - \sqrt{3} = 0$ **30.** $\csc^2 x = 4$

31. $\sec x - 2\cos x = 0$ **32.** $\cot x \csc x = 0$

■ *Solve each equation in the interval $0° \leq x < 360°$.*

33. $\cot^2 x - 2\csc^2 x + 5 = 0$ **34.** $\sec^2 x - 4\cos^2 x = 0$

[5.5] ■ *Solve each equation in the interval $0° \leq x < 360°$.*

35. $\cos 3x = \dfrac{1}{\sqrt{2}}$ **36.** $\sin 4x = \dfrac{\sqrt{3}}{2}$

37. $\sin 2x - \tan x = 0$ **38.** $\tan 2x = \tan x$

[5.6] ▪ *Find solutions of each system over the interval* $0 \leq x \leq 2\pi$.

39. $y = \cos x$
$\quad\ y = \frac{1}{3}$

40. $y = 2 \sin x$
$\quad\ y = \frac{3}{4}$

41. $y = \sin 2x$
$\quad\ y = \frac{1}{2} \cos x$

42. $y = \cos 2x$
$\quad\ y = \frac{1}{2} \sin x$

43. Find an equation in x and y for the parametric system

$$x = 4t - 1$$
$$y = 2t^2.$$

44. Find an equation in x and y for the parametric system

$$x = \cos t$$
$$y = 4 \sin t.$$

MORE ABOUT TRIANGLES; VECTORS

In Chapter 1 we used trigonometric ratios of acute angles to solve right triangles and solved a variety of applied problems that could be modeled by such triangles. In this chapter we derive special formulas that will enable us to solve any triangle.

6.1 Law of Sines

Consider $\triangle ABC$ in Figure 6.1 (β may be either acute or obtuse), and introduce the altitude CD. Thus, $\triangle ADC$ and $\triangle BDC$ are right triangles. In either figure, in $\triangle ADC$,

$$\sin \alpha = \frac{h}{b}, \qquad \text{from which} \qquad h = b \sin \alpha.$$

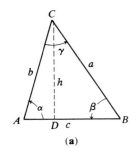

In $\triangle BDC$ in Figure 6.1a,

$$\sin \beta = \frac{h}{a}, \qquad \text{from which} \qquad h = a \sin \beta.$$

In $\triangle BDC$ in Figure 6.1b, $\angle CBD$ is equal to the reference angle for β; therefore,

$$\sin \beta = \sin \angle CBD = \frac{h}{a},$$

$$h = a \sin \beta.$$

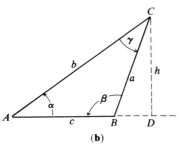

(a)

(b)

Figure 6.1

191

Hence, we have in each case

$$h = b \sin \alpha = a \sin \beta.$$

Dividing each member of $b \sin \alpha = a \sin \beta$ by $a \cdot b$, we have

$$\frac{\sin \alpha}{a} = \frac{\sin \beta}{b}.$$ (

If in Figure 6.1 we use the altitude from vertex A and a similar argumen
we will have

$$\frac{\sin \beta}{b} = \frac{\sin \gamma}{c}.$$ (.

Comparing Equations (1) and (2), we obtain the following relationshi

Law of Sines:

$$\frac{\sin \alpha}{a} = \frac{\sin \beta}{b} = \frac{\sin \gamma}{c}$$

In many applications this relationship is conveniently used in the for

$$\frac{a}{\sin \alpha} = \frac{b}{\sin \beta} = \frac{c}{\sin \gamma}.$$

Two Angles and a Side   The Law of Sines can be used to solve triangles when several kinds c
data are given. We consider first the case in which the measures of tw
angles and the length of one side of a triangle are given.

Example   Solve the triangle for which $\alpha = 30°$, $\beta = 45°$, and $a = 2$

Solution   A sketch of the triangle showing the given information
helpful. First, from the fact that $\alpha + \beta + \gamma = 180°$, it follows that

$$\gamma = 180° - \alpha - \beta$$
$$= 180° - 30° - 45° = 105°.$$

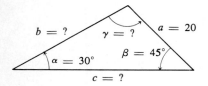

From the Law of Sines,

$$\frac{20}{\sin 30°} = \frac{b}{\sin 45°},$$

or, equivalently,

$$b = \frac{20 \sin 45°}{\sin 30°} \approx \frac{20(0.7071)}{0.5000} \approx 28.3.$$

Then, again from the Law of Sines,

$$\frac{20}{\sin 30°} = \frac{c}{\sin 105°},$$

or, equivalently,

$$c = \frac{20 \sin 105°}{\sin 30°} = \frac{20 \sin 75°}{\sin 30°} \approx \frac{20(0.9659)}{0.5000}$$

$$= 40(0.9659) \approx 38.6.$$

Hence,

$$b \approx 28.3, \qquad c \approx 38.6, \qquad \text{and} \qquad \gamma = 105°.$$

Two Sides and an Opposite Angle   A second case in which the Law of Sines is used is one in which the lengths of two sides of a triangle, say a and b, and the measure of an angle opposite one of them, say α, are given. This case is called the *ambiguous case* because ambiguity may arise, depending on a in relation to b and the measure of α. Figure 6.2, in which α is an *acute angle*

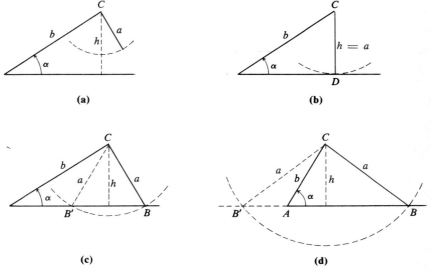

Figure 6.2

and h is the length of a perpendicular segment, illustrates the following possibilities that may occur.

1. If $a < h$, the side with length a cannot intersect the side opposite
   C and *no triangle is possible,* as shown in Figure 6.2a. In this case
   since $h = b \sin \alpha$, we have $a < b \sin \alpha$.
2. If $a = h = b \sin \alpha$, the side with length a intersects the side
   opposite C at D. In this case *exactly one right triangle is possible,*
   as shown in Figure 6.2b.
3. If $a > h$ (or $b \sin \alpha$) and $a < b$, or $b \sin \alpha < a < b$, *two triangles
   are possible,* as shown in Figure 6.2c.
4. If $a \geq b$, the triangle $AB'C$, shown in Figure 6.2d, does not contain the given angle α and therefore does not lead to a solution.
   Hence, in this case, *only one triangle is possible.* In particular, if
   $a = b$, the triangle is isosceles.

Figure 6.3, in which α is an *obtuse angle,* illustrates the following two
possibilities.

1. If $a > b$, then one triangle is possible, as shown in Figure 6.3a.
2. If $a \leq b$, then no triangle is possible as shown in Figure 6.3b.

While we have concerned ourselves only with the variables a, b, and
α in the above discussion, similar results are valid when other variables
are involved in the given information. A sketch of the triangle using the
given information will help to indicate the possibilities that exist. In
particular, sketches of triangles like those shown in Figures 6.2 and 6.3,
in which the given angle is drawn on the left, are usually the most helpful in deciding which possibility exists for given data.

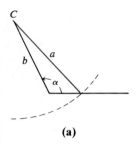

(a)

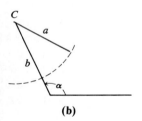

(b)

Figure 6.3

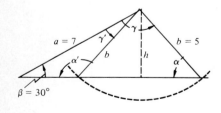

Example Solve the triangle(s) for which $a = 7$, $b = 5$, and $\beta = 30°$.

Solution Sketch the triangle(s) using the given information. Since the
lengths of two sides and the measure of an angle β, opposite the side
with measure b, are given, a check must be made to see how many triangles are possible. Since

$$h = a \sin \beta = 7 \sin 30°$$

$$= 7(0.5000) = 3.5,$$

it follows that $h < b < a$ and that two triangles are possible, as indicated in the figure.

From the Law of Sines,

$$\frac{\sin \alpha}{7} = \frac{\sin 30°}{5}$$

$$\sin \alpha = \frac{7 \sin 30°}{5} = \frac{7(0.5000)}{5} = 0.7000.$$

Hence,

$$\alpha = \text{Sin}^{-1} \, 0.7000 \approx 44.4°.$$

Since α is the base angle of an isosceles triangle, and α' is the supplement of the other base angle,

$$\alpha' \approx 180° - 44.4° = 135.6°.$$

For $\alpha \approx 44.4°$, the solution of one of the two possible triangles is completed as follows. Since the sum of the measures of the angles of a triangle equals $180°$,

$$\gamma = 180° - \alpha - \beta$$
$$\approx 180° - 44.4° - 30° = 105.6°.$$

From the Law of Sines

$$\frac{5}{\sin 30°} = \frac{c}{\sin 105.6°};$$

$$c = \frac{5 \sin 105.6°}{\sin 30°} \approx \frac{5(0.9632)}{0.5000} \approx 9.63.$$

For $\alpha' \approx 135.6°$, the solution of the second possible triangle can be completed. Since

$$\gamma' \approx 180° - 135.6° - 30° = 14.4°,$$

then by the Law of Sines

$$\frac{5}{\sin 30°} = \frac{c'}{\sin 14.4°};$$

$$c' \approx \frac{5(0.2487)}{0.5000} \approx 2.49.$$

Thus, the two solutions are

$$\alpha \approx 44.4°, \quad \gamma \approx 105.6°, \quad \text{and} \quad c \approx 9.63;$$
$$\alpha' \approx 135.6°, \quad \gamma' \approx 14.4°, \quad \text{and} \quad c' \approx 2.49.$$

EXERCISE SET 6.1

A

■ *Solve each triangle.*

1. $a = 5$, $\alpha = 20°$, and $\beta = 75°$

2. $a = 32$, $\alpha = 26.3°$, and $\gamma = 81.8°$

3. $a = 58.4$, $\beta = 37.2°$, and $\gamma = 100°$

4. $b = 1.9$, $\alpha = 111.7°$, and $\beta = 5.1°$

5. $b = 0.42$, $\alpha = 35° \ 36'$, and $\gamma = 91° \ 30'$

6. $b = 0.88$, $\beta = 63° \ 54'$, and $\gamma = 34° \ 12'$

7. $c = 13.6$, $\alpha = 30° \ 24'$, and $\beta = 72° \ 6'$

8. $c = 1.4$, $\alpha = 135° \ 12'$, and $\gamma = 34° \ 48'$

■ *Determine the number of triangles that satisfy each set of conditions an
solve each triangle.*

9. $a = 7$, $b = 5$, and $\alpha = 30°$

10. $a = 10$, $b = 9$, and $\beta = 60°$

11. $a = 4.2$, $c = 6.1$, and $\alpha = 32.2°$

12. $a = 38$, $c = 45$, and $\gamma = 35.6°$

13. $b = 16$, $c = 32$, and $\beta = 30°$

14. $b = 0.3$, $c = 0.4$, and $\gamma = 62°$

15. $b = 3.9$, $a = 5$, and $\beta = 42° \ 42'$

16. $b = 37$, $a = 51$, and $\alpha = 135° \ 30'$

■ *Find the area of each triangle.*

17. $a = 3.4$, $\beta = 13°$, and $\alpha = 50°$

18. $c = 3.8$, $\beta = 15° \ 6'$, and $\gamma = 50°$

19. To find the distance from an observer's point A on one side of a riv
    (see figure) to a point P on the other side, a line AP' measuring 254
    feet was laid out on one side and the angles $P' \ AP$ and $AP' \ P$ we
    measured and found to be $47° \ 18'$ and $78° \ 6'$, respectively. Find th
    distance fom A to P to the nearest foot.

20. Two men, 400 feet apart, observe a balloon between them that is in
    vertical plane with them. The respective angles of elevation of th

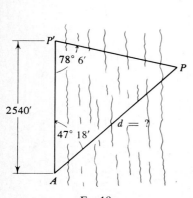

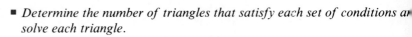

P'

$78° \ 6'$

P

2540'

$47° \ 18'$

$d = \ ?$

A

Ex. 19

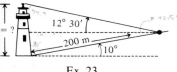

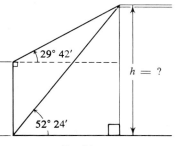

Ex. 23

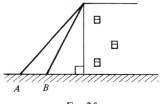

$h = ?$

Ex. 24

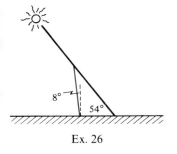

A B

Ex. 25

$8°$ $54°$

Ex. 26

balloon are observed by the men to be 75° 24′ and 49° 30′. Find the height of the balloon above the ground.

21. One diagonal of a parallelogram is 20 centimeters long and at one end forms angles of 20° and 40° with the sides of the parallelogram. Find the lengths of the sides.

22. A ship at sea is sighted from two points, A and B, on the shore. A and B are 5 kilometers apart, and the line of sight from A to the ship forms an angle of 30° 18′ with $\overline{AB}$, while the line of sight from B to the ship forms an angle of 41° 12′ with $\overline{AB}$. How far is the ship from point A?

23. A tower is built vertically on a hillside that slants upward 10° from the horizontal. From a point 200 meters uphill from the foot of the tower, the angle of elevation of the top of the tower is 12° 30′ (slant height). How high is the tower?

24. From a window 75 meters above the ground, the measure of the angle of elevation to the top of a nearby building is 29° 42′ (see figure). From a point on the ground directly below the window, the measure of the angle of elevation of the top of the same building is 52° 24′. Find the height of the building.

25. The angle of elevation of the building is 48° from A and 61° from B. If AB is 20 feet, find the height of the building.

26. The pole is tilted toward the sun at an angle of 8° from the vertical. When the angle of elevation of the sun is 54°, the shadow of the pole is 12 feet long. How long is the pole?

27. A hill makes an angle of 10° with the horizontal. From a point 50 meters downhill from a vertical tower, the angle of elevation of the tower is 56°. Find the height of the tower.

28. From two points on shore, the angles α and β to a light at C are 16.4° and 58.6°, respectively. If AB is 1500 feet, find the distance, d, of the light from the shore.

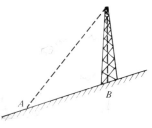

Ex. 27

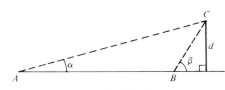

Ex. 28

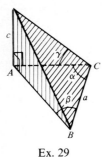

Ex. 29

B

29. Using the figure, show that

$$c = \frac{a \sin \beta \tan \gamma}{\sin(\alpha + \beta)}.$$

30. Using the figure and the results for Exercise 29, if $\alpha = 40°$, $\beta = 6$ $\gamma = 30°$, and $a = 100$ feet, find c.

■ *Show that, in any triangle ABC, each of the following statements true.*   *Hint: In Exercises 31 and 32 start with the Law of Sines.*

31. $\dfrac{a + b}{b} = \dfrac{\sin \alpha + \sin \beta}{\sin \beta}$ **32.** $\dfrac{a - b}{b} = \dfrac{\sin \alpha - \sin \beta}{\sin \beta}$

33. $\dfrac{a - b}{a + b} = \dfrac{\sin \alpha - \sin \beta}{\sin \alpha + \sin \beta}$

34. $\dfrac{a + b + c}{b} = \dfrac{\sin \alpha + \sin \beta + \sin \gamma}{\sin \beta}$

6.2 Law of Cosines

The Law of Sines considered in Section 6.1 cannot be used directly solve triangles in which the lengths of two sides and the measure of t angle formed by their respective rays are specified or in which only t lengths of the three sides are specified. Equations (1), (2), and (3) bel can be used directly in such cases.

Law of Cosines:

$$c^2 = a^2 + b^2 - 2ab \cos \gamma, \tag{1}$$

$$b^2 = a^2 + c^2 - 2ac \cos \beta, \tag{2}$$

$$a^2 = b^2 + c^2 - 2bc \cos \alpha. \tag{3}$$

These statements follow from a consideration of the distance form (see Appendix A.4). Each of the three formulas expresses the sa geometric fact as the other two. Hence, we will only prove Equation (

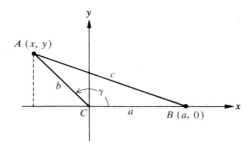

Figure 6.4

Orienting triangle ABC with γ at the origin, as shown in Figure 6.4, we find that

$$\cos \gamma = \frac{x}{b} \quad \text{and} \quad \sin \gamma = \frac{y}{b},$$

from which the coordinates of A are

$$b \cos \gamma \quad \text{and} \quad b \sin \gamma.$$

Of course, statements (1), (2), and (3) above would also be true if γ were an acute angle. From the distance formula,

$$c^2 = (y - 0)^2 + (x - a)^2$$
$$= (b \sin \gamma - 0)^2 + (b \cos \gamma - a)^2$$
$$= b^2 \sin^2 \gamma + b^2 \cos^2 \gamma - 2ab \cos \gamma + a^2$$
$$= b^2(\sin^2 \gamma + \cos^2 \gamma) - 2ab \cos \gamma + a^2.$$

Since $\sin^2 \gamma + \cos^2 \gamma = 1$, we have

$$c^2 = a^2 + b^2 - 2ab \cos \gamma.$$

If the triangle is oriented with A or B at the origin, Equations (2) and (3) in the Law of Cosines can be obtained in a similar way. The particular form of the equation that should be used will, of course, depend on the given information.

Notice that in Equation (1) if $\gamma = 90°$, then $\cos \gamma = \cos 90° = 0$ and the Law of Cosines reduces to

$$c^2 = a^2 + b^2 - 2ab(0),$$
$$c^2 = a^2 + b^2,$$

the familiar formula applicable to right triangles (the Pythagorean theorem).

Two Sides and the Included Angle

Now consider how the Law of Cosines can be used to solve a triangle when the measure of an angle and the lengths of the adjacent sides are known.

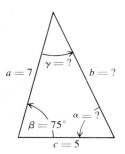

Example Given the measures $\beta = 75°$, $a = 7$, and $c = 5$ of an oblique triangle, find the length b.

Solution Sketch a triangle showing the given information. The length b can be found by using the Law of Cosines expressed in the form $b^2 = a^2 + c^2 - 2ac \cos \beta$. Substituting the known values gives

$$b^2 = 7^2 + 5^2 - 2(7)(5) \cos 75°$$

$$\approx 49 + 25 - 70(0.2588) \approx 55.9;$$

$$b \approx \sqrt{55.9} \approx 7.48.$$

Three Sides

The Law of Sines could be used to complete the solution of the triangle in the above example since the length b, opposite the given angle β, is known. However, we will use the Law of Cosines in the same example in order to show how it can be used to determine the measure of an angle if the lengths of three sides of a triangle are known.

Example To find the measure of α in the foregoing example, the equation

$$a^2 = b^2 + c^2 - 2bc \cos \alpha$$

is first written in the form

$$\cos \alpha = \frac{b^2 + c^2 - a^2}{2bc}.$$

From the previous calculations, $b^2 \approx 55.9$ and $b \approx 7.48$. Thus,

$$\cos \alpha \approx \frac{55.9 + 25 - 49}{2(7.48)(5)} \approx 0.4265,$$

from which

$$\alpha = \text{Cos}^{-1} 0.4265 \approx 64.8°.$$

Because $\alpha + \beta + \gamma = 180°$,

$$\gamma \approx 180° - 64.8° - 75° = 40.2°.$$

Thus, the solution to the triangle is

$$b \approx 7.48, \quad \alpha \approx 64.8°, \quad \text{and} \quad \gamma \approx 40.2°.$$

If, after first finding the length of the third side, $b = 7.48$ in the above example, the Law of Sines is used to find the measure of a second angle, the angle selected should be the smaller of the two remaining angles (the one opposite the shorter of the two given sides) and hence *an acute angle*. Thus, if the Law of Sines had been used to find the second angle in the above example, we would have found sin γ rather than sin α because, of the two given sides, $c < a$ and γ is opposite c.

Given three sides of a triangle, any angle can be obtained by using one of the Equations (1), (2), or (3) on page 198. If an equation is used that gives the measure of either of the smaller angles, which will necessarily be acute angles, then the cosine of these angles will be positive. If the equation is used that gives the measure of the largest angle, the angle may be acute or obtuse, and the cosine of the angle may be positive or negative.

Example Solve the triangle for which the lengths of the three sides are 3.4, 2.7, and 1.3.

Solution Draw a sketch showing the given information. We will first find the measure of the smallest angle. From the Law of Cosines,

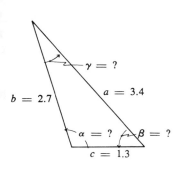

$$\cos \gamma = \frac{a^2 + b^2 - c^2}{2ab} = \frac{(3.4)^2 + (2.7)^2 - (1.3)^2}{2(3.4)(2.7)}$$

$$\approx 0.9346.$$

Hence,

$$\gamma = \text{Cos}^{-1}\, 0.9346 \approx 20.8°.$$

From the Law of Sines,

$$\frac{\sin \beta}{b} = \frac{\sin \gamma}{c},$$

$$\frac{\sin \beta}{2.7} = \frac{\sin 20.8°}{1.3},$$

from which

$$\sin \beta = \frac{2.7 \sin 20.8°}{1.3} \approx \frac{2.7(0.3551)}{1.3} \approx 0.7375.$$

Thus,

$$\beta = \text{Sin}^{-1}\, 0.7375 \approx 47.5°. \qquad \text{(continued)}$$

From the fact that $\alpha = 180° - \beta - \gamma$,

$$\alpha \approx 180° - 47.5° - 20.8°$$

$$\approx 111.7°.$$

Additional applications that require solutions of triangles occur in work with geometric vectors which are introduced in Sections 6.3 and 6.4.

EXERCISE SET 6.2

A

■ *Solve each triangle.*

1. $b = 4$, $c = 3.5$, and $\alpha = 71°$

2. $c = 0.3$, $a = 0.1$, and $\beta = 70°$

3. $a = 3.2$, $b = 2.2$, and $\gamma = 75.3°$

4. $b = 6.0$, $a = 5.1$, and $\gamma = 83.5°$

5. $a = 0.7$, $c = 0.8$, and $\beta = 141.5°$

6. $b = 3.4$, $a = 2.1$, and $\gamma = 122.2°$

7. $b = 1.6$, $c = 3.2$, and $\alpha = 100.4°$

8. $c = 2.1$, $b = 4.3$, and $\alpha = 130.6°$

■ *Solve the triangles for which the lengths of the three sides are given.*

9. $a = 9$, $b = 7$, and $c = 5$ **10.** $a = 2.7$, $b = 5.1$, and $c = 4.$

11. $a = 1.2$, $b = 9$, and $c = 10$ **12.** $a = 4.5$, $b = 7.5$, and $c = 5.$

13. $a = 6$, $b = 8$, and $c = 12$ **14.** $a = 6$, $b = 12$, and $c = 13$

15. $a = 4$, $b = 5$, and $c = 6$ **16.** $a = 5$, $b = 7$, and $c = 8$

17. Find the measure of the smallest angle of the triangle whose sides have lengths 4.3, 5.1, and 6.3.

18. Find the measure of the smallest angle of the triangle whose sides have lengths 3.0, 4.2, and 3.8.

19. Find the measure of the largest angle of the triangle whose sides have lengths 2.9, 3.3, and 4.1.

20. Find the measure of the largest angle of the triangle whose sides have lengths 6.0, 8.2, and 9.4.

21. Find the area of the triangle in Exercise 17.

22. Find the area of the triangle in Exercise 18.

23. A ship is supposed to travel directly from port A to port B, a distance of 10 kilometers. After traveling a distance of 5 kilometers, the captain discovers he has been traveling 15° off course. At this point, how far is the ship from port B?

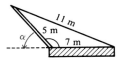

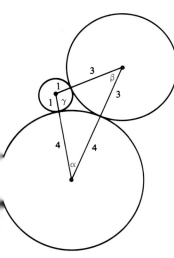

Ex. 27

24. Two ships leave the same port on courses that form an angle of 37° 12′. When one ship has traveled 12 kilometers, the other has traveled 18 kilometers. How far apart are the ships at that time?

25. The lengths of two adjacent sides of a parallelogram are 6 centimeters and 8 centimeters. The included angle is 67°. Find the length of the longer diagonal.

26. A tower 40 meters high is on top of a hill. A cable 50 meters long runs from the top of the tower to a point 20 meters down the hill. Find the angle the hillside forms with a horizontal line.

27. A crane at the edge of a dock is supported by a cable 11 meters long, attached to the dock 7 meters from the base of the crane (see figure). If the crane arm is 5 meters long, find the acute angle that the crane arm forms with the horizontal.

28. Three circles with radii 1, 3, and 4 centimeters are tangent to each other (see figure). To the nearest 6′, find the three angles formed by the lines joining their centers.

Ex. 28

29. Points P and Q are on opposite ends of a lake. If point R is 1200 feet from P and 1600 feet from Q, and the angle at R is 50°, how long is the lake?

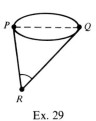

Ex. 29

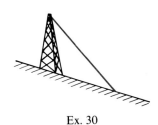

Ex. 30

30. A vertical tower 30 feet high is on a hill that forms an angle of 15° with the horizontal. How long is a guy wire that runs from the top of the tower to a point 20 feet downhill from the base of the tower?

31. Two airplanes leave a field at the same time. One flies 30° east of north at 250 kilometers per hour and the other 45° east of south at 300 kilometers per hour. How far apart are they at the end of 2 hours?

Ex. 31

32. A pilot is flying from Salt Lake City to Denver, a distance of abou 500 miles. When she is 300 miles from Salt Lake City, she discove she has been flying 16° off course. How far is she from Denver at tha time?

B

■ *The formulas in Exercises 33–36 where* $s = \frac{1}{2}(a + b + c)$ *are some times used in the solution of triangles. Show that each statement is tru for any* α *such that* $0° < \alpha < 180°$.

Hint: *Use the fact that* $\cos \alpha = \dfrac{b^2 + c^2 - a^2}{2bc}$.

33. $1 + \cos \alpha = \dfrac{(b + c + a)(b + c - a)}{2bc}$

34. $1 - \cos \alpha = \dfrac{(a - b + c)(a + b - c)}{2bc}$

35. $\dfrac{1 + \cos \alpha}{2} = \dfrac{s(s - a)}{bc}$.

36. $\dfrac{1 - \cos \alpha}{2} = \dfrac{(s - b)(s - c)}{bc}$.

37. Use the results of <u>Exercises 35 and 36</u> to prove Hero's (or Heron' formula, $\mathscr{A} = \sqrt{s(s - a)(s - b)(s - c)}$, which can be used to fin the area of any triangle directly from the lengths of its three side

38. Use Hero's formula to find the area of the triangle in Exercise above.

39. Use Hero's formula to find the area of the triangle in Exercise 1 above.

40. Two sides of a parallelogram are of lengths 6.8 and 8.3 inches, an one of the diagonals is of length 4.2 inches (see figure). Find the are of the parallelogram.

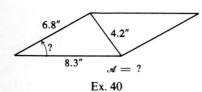

6.8"

4.2"

?

8.3"

$\mathscr{A} = ?$

Ex. 40

41. Show that the radius r of an inscribed circle of a triangle (see figure is given by

$$r = \sqrt{\dfrac{(s - a)(s - b)(s - c)}{s}}$$

where $s = \frac{1}{2}(a + b + c)$.

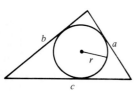

b

a

r

c

Ex. 41

42. Show that the radius r of a circumscribed circle of a triangle (se figure) is given by

$$r = \dfrac{a}{2 \sin \alpha} = \dfrac{b}{2 \sin \beta} = \dfrac{c}{2 \sin \gamma}.$$

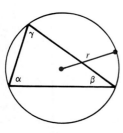

γ

r

α

β

Ex. 42

6.3 Geometric Vectors

Some physical quantities, such as length and mass, are characterized by *magnitude alone*. If units of measure are specified, the quantities can be represented by real numbers and associated with points in a number line. In the description of other physical quantities, such as force, velocity, and acceleration, *direction as well as magnitude* is required. Such a quantity can be represented by a pair of real numbers and associated with a line segment in the form of an arrow with length representing the magnitude and pointing in an assigned direction.

A two-dimensional **geometric vector*** *designated by $\vec{v}$ is a line segment with a specified direction.*

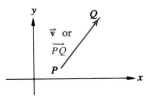

Figure 6.5

If the end points of a geometric vector are named, it is sometimes convenient to specify the vector by using these names. For example, in Figure 6.5 the geometric vector can also be designated by $\overrightarrow{PQ}$, where the point named P is called the **initial point** and the point named Q is called the **terminal point.** For discussion purposes, the geometric vector in Figure 6.5 is shown in the framework of a rectangular coordinate system.

Sometimes a geometric vector is simply represented by **v** without using the arrow above the symbol. We are using the arrow to differentiate this symbol for a geometric vector from the symbol v as a variable for a real number. In addition, the arrow emphasizes the fact that a geometric vector has *magnitude* and *direction.*

Magnitude and Direction

We will assume that each geometric vector in the plane has a length that is a real number that depends on the unit of measurement.

For all geometric vectors $\vec{v}$, the **norm** (*or* **magnitude**) *of $\vec{v}$ is the length of $\vec{v}$ and is designated by $\|\vec{v}\|$.*

When a geometric vector is considered in relation to a rectangular coordinate system, the following concept is useful.

The **direction angle** α *of a geometric vector $\vec{v}$, $-180° < \alpha \le 180°$, is an angle such that the initial side is the ray from the initial point of $\vec{v}$ parallel to the x-axis and directed in the positive x direction; the terminal side is the ray from the initial point of $\vec{v}$ and containing $\vec{v}$.*

Figure 6.6 shows $\|\vec{v}\|$ and the direction angle α associated with a vector $\vec{v}$. From a vector viewpoint, the real number $\|\vec{v}\|$ is called a **scalar.**

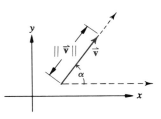

Figure 6.6

---

* The term *geometric vector* is used to differentiate this notion of a directed line segment from a vector as an ordered pair of real numbers introduced in Section 6.5.

For all geometric vectors $\vec{v}$, the geometric vector $-\vec{v}$ has the same magnitude as $\vec{v}$ and has a direction angle whose measure differs from the measure of the direction angle of $\vec{v}$ by 180°.

Figure 6.7 shows two vectors $\vec{v}$ and $-\vec{v}$.

Two geometric vectors $\vec{v}_1$ and $\vec{v}_2$ are **equivalent** *if they have the same magnitude and the same direction angle.*

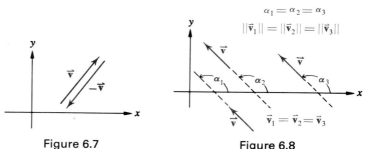

Figure 6.7   Figure 6.8

Figure 6.8 illustrates three equivalent vectors. We use the symbol $=$ to indicate that two vectors are equivalent.

If we can imagine "sliding" a geometric vector so that it remains parallel to its original position in the plane, then any geometric vector can "slide" onto any vector equivalent to it.

Operations   Geometric vectors are very useful in setting up mathematical models of practical problems. However, before this can be done, certain operations must be defined.

For all geometric vectors $\vec{v}_1$ and $\vec{v}_2$, such that the terminal point of $\vec{v}_1$ is the initial point of $\vec{v}_2$, $\vec{v}_1 + \vec{v}_2$ is the geometric vector having the same initial point as $\vec{v}_1$ and the same terminal point as $\vec{v}_2$. $\vec{v}_1 + \vec{v}_2$ is called the **sum** *(or* **resultant***) of $\vec{v}_1$ and $\vec{v}_2$.*

As indicated in Figure 6.9a, if $\vec{v}_1$ and $\vec{v}_2$ are not in the relative locations required by the definition, we "slide" one of them, say $\vec{v}_2$, parallel to its original position into the correct location and then find the sum $\vec{v}_1 + \vec{v}_2$.

Because a triangle is formed by $\vec{v}_1, \vec{v}_2$, and $\vec{v}_1 + \vec{v}_2$, we sometimes say that geometric vectors are added according to "the triangle law." From the properties of a parallelogram, we have that, if $\vec{v}_1$ and $\vec{v}_2$ have the same initial point, as shown in Figure 6.9b, the sum $\vec{v}_1 + \vec{v}_2$ is a diagonal of the parallelogram with adjacent sides $\vec{v}_1$ and $\vec{v}_2$. Therefore, we sometimes also say that geometric vectors are added according to "the parallelogram law." Figure 6.9c shows the sum $\vec{v}_3 + \vec{v}_4$ if $\vec{v}_3$ and $\vec{v}_4$ are collinear (vectors lie on the same line).

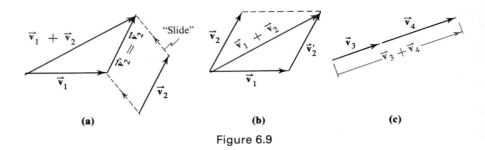

Figure 6.9

Note that the symbol "+" in a vector sum is being used in a different sense from that of its ordinary usage to pair two real numbers in a sum, although the operation is commonly referred to as the "addition" of geometric vectors.

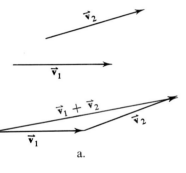

a.

b.

Example   Construct the sum $\vec{v}_1 + \vec{v}_2$.

Solution   "Slide" the initial point of $\vec{v}_2$ onto the terminal point of $\vec{v}_1$. The sum $\vec{v}_1 + \vec{v}_2$ is the geometric vector from the initial point of $\vec{v}_1$ to the terminal point of $\vec{v}_2$ (figure a).

Alternate Solution   "Slide" the initial point of $\vec{v}_2$ onto the initial point of $\vec{v}_1$ and form a parallelogram. The sum $\vec{v}_1 + \vec{v}_2$ is the diagonal of the parallelogram (figure b).

*For all geometric vectors $\vec{v}$ and any real number c, $c\vec{v}$ is a geometric vector with magnitude $\|c\vec{v}\| = |c| \cdot \|\vec{v}\|$, having the same direction angle as $\vec{v}$ if $c > 0$ and the same direction angle as $-\vec{v}$ if $c < 0$. The geometric vector $c\vec{v}$ is called a **scalar multiple** of $\vec{v}$.*

Example   As shown in the figure below, $2\vec{v}$ and $\frac{1}{2}\vec{v}$ are geometric vectors in the same direction as $\vec{v}$ such that

$$\|2\vec{v}\| = 2\|\vec{v}\| \qquad \text{and} \qquad \|\tfrac{1}{2}\vec{v}\| = \tfrac{1}{2}\|\vec{v}\|,$$

while $-2\vec{v}$ is in the opposite direction such that

$$\|-2\vec{v}\| = |-2|\|\vec{v}\| = 2\|\vec{v}\|.$$

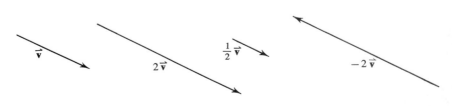

EXERCISE SET 6.3

A

■ *In Exercises 1–40, use the geometric vectors below.*

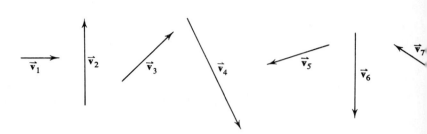

■ *If $\|\vec{v}_1\| = 1$, estimate the magnitude and the measure of the directio[n] angle for each of the following geometric vectors, where $\vec{v}_1$ is in th[e] direction of the positive x-axis.*

1. $\vec{v}_2$ **2.** $\vec{v}_3$ **3.** $\vec{v}_4$ **4.** $\vec{v}_5$ **5.** $\vec{v}_6$ **6.** $\vec{v}_7$

■ *Using freehand methods, draw a single vector representing each of th[e] following sums.*

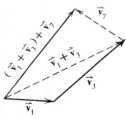

Example $(\vec{v}_1 + \vec{v}_3) + \vec{v}_7$

Solution   See the figure.

7. $\vec{v}_1 + \vec{v}_2$ **8.** $\vec{v}_1 + \vec{v}_3$ **9.** $\vec{v}_1 + \vec{v}_4$

10. $\vec{v}_1 + \vec{v}_5$ **11.** $\vec{v}_1 + \vec{v}_6$ **12.** $\vec{v}_1 + \vec{v}_7$

13. $\vec{v}_2 + \vec{v}_3$ **14.** $\vec{v}_2 + \vec{v}_4$ **15.** $(\vec{v}_2 + \vec{v}_5) + \vec{v}_6$

16. $(\vec{v}_2 + \vec{v}_6) + \vec{v}_7$ **17.** $(\vec{v}_2 + \vec{v}_3) + \vec{v}_5$ **18.** $\vec{v}_3 + (\vec{v}_4 + \vec{v}_7)$

■ *Using freehand methods, draw each of the following.*

19. $-\vec{v}_1$ **20.** $-\vec{v}_3$ **21.** $2\vec{v}_7$

22. $4\vec{v}_1$ **23.** $-2\vec{v}_5$ **24.** $-\vec{v}_6$

25. $\frac{1}{2}\vec{v}_3$ **26.** $-\frac{1}{3}\vec{v}_4$ **27.** $2\vec{v}_1 + 3\vec{v}_5$

28. $4\vec{v}_7 + 2\vec{v}_6$ **29.** $\frac{1}{2}\vec{v}_4 + \vec{v}_7$ **30.** $\vec{v}_2 + \frac{3}{4}\vec{v}_3$

31. $2\vec{v}_3 + (-3\vec{v}_1)$ **32.** $\vec{v}_7 + (-2\vec{v}_2)$

B

■ *By appropriate freehand sketches show that each statement is true.*

33. $\vec{v}_2 + \vec{v}_3 = \vec{v}_3 + \vec{v}_2$ **34.** $\vec{v}_5 + \vec{v}_4 = \vec{v}_4 + \vec{v}_5$

35. $(\vec{v}_6 + \vec{v}_7) + \vec{v}_2 = \vec{v}_6 + (\vec{v}_7 + \vec{v}_2)$ **36.** $\vec{v}_3 + (\vec{v}_7 + \vec{v}_5) = (\vec{v}_3 + \vec{v}_7) + \vec{v}_5$

37. $2(\vec{v}_3 + \vec{v}_5) = 2\vec{v}_3 + 2\vec{v}_5$ **38.** $(2 + 3)\vec{v}_7 = 2\vec{v}_7 + 3\vec{v}_7$

39. $\frac{1}{3}\vec{v}_1 + 2\vec{v}_1 = \frac{7}{3}\vec{v}_1$ **40.** $\vec{v}_2 + (-2\vec{v}_2) = -\vec{v}_2$

6.4 Applications

The solution to many kinds of practical problems involving such quantities as velocities, accelerations, forces, and directed line segments can be found by means of geometric vectors.

Projections   First, consider the *projections* of a geometric vector on the *x*- and *y*-axes of a Cartesian coordinate system. Figure 6.10 depicts a geometric vector $\vec{v}$ and perpendicular lines drawn to the *x*- and *y*-axes from the end points P_1 and P_2 of $\vec{v}$. The geometric vector projections $\vec{v}_x$ and $\vec{v}_y$ of $\vec{v}$ on the *x*-axis and *y*-axis, respectively, are called the **rectangular components** of $\vec{v}$; $\vec{v}_x$ is called the **horizontal component** of $\vec{v}$, and $\vec{v}_y$ is called the **vertical component**. These components are determined by the magnitude of $\vec{v}$ and its direction. Furthermore,

$$\vec{v} = \vec{v}_x + \vec{v}_y.$$

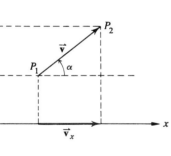

Figure 6.10

In Figure 6.11, $\overline{P_1 P_3}$ is drawn parallel to the *x*-axis, and $\overline{P_2 P_3}$ is drawn parallel to the *y*-axis. Since

$$|\sin \alpha| = \frac{\|\vec{v}_y\|}{\|\vec{v}\|},$$

$$\|\vec{v}_y\| = \|\vec{v}\| \cdot |\sin \alpha|. \tag{1}$$

Also, since

$$|\cos \alpha| = \frac{\|\vec{v}_x\|}{\|\vec{v}\|},$$

$$\|\vec{v}_x\| = \|\vec{v}\| \cdot |\cos \alpha|. \tag{2}$$

In Figure 6.11, α is shown as a positive acute angle. However, α may be anywhere between $-180°$ and $180°$; hence, $\cos \alpha$ or $\sin \alpha$ may be negative. Therefore, because $\|\vec{v}_x\|$, $\|\vec{v}_y\|$, and $\|\vec{v}\|$ are all positive, we use $|\sin \alpha|$ and $|\cos \alpha|$ in Equations (1) and (2).

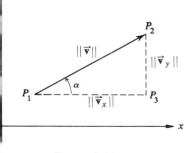

Figure 6.11

If the geometric vectors $\vec{v}_x$ and $\vec{v}_y$ are known, the magnitude $\|\vec{v}\|$ a■ direction angle α of $\vec{v}$ can be determined from the relationships in a rig■ triangle, as shown in Figure 6.11

$$\|\vec{v}\| = \sqrt{\|\vec{v}_x\|^2 + \|\vec{v}_y\|^2},$$

and

$$|\sin \alpha| = \frac{\|\vec{v}_y\|}{\|\vec{v}\|} \quad \text{or} \quad |\cos \alpha| = \frac{\|\vec{v}_x\|}{\|\vec{v}\|}.$$

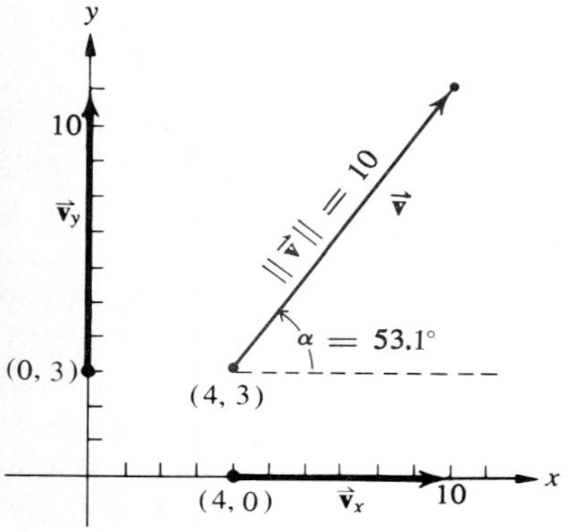

Example The horizontal component, $\vec{v}_x$, has its initial point at (4, with $\|\vec{v}_x\| = 6$. The vertical component, $\vec{v}_y$, has its initial point at (0, with $\|\vec{v}_y\| = 8$. Find α and $\|\vec{v}\|$ and sketch $\vec{v}$. Assume $\vec{v}_x$ and $\vec{v}_y$ a■ directed in a positive direction.

Solution The magnitude

$$\|\vec{v}\| = \sqrt{\|\vec{v}_x\|^2 + \|\vec{v}_y\|^2}$$
$$= \sqrt{6^2 + 8^2} = \sqrt{100} = 10.$$

Because

$$\cos \alpha = \frac{6}{10} = 0.6000,$$

$$\alpha \approx 53.1°.$$

Thus, the geometric vector $\vec{v}$ has its initial point at (4, 3), a direction ang■ whose measure is approximately 53.1°, and a magnitude of 10.

Force Consider two forces, F_1 and F_2, acting on a particle at the point ■ as in Figure 6.12. Since a force has magnitude and direction, we c■ represent it with a geometric vector, with magnitude given in term■ of the units of force and with direction angle given by the directic■ of the force. It is shown in physics that the two forces represente■ by $\vec{v}_1$ and $\vec{v}_2$ can be replaced by a single equivalent force, F_3, ass■ ciated with the vector $\vec{v}_1 + \vec{v}_2$, the resultant of the forces represente■ by $\vec{v}_1$ and $\vec{v}_2$. The resultant $\vec{v}_1 + \vec{v}_2$ can be obtained, as discussed ■ Section 6.3, by constructing the parallelogram with adjacent sides ■ and $\vec{v}_2$ and drawing the diagonal; $\vec{v}_1 + \vec{v}_2$ is then associated with th■ resultant force F_3.

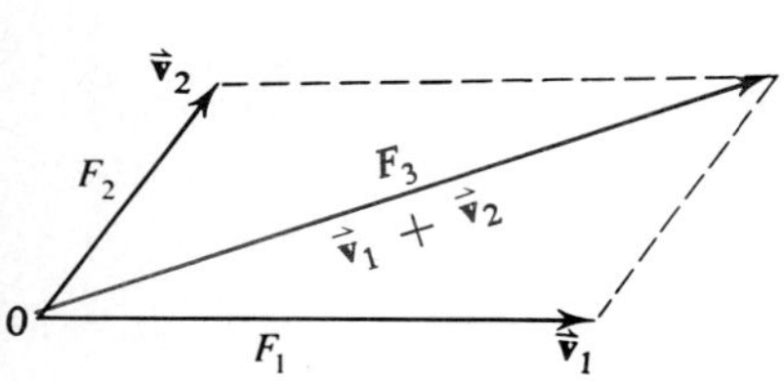

Figure 6.12

Example Two forces, the first one of 5 pounds and the second ■ 12 pounds, act on a body at right angles to each other. Find th■ magnitude of the resultant force and the measure of the angle that th■ resultant force makes with the 5-pound force.

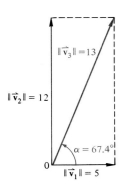

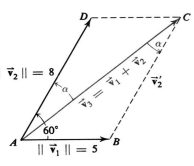

Solution   Because $\|\vec{v}_1\| = 5$ and $\|\vec{v}_2\| = 12$,

$$\|\vec{v}\| = \sqrt{5^2 + 12^2} = \sqrt{169} = 13,$$

and

$$\cos \alpha = \frac{5}{13} \approx 0.3846.$$

Hence, $\alpha \approx 67.4°$. Thus, the magnitude of the resultant force is 13 pounds, and the angle it makes with the 5-pound force has a measure of approximately 67.4°.

In cases where the measure of an angle between two forces is not 90°, it is sometimes convenient to use the Law of Cosines and the Law of Sines in order to find the resultant force.

Example   A force of 5 pounds and another force of 8 pounds act on an object at an angle of 60° with respect to each other. Find the magnitude of the resultant force and the angle it forms with respect to the 8-pound force.

Solution   The figure shows the forces of 5 pounds and 8 pounds represented by the geometric vectors $\vec{v}_1$ and $\vec{v}_2$, respectively, and the resultant $\vec{v}_3 = \vec{v}_1 + \vec{v}_2$. Since $\vec{v}_2'$ is equivalent to $\vec{v}_2$, and since the consecutive angles of a parallelogram are supplementary.

$$\angle ABC = 180° - 60° = 120°,$$

and it follows from the Law of Cosines that

$$\|\vec{v}_3\|^2 = \|\vec{v}_1\|^2 + \|\vec{v}_2\|^2 - 2\|\vec{v}_1\| \cdot \|\vec{v}_2\| \cos 120°$$
$$= 5^2 + 8^2 - 2(5)(8)(-\cos 60°)$$
$$= 25 + 64 + 80(0.5000) = 129.$$

Therefore, $\|\vec{v}_3\| = \sqrt{129} \approx 11.4$. From the Law of Sines,

$$\frac{\sin \alpha}{\|\vec{v}_1\|} = \frac{\sin 120°}{\|\vec{v}_3\|},$$

or, equivalently,

$$\sin \alpha = \frac{\|\vec{v}_1\| \sin 120°}{\|\vec{v}_3\|}$$

$$\approx \frac{5(0.8660)}{11.4} \approx 0.3798.$$

(*continued*)

Hence, $\alpha \approx 22.3°$. Thus, the forces of 5 and 8 pounds, acting at a angle of $60°$ with respect to each other, can be represented by the singl resultant force of 11.4 pounds. The angle it makes with the 8-pound forc has a measure of approximately $22.3°$.

Velocity   Geometric vectors also can facilitate the solution of problems involv ing velocities and accelerations. As examples, we will look at severa air navigation problems. First, however, we consider some terminolog that is commonly used. The **heading** of an airplane is the direction i which it is pointed, and the **course** is the direction in which it is actuall moving over the ground. Its **air speed** is the speed relative to the ai and its **ground speed** is the speed relative to the ground. The two direc tions and two speeds may differ because of wind effect.

If we use one geometric vector, say $\vec{v}_1$, to represent the heading and ai speed of an airplane (see Figure 6.13) and another, say $\vec{v}_2$, to represen the wind direction and speed, then $\vec{v}_3 = \vec{v}_1 + \vec{v}_2$ represents the cours and ground speed. The angle α between the geometric vectors $\vec{v}_1$ and $\vec{v}_3$ i called the **drift angle**. The measures of the angles for the heading, wind and course directions in air navigation are generally taken clockwis from the ray directed toward true north. The angle is called the **bearin** and has a positive measure. In graphical representations, we will use th positive y-axis as the true north direction and the positive x-axis a the true east direction. For example, in Figure 6.14, the bearing of th geometric vector $\vec{v}_1$ is $120°$, that of $\vec{v}_2$ is $285°$, and that of $\vec{v}_3$ is $354°$.

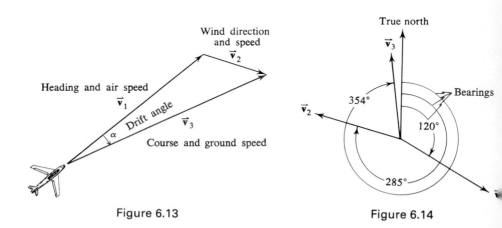

Figure 6.13   Figure 6.14

* Other types of bearings, each of which has its own reference, or initial, ray from which th angle is measured, are also defined. A bearing used in surveying is designated by an acu angle measured to the east or west of a ray pointing toward true north or true south, fo example, N 35° E, N 38° W, S 22° E, and S 47° W.

Example   An airplane is headed northeast (bearing 45°) with an air speed of 500 miles per hour, with a wind blowing from the southeast (bearing 315°) at a speed of 75 miles per hour. Find the drift angle, the ground speed, and the course of the plane.

Solution   The figure below shows three vectors: $\vec{v}_1$ representing the heading and air speed, $\vec{v}_2$ representing the wind direction and speed, and $\vec{v}_3$ representing the course and ground speed. From geometry we have that

$$\angle ABC = 315° - 45° - 180° = 90°.$$

Because $\angle ABC$ is a right angle,

$$\tan \alpha = \frac{75}{500} = 0.1500,$$

and the drift angle $\alpha \approx 8.5°$. The bearing, or direction, of the course is

$$45° - \alpha \approx 45° - 8.5° = 36.5°.$$

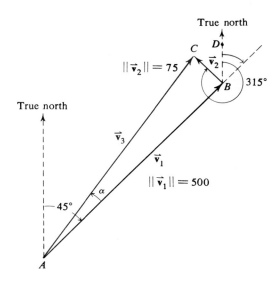

Now, because $\cos \alpha = 500/\|\vec{v}_3\|$,

$$\|\vec{v}_3\| \approx \frac{500}{\cos 8.5°} \approx \frac{500}{0.9890} \approx 506.$$

Thus, the ground speed (magnitude of the course vector) is approximately 506 miles per hour.

EXERCISE SET 6.4

A

■ *The direction angle α and $\|\vec{v}\|$ are given. Find $\|\vec{v}_x\|$ and $\|\vec{v}_y\|$.*

Example $\alpha = 60°$ and $\|\vec{v}\| = 8$

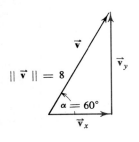

Solution   The figure shows the geometric vector $\vec{v}$ and its rectangular components $\vec{v}_x$ and $\vec{v}_y$. From Equations (1) and (2) on page 209,

$$\|\vec{v}_x\| = \|\vec{v}\||\cos 60°| = 8\left(\frac{1}{2}\right) = 4,$$

$$\|\vec{v}_y\| = \|\vec{v}\||\sin 60°| = 8\left(\frac{\sqrt{3}}{2}\right) = 4\sqrt{3}.$$

1. $\alpha = 45°$ and $\|\vec{v}\| = 5\sqrt{2}$ **2.** $\alpha = 30°$ and $\|\vec{v}\| = 6$

3. $\alpha = 90°$ and $\|\vec{v}\| = 15$ **4.** $\alpha = 120°$ and $\|\vec{v}\| = 2$

5. $\alpha = -30°$ and $\|\vec{v}\| = 12$ **6.** $\alpha = -45°$ and $\|\vec{v}\| = 2\sqrt{2}$

7. $\alpha = -120°$ and $\|\vec{v}\| = 8$ **8.** $\alpha = -150°$ and $\|\vec{v}\| = 4$

■ *In Exercises 9–14, the two given forces act on a point in a plane at an angle with the specified measure. Find the magnitude of the resultant force and the angles the resultant force makes with the given forces.*

Example   Forces of 10 and 20 newtons act on a point at an angle of 45° with respect to each other.

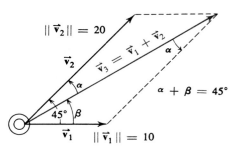

Solution   The figure shows the given forces $\vec{v}_1$ and $\vec{v}_2$ and the resultant $\vec{v}_3 = \vec{v}_1 + \vec{v}_2$. Since $\alpha + \beta = 45°$, the measure of the angle opposite is 135°. From the Law of Cosines,

$$\|\vec{v}_3\|^2 = 10^2 + 20^2 - 2(10)(20)(\cos 135°)$$

$$\approx 100 + 400 + 283 \approx 783;$$

$$\|\vec{v}_3\| \approx 28.0.$$

From the Law of Sines,

$$\frac{\sin \alpha}{\|\vec{v}_1\|} = \frac{\sin 135°}{\|\vec{v}_3\|},$$

so

$$\sin \alpha = \frac{\|\vec{v}_1\| \sin 135°}{\|\vec{v}_3\|} \approx \frac{10(0.7071)}{28.0} \approx 0.2525.$$

Hence, $\alpha \approx 14.6°$. If $\alpha + \beta = 45°$, then

$$\beta \approx 45° - 14.6° = 30.4°.$$

Thus, the resultant force is approximately 28.0 newtons. The angles this force makes with the force of 10 newtons and with the force of 20 newtons have measures of approximately 30.4° and 14.6°, respectively.

9. Forces of 3 and 4 pounds, acting at an angle of 90° with respect to each other.

10. Forces of 5 and 7 pounds, acting at an angle of 30° with respect to each other.

11. Forces of 15 and 20 pounds, acting at an angle of 75° with respect to each other.

12. Forces of 8 and 11 pounds, acting at an angle of 132° 48′ with respect to each other.

13. Forces of 6 and 10 newtons acting at an angle of 120° with respect to each other.

14. Forces of 20 and 30 newtons, acting at an angle of 100° with respect to each other.

15. The resultant of two forces acting at an angle of 90° with respect to each other is 50 newtons. If one of the forces is 30 newtons, find the other force and the angle it makes with the resultant.

16. Two equal forces acting at an angle of 60° with respect to each other have a resultant of 80 newtons. Find the magnitude of each force.

17. Two forces acting at an angle of 60° with respect to each other have a resultant of 50 pounds. If one force acts at an angle of 20° with respect to the resultant, find the magnitude of each force.

18. Two forces, one of 100 pounds and the other 200 pounds, have a resultant of 250 pounds. Find the angle formed by the two forces.

Example An auto weighing 3000 pounds is on a slope that makes an angle of 9° 30′ with the horizontal. Find the force that pulls the auto down the hill.

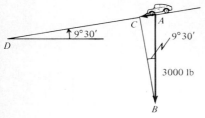

Solution   The force due to gravity can be resolved into two forces, on force in the direction of the incline and one force perpendicular to th plane. See the figure. In this case, $\overrightarrow{AB}$, the force due to gravity, is resolve into $\overrightarrow{AC}$ along the incline and $\overrightarrow{CB}$, perpendicular to $\overrightarrow{AC}$. From geometry $\angle B = \angle D$ because the sides of $\angle B$ are perpendicular to the sides o $\angle D$. Thus,

$$\sin 9° \; 30' = \frac{\|\overrightarrow{AC}\|}{3000},$$

from which

$$\|\overrightarrow{AC}\| = 3000 \sin 9° \; 30' = 3000 \sin 9.5° \approx 495.$$

Hence, a force of 495 pounds pulls the auto down the hill.

19. An auto, weighing 3500 pounds, is on a slope that makes an angle c 8° 24′ with the horizontal. Find the force that pulls the auto dow the hill.

20. A steel ball, weighing 50 pounds, is on an inclined surface. Wha angle does the plane make with the horizontal if a force of 10 pound is pulling the ball down the surface?

21. A force of 110 pounds is needed to keep a weight of 200 pounds fror sliding down an inclined plane. What angle does the plane mak with the horizontal?   *Note:* The force that keeps it from sliding opposite to the force pulling it down the plane and has the sam magnitude.

22. A force of 200 pounds is needed to keep a weight of 320 pounds fror sliding down an inclined ramp. What angle does the plane mak with the horizontal?

23. A bullet has an acceleration of 2 meters per second² in a directio making an angle of 35° with the horizontal. Find the horizontal an vertical components ($\|\vec{v}_x\|$ and $\|\vec{v}_y\|$) of its acceleration.

24. A space rocket has an acceleration of 4.2 meters per second² in direction making an angle of 73° with the horizontal. Find th horizontal and vertical components of the rocket's acceleration

25. A crew can row a boat at a speed of 9.2 kilometers per hour in sti water. If the crew heads across a river at right angles to the currer and finds its "drift angle" to be 6°, find the speed of the current.

26. A person can swim at a speed of 2 kilometers per hour in still wate If he heads across a river at right angles to a current of 5 kilometer per hour, find his speed in relation to the land and the direction i which he actually moves.

27. An airplane is headed southeast (bearing 135°) with an air speed of 600 miles per hour, with the wind blowing from the northeast (bearing 225°) at a speed of 120 miles per hour. Find the drift angle, the ground speed, and the course of the airplane.

28. Solve Exercise 27 if the wind is from the southwest (bearing 45°) at a speed of 60 miles per hour.

29. An airplane is headed on a bearing of 250° with an air speed of 425 miles per hour. The course has a bearing of 262°. The ground speed is 475 miles per hour. Find the drift angle, the wind direction, and the wind speed.

30. Solve Exercise 29 if the course has a bearing of 241°.

31. A pilot wishes to fly on a course bearing 90° and with a ground speed of 600 kilometers per hour. If a wind is blowing from the north (bearing 180°) with a speed of 50 kilometers an hour, what must be the heading and air speed of the aircraft?

32. Solve Exercise 31 if the wind is from the northwest (bearing 135°).

B

33. Two ships leave a harbor at the same time, the first traveling at 20 knots (nautical miles per hour) on a course bearing 40° and the second at 25 knots on a course bearing 300°. After one hour, how far apart are they and what is the bearing of a course from the first ship to the second?

34. Three forces of 25, 35, and 45 pounds act in directions having bearings of 240°, 290°, and 340°, respectively. Find a force, acting due east, and a force, acting due south, that would counterbalance these three forces.

35. A 100-pound weight hangs from a rope that makes an angle of 90° where it is attached to the weight (see figure). What is the tension (force) on each half of the rope?

36. A 50-pound weight hangs from a rope whose two ends are tied at 60° angles to a support (see figure). What is the tension on each half of the rope?

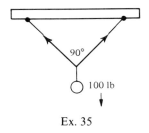

Ex. 35

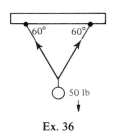

Ex. 36

6.5 Vectors as Ordered Pairs

In Sections 6.3 and 6.4 we considered geometric vectors. In this section we will consider the more general notion of a two-dimensional vector as an ordered pair of real numbers, some of its properties, and the close

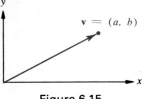

Figure 6.15

relationship that exists between such a vector and a geometric vector i̇ the plane.

Geometric vectors whose initial points are located at a fixed positio in the plane are called **bound geometric vectors, or position vectors.** Th terminal point of such a geometric vector with initial point at the origi̇ has coordinates that are the components of exactly one ordered pȧ (a, b) of real numbers. This relationship, shown in Figure 6.15, suggest̄ the following definition.

*A **two-dimensional vector v** is an ordered pair of real numbers*

$$\mathbf{v} = (a, b), \quad where \ a, b \in R.$$

The set of all such vectors is designated by **V.**

For a vector $\mathbf{v} = (a, b)$, the numbers a and b are called the **scalȧ components** of **v**.

Notice that boldface type *without* an arrow above the symbol is bein̄ used to differentiate a vector **v**, an ordered pair of real numbers, from geometric vector $\vec{\mathbf{v}}$.

Example   The figure shows the geometric vectors associated with th̄ vectors $\mathbf{v_1} = (2, 3), \ \mathbf{v_2} = (-3, 5), \ \mathbf{v_3} = (-5, -1), \ and \ \mathbf{v_4} = (0, -4)$ respectively.

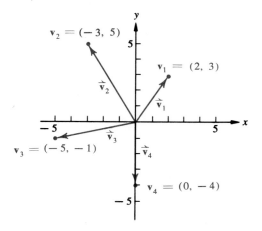

The following definitions are suggested by the close association of . vector with a geometric vector.

For all $\mathbf{v} \in \mathbf{V}$, *where* $\mathbf{v} = (a, b)$,

$$-\mathbf{v} = -(a, b) = (-a, -b);$$

the **norm** *or* **magnitude** *of* **v** *is*

$$\|\mathbf{v}\| = \|(a, b)\| = \sqrt{a^2 + b^2};$$

the **direction angle** *of* **v** *is the angle* α *such that*

$$\cos \alpha = \frac{a}{\|\mathbf{v}\|} \quad \text{and} \quad \sin \alpha = \frac{b}{\|\mathbf{v}\|}, \ \|\mathbf{v}\| \neq 0,$$

where $-180° < \alpha \le 180°$.

Although the definition defines the direction angle of $\mathbf{v} = (a, b)$ in terms of both $\cos \alpha$ and $\sin \alpha$, in practice we generally determine only one function value in order to find α. This is possible because the quadrant in which **v** lies can readily be determined from the ordered pair (a, b).

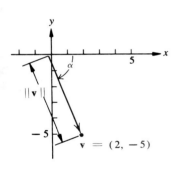

Example   The norm of the vector $\mathbf{v} = (2, -5)$ is

$$\|\mathbf{v}\| = \sqrt{2^2 + (-5)^2} = \sqrt{29}.$$

Then,

$$\sin \alpha = \frac{-5}{\sqrt{29}} \approx -0.9285,$$

and because $(2, -5)$ is in Quadrant IV $(-90° < \alpha < 0°)$,

$$\alpha \approx -68.2° = -68° \ 12'.$$

Notice that the norm of a vector **v** is the length of the corresponding geometric vector $\vec{v}$, and its direction angle is the angle α such that $-180° < \alpha \le 180°$, associated with the geometric vector.

Several basic operations on vectors are closely related to the operations on geometric vectors that have been defined in Section 6.3.

Vector Addition   As was the case for geometric vectors, the operation of forming the sum of two vectors is called **vector addition.**

For all vectors $\mathbf{v}_1$ *and* $\mathbf{v}_2$, *where* $\mathbf{v}_1 = (a_1, b_1)$ *and* $\mathbf{v}_2 = (a_2, b_2)$, *the* **sum** $\mathbf{v}_1 + \mathbf{v}_2$ *is a vector such that*

$$\mathbf{v}_1 + \mathbf{v}_2 = (a_1, b_1) + (a_2, b_2) = (a_1 + a_2, b_1 + b_2).$$

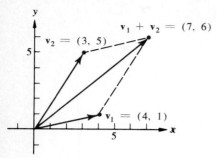

Example   Given that $v_1 = (4, 1)$ and $v_2 = (3, 5)$,

$$v_1 + v_2 = (4, 1) + (3, 5)$$
$$= (4 + 3, 1 + 5)$$
$$= (7, 6).$$

The corresponding geometric vectors for v_1, v_2, and their sum $v_1 +$
are shown in the figure.

Note that, in the above example, the resultant $v_1 + v_2$ is associate
with the geometric vector that is one of the diagonals of a parallelogram

Difference of Two Vectors   The difference of two vectors, $v_2 - v_1$, is defined similarly to th
difference of two real numbers.

For all vectors v_1 and v_2, the **difference** *$v_2 - v_1$ is a vector such that*

$$v_2 - v_1 = v_2 + (-v_1).$$

Example   Given that $v_1 = (2, 5)$ and $v_2 = (7, 1)$,

$$v_2 - v_1 = (7, 1) - (2, 5)$$
$$= (7, 1) + (-2, -5)$$
$$= (7 - 2, 1 - 5)$$
$$= (5, -4).$$

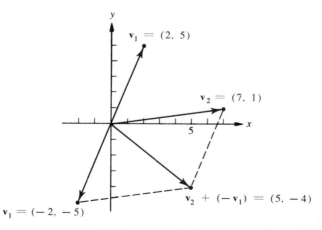

Multiplication of a Vector by a Scalar   The second basic operation on vectors concerns multiplication of
scalar (real number) and a vector and is related to the similar operation
on geometric vectors.

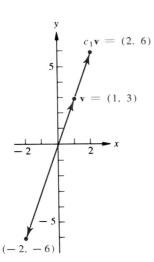

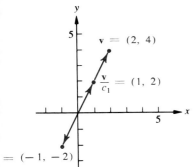

For all $\mathbf{v} \in \mathbf{V}$, *where* $\mathbf{v} = (a, b)$ *and* $c \in R$,

$$c\mathbf{v} = c(a, b) = (ca, cb).$$

Example   If $\mathbf{v} = (1, 3)$, $c_1 = 2$, and $c_2 = -2$, then

$$c_1\mathbf{v} = 2(1, 3) = (2, 6)$$

and

$$c_2\mathbf{v} = -2(1, 3) = (-2, -6).$$

Notice from the above example that $c\mathbf{v}$ has a magnitude $|c| \cdot \|\mathbf{v}\|$ if c is a positive or negative real number. The direction angles of $\mathbf{v}$ and $c\mathbf{v}$ are the same if c is positive, and they differ by $180°$ if c is negative.

The quotient of a vector divided by a scalar is defined in terms of a product in much the same way as is the quotient of two real numbers.

Example   If $\mathbf{v} = (2, 4)$, $c_1 = 2$, and $c_2 = -2$, then

$$\frac{\mathbf{v}}{c_1} = \frac{(2, 4)}{2} = \frac{1}{2}(2, 4) = (1, 2),$$

and

$$\frac{\mathbf{v}}{c_2} = \frac{(2, 4)}{-2} = -\frac{1}{2}(2, 4) = (-1, -2).$$

$a\mathbf{i} + b\mathbf{j}$ Form   Vectors $\mathbf{v}_1 = (1, 0)$ and $\mathbf{v}_2 = (0, 1)$ are called **unit vectors.** If they are designated $\mathbf{i}$ and $\mathbf{j}$, respectively, then each vector

$$\mathbf{v} = (a, b)$$
$$= (a, 0) + (0, b)$$
$$= a(1, 0) + b(0, 1)$$
$$= a\mathbf{i} + b\mathbf{j}.$$

Here again the real numbers a and b are called the **scalar components** of the vector.

Examples   Express each vector as the sum of scalar multiples of $\mathbf{i}$ and $\mathbf{j}$ and sketch the corresponding geometric vectors.

a. $(-4, 0)$ b. $(-4, 3)$ *(continued)*

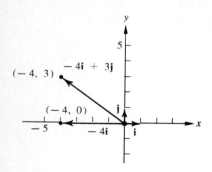

Solutions

a. $(-4, 0) = -4(1, 0)$ b. $(-4, 3) = (-4, 0) + (0, 3)$

$\qquad\qquad = -4\mathbf{i}$ $= -4(1, 0) + 3(0, 1)$

$\qquad\qquad\qquad\qquad\qquad\qquad\qquad = -4\mathbf{i} + 3\mathbf{j}$

Example   Find the norm and the direction angle of $5\mathbf{i} - 3\mathbf{j}$.

Solution $5\mathbf{i} - 3\mathbf{j} = 5(1, 0) - 3(0, 1) = (5, -3)$. Thus,

$$\|5\mathbf{i} - 3\mathbf{j}\| = \|(5, -3)\|,$$

from which we have

$$\|5\mathbf{i} - 3\mathbf{j}\| = \sqrt{5^2 + (-3)^2} = \sqrt{34}.$$

Then,

$$\sin \alpha = \frac{-3}{\sqrt{34}} \approx -0.5145.$$

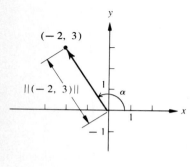

Furthermore, $(5, -3)$ corresponds to a point in Quadrant IV. Hence

$$\alpha \approx -31.0°.$$

EXERCISE SET 6.5

A

▪ *For each ordered pair* (**a**) *sketch each corresponding geometric vector with initial point at the origin and* (**b**) *find the norm and the direction angle (to the nearest six minutes) of each vector.*

Example $(-2, 3)$

Solution
   a. See the figure.
   b. The norm is

$$\|(-2, 3)\| = \sqrt{(-2)^2 + (3)^2} = \sqrt{13}.$$

Since

$$\sin \alpha = \frac{3}{\sqrt{13}} \approx 0.8321,$$

and because $(-2, 3)$ is in Quadrant II $(90° < \alpha < 180°)$,

$$\alpha \approx 123.7° = 123°\ 42'.$$

1. $(3, 4)$ **2.** $(-3, 4)$ **3.** $(1, 1)$ **4.** $(-1, 1)$

5. $(-3, -3)$ **6.** $(0, -5)$ **7.** $(4, 0)$ **8.** $(\sqrt{3}, 1)$

■ *Given the following norms and direction angles, find the vector* **v**, *and sketch the corresponding geometric vector.*

Example $\|\mathbf{v}\| = 5$ and $\alpha = 30°$

Solution   If $\mathbf{v} = (a, b)$, then

$$a = \|\mathbf{v}\| \cos \alpha,$$
$$b = \|\mathbf{v}\| \sin \alpha.$$

Thus,

$$\mathbf{v} = (5 \cos 30°, 5 \sin 30°) = \left(\frac{5\sqrt{3}}{2}, \frac{5}{2}\right).$$

9. $\|\mathbf{v}\| = 2\sqrt{2}$ and $\alpha = 45°$ **10.** $\|\mathbf{v}\| = 2$ and $\alpha = 60°$
11. $\|\mathbf{v}\| = 1$ and $\alpha = 90°$ **12.** $\|\mathbf{v}\| = 4$ and $\alpha = 150°$
13. $\|\mathbf{v}\| = 5$ and $\alpha = -150°$ **14.** $\|\mathbf{v}\| = 3\sqrt{2}$ and $\alpha = -30°$

■ *Write the sum of each pair of vectors as an ordered pair and show the corresponding geometric vectors.*

Example

$\mathbf{v}_1 = (2, -5)$ and $\mathbf{v}_2 = (-4, 2)$

Solution

$$\mathbf{v}_1 + \mathbf{v}_2 = (2, -5) + (-4, 2)$$
$$= (2 + (-4), -5 + 2)$$
$$= (-2, -3)$$

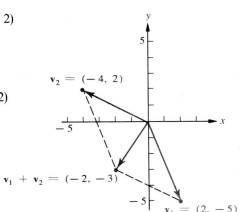

15. $\mathbf{v}_1 = (-4, 0)$ and $\mathbf{v}_2 = (0, 6)$

16. $\mathbf{v}_1 = (-6, 0)$ and $\mathbf{v}_2 = (0, 1)$

17. $\mathbf{v}_1 = (1, -4)$ and $\mathbf{v}_2 = (2, 5)$

18. $\mathbf{v}_1 = (7, -8)$ and $\mathbf{v}_2 = (-3, -7)$

19. $\mathbf{v}_1 = (-5, 3)$ and $\mathbf{v}_2 = (1, -2)$

20. $\mathbf{v}_1 = (-6, 6)$ and $\mathbf{v}_2 = (3, -3)$

■ *Let* $\mathbf{v}_1 = (3, 6)$, $\mathbf{v}_2 = (5, -1)$, $\mathbf{v}_3 = (-4, 2)$, $c_1 = 2$, *and* $c_2 = -$
Write each expression as an ordered pair and show the correspondi
geometric vectors.

Example $c_1\mathbf{v}_2 + c_1\mathbf{v}_3$

Solution $c_1\mathbf{v}_2 + c_1\mathbf{v}_3 = 2(5, -1) + 2(-4, 2)$

$$= (10, -2) + (-8, 4)$$
$$= (10 + (-8), -2 + 4)$$
$$= (2, 2)$$

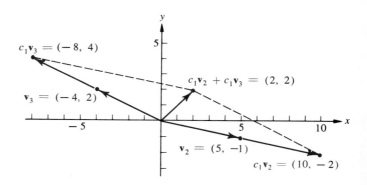

21. $-\mathbf{v}_1$ **22.** $-\mathbf{v}_2$ **23.** $c_1\mathbf{v}_1$

24. $c_2\mathbf{v}_2$ **25.** $c_2\mathbf{v}_1 + \mathbf{v}_3$ **26.** $c_1\mathbf{v}_2 + \mathbf{v}_1$

27. $c_1\mathbf{v}_1 + c_2\mathbf{v}_2$ **28.** $c_1\mathbf{v}_2 + c_2\mathbf{v}_3$ **29.** $c_1\mathbf{v}_2 - \mathbf{v}_1$

30. $c_2\mathbf{v}_3 - \mathbf{v}_2$ **31.** $c_2\mathbf{v}_2 - c_1\mathbf{v}_3$ **32.** $c_2\mathbf{v}_3 - c_1\mathbf{v}_1$

33. $\dfrac{\mathbf{v}_2}{c_1}$ **34.** $\dfrac{\mathbf{v}_3}{c_2}$

■ *Express each vector as a sum of scalar multiples of* **i** *and* **j.**

Example $(2, 9)$

Solution $(2, 9) = (2, 0) + (0, 9)$

$$= 2(1, 0) + 9(0, 1)$$
$$= 2\mathbf{i} + 9\mathbf{j}$$

35. $(3, 5)$ **36.** $(-2, 1)$ **37.** $(1, -4)$

38. $(-2, -3)$ **39.** $(-3, 0)$ **40.** $(0, 4)$

■ *Find the norm and the direction angle, to the nearest six minutes, of each vector. Sketch the corresponding geometric vector.*

Example $2\mathbf{i} - 3\mathbf{j}$

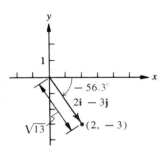

Solution $2\mathbf{i} - 3\mathbf{j} = 2(1, 0) - 3(0, 1)$

$$= (2, 0) + (0, -3)$$

$$= (2, -3)$$

Hence,

$$\|2\mathbf{i} - 3\mathbf{j}\| = \|(2, -3)\|$$

$$= \sqrt{2^2 + (-3)^2}$$

$$= \sqrt{13},$$

from which

$$\sin \alpha = \frac{-3}{\sqrt{13}} \approx -0.8321.$$

Since $(2, -3)$ is in Quadrant IV, $-90° < \alpha < 0°$. Thus,

$$\alpha \approx -56.3° = -56° \ 18'.$$

41. $4\mathbf{i} - 4\mathbf{j}$ **42.** $\sqrt{3}\mathbf{i} + \mathbf{j}$ **43.** $2\mathbf{i} + \mathbf{j}$ **44.** $-\mathbf{i} - 5\mathbf{j}$

45. $-\mathbf{i} - 3\mathbf{j}$ **46.** $-6\mathbf{i} + \mathbf{j}$ **47.** $-\mathbf{i} + \sqrt{3}\mathbf{j}$ **48.** $3\mathbf{i} - 2\mathbf{j}$

6.6 Inner Products

In Section 6.5 we considered two basic operations, vector addition and the multiplication of a vector by a scalar. The result of either operation is a vector. Another operation on two vectors is defined in such a way as to be useful in a variety of problems related to physical and geometric concepts. The result of this operation is a *scalar*.

For all $\mathbf{v}_1, \mathbf{v}_2 \in V,$ $\mathbf{v}_1 = (a_1, b_1)$ *and* $\mathbf{v}_2 = (a_2, b_2),$ *the* **inner product**

$$\mathbf{v}_1 \cdot \mathbf{v}_2 = (a_1, b_1) \cdot (a_2, b_2) = a_1 a_2 + b_1 b_2. \tag{1}$$

Since the symbol for the operation is a dot, the inner product is also called a **dot product.** Notice that the dot as used here has a different

meaning than when it is used for the operation of multiplication in t**
set of real numbers.

Examples

a. If $\mathbf{v}_1 = (2, 3)$ and $\mathbf{v}_2 = (-5, 4)$, then

$$\mathbf{v}_1 \cdot \mathbf{v}_2 = (2, 3) \cdot (-5, 4)$$
$$= (2)(-5) + (3)(4) = 2.$$

b. If $\mathbf{v}_1 = 3\mathbf{i} - 4\mathbf{j}$ and $\mathbf{v}_2 = \mathbf{i} + 2\mathbf{j}$, then

$$\mathbf{v}_1 \cdot \mathbf{v}_2 = (3\mathbf{i} - 4\mathbf{j}) \cdot (\mathbf{i} + 2\mathbf{j})$$
$$= (3)(1) + (-4)2 = -5.$$

Because the inner product of two vectors is a scalar, the inner produ**
is also called the **scalar product.**

Another useful form of the inner product can be obtained from th**
Law of Cosines:

If $\mathbf{v}_1 = (a_1, b_1)$ *and* $\mathbf{v}_2 = (a_2, b_2)$, *where* $\mathbf{v}_1$ *and* $\mathbf{v}_2$ *are nonzero vectors, and* γ *is the angle formed by the geometric vectors corresponding to* $\mathbf{v}_1$ *and* $\mathbf{v}_2$, *then*

$$\mathbf{v}_1 \cdot \mathbf{v}_2 = \|\mathbf{v}_1\| \|\mathbf{v}_2\| \cos \gamma. \qquad (2)$$

The above property follows from the following argument. Conside**
the line segment with length c drawn between the points correspondin**
to (a_1, b_1) and (a_2, b_2), as shown in Figure 6.16. From the Law c**
Cosines,

$$c^2 = \|\mathbf{v}_1\|^2 + \|\mathbf{v}_2\|^2 - 2\|\mathbf{v}_1\| \|\mathbf{v}_2\| \cos \gamma,$$

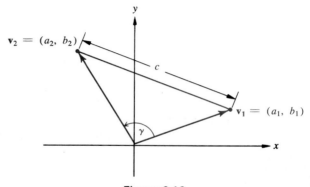

Figure 6.16

from which, for nonzero vectors $\mathbf{v}_1$ and $\mathbf{v}_2$,

$$\cos \gamma = \frac{\|\mathbf{v}_1\|^2 + \|\mathbf{v}_2\|^2 - c^2}{2\|\mathbf{v}_1\|\|\mathbf{v}_2\|}.$$

Substituting $\sqrt{a_1^2 + b_1^2}$ for $\|\mathbf{v}_1\|$, $\sqrt{a_2^2 + b_2^2}$ for $\|\mathbf{v}_2\|$, and

$$\sqrt{(a_2 - a_1)^2 + (b_2 - b_1)^2}$$

for c and simplifying, we obtain

$$\cos \gamma = \frac{a_1 a_2 + b_1 b_2}{\|\mathbf{v}_1\|\|\mathbf{v}_2\|}.$$

Substituting $\mathbf{v}_1 \cdot \mathbf{v}_2$ for $a_1 a_2 + b_1 b_2$ from the definition of the inner product on page 225 yields

$$\cos \gamma = \frac{\mathbf{v}_1 \cdot \mathbf{v}_2}{\|\mathbf{v}_1\|\|\mathbf{v}_2\|}$$

from which

$$\mathbf{v}_1 \cdot \mathbf{v}_2 = \|\mathbf{v}_1\|\|\mathbf{v}_2\| \cos \gamma.$$

We can now express the inner product of two vectors $\mathbf{v}_1 = (a_1, b_1)$ and $\mathbf{v}_2 = (a_2, b_2)$ either, as shown in Equation (1), in terms of their scalar components or, as in Equation (2), in terms of the norms of the two vectors and the cosine of the angle γ formed by their corresponding geometric vectors.

Example   Given that $\|\mathbf{v}_1\| = 3$, and $\mathbf{v}_1$ has a direction angle, $\alpha_1 = 20°$, $\|\mathbf{v}_2\| = 8$, and $\mathbf{v}_2$ has a direction angle, $\alpha_2 = 170°$, find $\mathbf{v}_1 \cdot \mathbf{v}_2$.

Solution   From Equation (2) on page 226,

$$\mathbf{v}_1 \cdot \mathbf{v}_2 = (3)(8) \cos(170° - 20°) = (3)(8) \cos 150°$$

$$= 24\left(-\frac{\sqrt{3}}{2}\right) = -12\sqrt{3}.$$

Scalar Projections   The property of inner products exhibited for two nonzero vectors in Equation (2) is particularly useful in several equivalent forms:

$$\cos \gamma = \frac{\mathbf{v}_1 \cdot \mathbf{v}_2}{\|\mathbf{v}_1\|\|\mathbf{v}_2\|}, \tag{3}$$

$$\|v_1\| \cos \gamma = \frac{v_1 \cdot v_2}{\|v_2\|}, \tag{4}$$

$$\|v_2\| \cos \gamma = \frac{v_1 \cdot v_2}{\|v_1\|}. \tag{5}$$

Equation (3) enables us to find the angle γ between any two nonzero vectors.

Example The angle γ between the geometric vectors corresponding to (2, 3) and (0, 2) is given by

$$\cos \gamma = \frac{(2, 3) \cdot (0, 2)}{\|(2, 3)\| \, \|(0, 2)\|}$$

$$= \frac{2 \cdot 0 + 3 \cdot 2}{\sqrt{4 + 9}\sqrt{0 + 4}}$$

$$= \frac{3}{\sqrt{13}} \approx 0.8321.$$

Thus, $\gamma \approx 33.7° = 33° \, 42'$.

In Section 6.4 we obtained projections of a vector on the x- and y-axes of a Cartesian coordinate system. Equations (4) and (5) above now give us a way to find the projection of a vector on any other vector.

The product $\|v_1\| \cos \gamma$ is called the **scalar projection** of v_1 on v_2, and the product $\|v_2\| \cos \gamma$ is called the scalar projection of v_2 on v_1. In a geometric sense, these can be interpreted as the lengths of the line segments shown in Figure 6.17 (a and b).

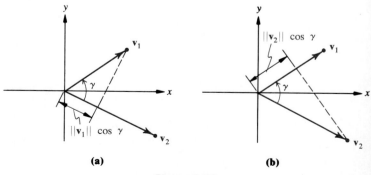

(a)   (b)

Figure 6.17

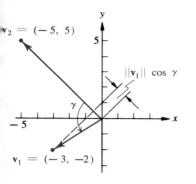

Example   Given that $v_1 = (-3, -2)$ and $v_2 = (-5, 5)$, the scalar projection of v_1 on v_2 [Equation (4) on page 228] is

$$\|v_1\| \cos \gamma = \frac{v_1 \cdot v_2}{\|v_2\|} = \frac{(-3, -2) \cdot (-5, 5)}{\|(-5, 5)\|}$$

$$= \frac{15 - 10}{\sqrt{(-5)^2 + 5^2}} = \frac{5}{\sqrt{50}}$$

$$= \frac{5}{5\sqrt{2}} = \frac{1}{\sqrt{2}}.$$

The scalar projection of v_2 on v_1 [Equation (5) on page 228] is

$$\|v_2\| \cos \gamma = \frac{v_1 \cdot v_2}{\|v_1\|} = \frac{(-3, -2) \cdot (-5, 5)}{\|(-3, -2)\|}$$

$$= \frac{15 - 10}{\sqrt{(-3)^2 + (-2)^2}}$$

$$= \frac{5}{\sqrt{13}}.$$

Orthogonal Vectors   If v_1 and v_2 are nonzero vectors and orthogonal (perpendicular), then $\|v_1\| \neq 0$, $\|v_2\| \neq 0$, $\cos \gamma = 0$, and

$$v_1 \cdot v_2 = \|v_1\| \|v_2\| \cos \gamma = 0.$$

We also note that for $0° \leq \gamma \leq 180°$, $\cos \gamma = 0$ only if $\gamma = 90°$. Since

$$v_1 \cdot v_2 = a_1 a_2 + b_1 b_2,$$

we can test for the orthogonality (perpendicularity) of two vectors directly from their scalar components using the following property:

The vectors $v_1 = (a_1, b_1)$ and $v_2 = (a_2, b_2)$ are orthogonal if and only if

$$v_1 \cdot v_2 = a_1 a_2 + b_1 b_2 = 0. \tag{6}$$

Examples
   a. $(2, -4)$ and $(6, 3)$ are orthogonal because

$$(2, -4) \cdot (6, 3) = (2)(6) + (-4)(3) = 0.$$

(*continued*)

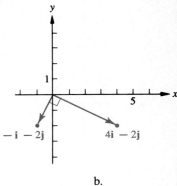

a.   b.

b. $-\mathbf{i} - 2\mathbf{j}$ and $4\mathbf{i} - 2\mathbf{j}$ are orthogonal because

$$(-\mathbf{i} - 2\mathbf{j}) \cdot (4\mathbf{i} - 2\mathbf{j}) = (-1)(4) + (-2)(-2) = 0.$$

Applications   If a force is applied to an object so as to cause motion in a particula
direction, and if the force is applied throughout the motion and in th
same direction as the motion, then the product of the magnitude of th
force (F) and the distance (d) that the object moves is called **work.**
many situations, the direction of an applied force is not the same as th
direction in which the object moves. In this case, the work is the produ
of the magnitude of the *component of the force in the direction of moti*
and the distance the object moves. Since the inner product of two vecto
can be used to find such a product whether or not the two vectors ha
the same direction, we provide a more general definition of work.

*If **F** is a force vector and **d** a displacement vector, then the work W is
given by*

$$W = \mathbf{F} \cdot \mathbf{d}.$$

Example   A force of 40 pounds at an angle of 30° to the horizontal
used to move a box 8 feet on a level floor. Find the work done.

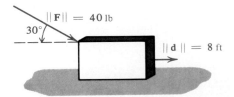

Solution   The figure shows the vectors in a Cartesian system. Since $\|\mathbf{F}\| = 40$ pounds, $\|\mathbf{d}\| = 8$ feet, and $\gamma = 30°$,

$$W = \mathbf{F} \cdot \mathbf{d} = \|\mathbf{F}\|\|\mathbf{d}\| \cos \gamma$$

$$= (40)(8) \cos 30°$$

$$= (40)(8)\left(\frac{\sqrt{3}}{2}\right)$$

$$= 160\sqrt{3} \approx 277.$$

If the force magnitude is measured in pounds and the displacement magnitude in feet, as in this case, the work is measured in foot-pounds. Hence, the work done is approximately 277 foot-pounds.

Example   How much work is done in lifting an 80-pound weight to a height of 3 feet?

Solution   Since the force and displacement vectors are in the same direction, the angle $\gamma = 0°$ and $\cos \gamma = 1$. Hence,

$$W = \mathbf{F} \cdot \mathbf{d} = \|\mathbf{F}\|\|\mathbf{d}\| \cos \gamma = (80)(3)(1) = 240.$$

Thus, the work done is 240 foot-pounds. In this case the work is simply the product of the force times the distance.

EXERCISE SET 6.6

A

■ *Find the inner product, $\mathbf{v}_1 \cdot \mathbf{v}_2$, of the given vectors.*

Examples
   a. $\mathbf{v}_1 = (7, -9)$ and $\mathbf{v}_2 = (-6, -5)$
   b. $\mathbf{v}_1 = -9\mathbf{i} + 8\mathbf{j}$ and $\mathbf{v}_2 = 3\mathbf{i} - 2\mathbf{j}$

Solutions $\mathbf{v}_1 \cdot \mathbf{v}_2 = a_1 a_2 + b_1 b_2$. Hence,

a. $\mathbf{v}_1 \cdot \mathbf{v}_2 = (7, -9) \cdot (-6, -5)$ b. $\mathbf{v}_1 \cdot \mathbf{v}_2 = (-9\mathbf{i} + 8\mathbf{j}) \cdot (3\mathbf{i} - 2\mathbf{j})$

$\qquad = (7)(-6) + (-9)(-5)$ $\qquad = (-9)(3) + (8)(-2)$

$\qquad = 3$ $\qquad = -43$

1. $\mathbf{v}_1 = (6, 0)$ and $\mathbf{v}_2 = (8, 10)$ 2. $\mathbf{v}_1 = (9, 3)$ and $\mathbf{v}_2 = (8, -5)$

3. $\mathbf{v}_1 = (7, -8)$ and $\mathbf{v}_2 = (2, 2)$ 4. $\mathbf{v}_1 = (3, 0)$ and $\mathbf{v}_2 = (0, -5)$

5. $\mathbf{v}_1 = 9\mathbf{i} - 4\mathbf{j}$ and $\mathbf{v}_2 = 5\mathbf{i}$ 6. $\mathbf{v}_1 = 8\mathbf{j}$ and $\mathbf{v}_2 = 7\mathbf{i} + 7\mathbf{j}$

7. $\mathbf{v}_1 = \mathbf{i} - 2\mathbf{j}$ and $\mathbf{v}_2 = 3\mathbf{i} - 5\mathbf{j}$ 8. $\mathbf{v}_1 = -4\mathbf{i} + 3\mathbf{j}$ and $\mathbf{v}_2 = 10\mathbf{j}$

■ *Find the exact value of* $\mathbf{v}_1 \cdot \mathbf{v}_2$ *given the norms and direction angles for* $\mathbf{v}_1$ *and* $\mathbf{v}_2$.

Example $\|\mathbf{v}_1\| = 10, \alpha_1 = 75°$ and $\|\mathbf{v}_2\| = 3, \alpha_2 = 45°$

Solution From Equation (2) on page 226,

$$\mathbf{v}_1 \cdot \mathbf{v}_2 = (10)(3) \cos(75° - 45°) = 30 \cos 30°$$

$$= 30\left(\frac{\sqrt{3}}{2}\right) = 15\sqrt{3}.$$

9. $\|\mathbf{v}_1\| = 8, \alpha_1 = 60°$ and $\|\mathbf{v}_2\| = 9, \alpha_2 = 15°$

10. $\|\mathbf{v}_1\| = 7, \alpha_1 = 160°$ and $\|\mathbf{v}_2\| = 6, \alpha_2 = -170°$

11. $\|\mathbf{v}_1\| = 5, \alpha_1 = -155°$ and $\|\mathbf{v}_2\| = \dfrac{3}{10}, \alpha_2 = 145°$

12. $\|\mathbf{v}_1\| = \dfrac{1}{2}, \alpha_1 = -60°$ and $\|\mathbf{v}_2\| = 10, \alpha_2 = -150°$

■ *Determine the angle* γ, *where* $0° \leq \gamma \leq 180°$, *to the nearest six minutes, formed by each pair of vectors.*

Example $\mathbf{v}_1 = (1, -7)$ and $\mathbf{v}_2 = (9, -1)$

Solution From Equation (3),

$$\cos \gamma = \frac{(1, -7) \cdot (9, -1)}{\sqrt{1^2 + (-7)^2}\sqrt{9^2 + (-1)^2}}$$

$$= \frac{16}{\sqrt{50}\sqrt{82}} = \frac{16}{5\sqrt{2} \cdot \sqrt{2} \cdot \sqrt{41}} = \frac{8}{5\sqrt{41}} \approx 0.2499,$$

from which $\gamma \approx 75.5° = 75° 30'$.

13. $\mathbf{v}_1 = (3, 0)$ and $\mathbf{v}_2 = (-7, 3)$

14. $\mathbf{v}_1 = (1, -3)$ and $\mathbf{v}_2 = (8, -2)$

15. $\mathbf{v}_1 = (4, 0)$ and $\mathbf{v}_2 = (0, -2)$

16. $\mathbf{v}_1 = (-2, 0)$ and $\mathbf{v}_2 = (0, -3)$

17. $v_1 = 9i + 4j$ and $v_2 = 2i$

18. $v_1 = -4i - 8j$ and $v_2 = i - 5j$

■ *Find the scalar projection of v_1 on v_2 and of v_2 on v_1 for the given vectors. Sketch the corresponding geometric vectors and the scalar projections.*

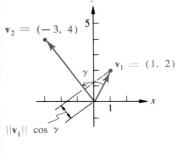

Example $v_1 = (1, 2)$ and $v_2 = (-3, 4)$

Solution   From Equation (4) on page 228, the scalar projection of v_1 on v_2 is

$$\|v_1\| \cos \gamma = \frac{v_1 \cdot v_2}{\|v_2\|} = \frac{(1)(-3) + (2)(4)}{\sqrt{(-3)^2 + (4)^2}}$$

$$= \frac{-3 + 8}{\sqrt{25}} = \frac{5}{5} = 1.$$

From Equation (5), the scalar projection of v_2 on v_1 is

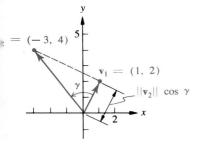

$$\|v_2\| \cos \gamma = \frac{v_1 \cdot v_2}{\|v_1\|} = \frac{5}{\sqrt{1^2 + 2^2}}$$

$$= \frac{5}{\sqrt{5}} = \sqrt{5}.$$

19. $v_1 = -4i$ and $v_2 = -5i - 2j$

20. $v_1 = i - 6j$ and $v_2 = -5j$

21. $v_1 = (2, 3)$ and $v_2 = (-4, 5)$

22. $v_1 = (-9, 10)$ and $v_2 = (2, 4)$

23. $v_1 = (-3, 5)$ and $v_2 = (3, -1)$

24. $v_1 = (-7, 4)$ and $v_2 = (3, -6)$

25. $v_1 = 3i$ and $v_2 = 2j$

26. $v_1 = 2i$ and $v_2 = -2j$

■ *Verify that the following pairs of vectors are orthogonal.*

Example $i + 4j$ and $8i - 2j$

Solution $(i + 4j) \cdot (8i - 2j) = (1)(8) + (4)(-2)$

$$= 8 - 8 = 0.$$

Hence, by Equation (6), the vectors are orthogonal.

27. $(0, 5)$ and $(-3, 0)$

28. $(-2, 2)$ and $(5, 5)$

29. $(-4, 1)$ and $(1, 4)$

30. $(9, 1)$ and $(-1, 9)$

31. $6\mathbf{i} - 9\mathbf{j}$ and $3\mathbf{i} + 2\mathbf{j}$

32. $-3\mathbf{i} + \sqrt{3}\,\mathbf{j}$ and $\mathbf{i} + \sqrt{3}\,\mathbf{j}$

■ *Find the work done, given each force vector and displacement vector.*

Example $\mathbf{F} = (1, 5)$ and $\mathbf{d} = (-2, 3)$

Solution $W = \mathbf{F} \cdot \mathbf{d} = (1, 5) \cdot (-2, 3)$

$$= 1 \cdot (-2) + 5 \cdot 3 = 13$$

33. $\mathbf{F} = (3, 6)$ and $\mathbf{d} = (4, 2)$

34. $\mathbf{F} = (5, -2)$ and $\mathbf{d} = (-1, -7)$

35. $\mathbf{F} = 7\mathbf{i} - 2\mathbf{j}$ and $\mathbf{d} = 4\mathbf{i} - 4\mathbf{j}$

36. $\mathbf{F} = -12\mathbf{i} + 3\mathbf{j}$ and $\mathbf{d} = -3\mathbf{i} + 12\mathbf{j}$

37. A small wagon is pulled horizontally a distance of 100 meters by short rope. If the rope is at an angle of 30° to the horizontal and constant force of 20 newtons is used, find the work done.

38. How much work is done in moving a string of cars 500 meters along a track by a truck exerting a force of 2000 newtons at an angle of 20 to the direction of the track?

39. In Exercise 38 how much work is done if the force is applied in the direction of motion?

40. How much work is done by a 175-pound man climbing 12 feet up vertical ladder?

Chapter Summary

[6.1]   The **Law of Sines:**

If α, β, and γ are the angles of any triangle, and if a, b, and c are the lengths of the sides opposite α, β, and γ, respectively, then

$$\frac{\sin \alpha}{a} = \frac{\sin \beta}{b} = \frac{\sin \gamma}{c}.$$

This law is used if the measures of two angles and the length of an side, or if the lengths of two sides and the measure of an angle opposit one of these sides, are given. In the latter case, certain ambiguities may arise. Two, one, or no triangles may exist.

[6.2]   The **Law of Cosines:**

If α, β, and γ are the angles of any triangle, and a, b, and c are the lengths of the sides opposite α, β, and γ, respectively, then

$$c^2 = a^2 + b^2 - 2ab \cos \gamma,$$
$$b^2 = a^2 + c^2 - 2ac \cos \beta,$$
$$a^2 = b^2 + c^2 - 2bc \cos \alpha.$$

This law is used if the lengths of two sides and the measure of the included angle, or if the lengths of three sides, are given.

[6.3]   A **geometric vector** $\vec{v}$ is a line segment with a specified direction. Its **norm** or **magnitude** is designated by $\|\vec{v}\|$. The **direction angle** is the angle α, where $-180° < \alpha \leq 180°$, such that

The initial side of α is the ray from the initial point of $\vec{v}$ parallel to the x-axis and directed in the positive x direction; the terminal side of α is the ray from the initial point of $\vec{v}$ and containing $\vec{v}$.

The **sum** $\vec{v}_1 + \vec{v}_2$ is defined as the geometric vector having as its initial point the initial point of $\vec{v}_1$ and as its terminal point the terminal point of $\vec{v}_2$, and where the terminal point of $\vec{v}_1$ is the initial point of $\vec{v}_2$. When viewed in this way, geometric vectors are said to be added according to "the triangle law." When viewed as the diagonal of a parallelogram with adjacent sides $\vec{v}_1$ and $\vec{v}_2$, geometric vectors are said to be added according to "the parallelogram law."

If $\vec{v}$ is any geometric vector and c is any real number (scalar), then $c\vec{v}$ is a geometric vector collinear with $\vec{v}$ with magnitude $|c| \cdot \|\vec{v}\|$.

[6.4]   The magnitudes of the **geometric vector projections,** $\vec{v}_x$ and $\vec{v}_y$, of $\vec{v}$ on the x-axis and y-axis, respectively, are given by

$$\|\vec{v}_x\| = \|\vec{v}\| \cdot |\cos \alpha| \quad \text{and} \quad \|\vec{v}_y\| = \|\vec{v}\| \cdot |\sin \alpha|.$$

Geometric vectors can help to visualize and thus facilitate the solution of problems involving such quantities as force, velocity, and acceleration, since these quantities have both magnitude and direction.

[6.5]   A **two-dimensional vector** $\mathbf{v}$ is an ordered pair of real numbers. The set of all such vectors is designated by $\mathbf{V}$. Each vector in the set corresponds to a unique bound geometric vector with initial point at the origin.

For all $\mathbf{v}_1, \mathbf{v}_2 \in \mathbf{V}$, where $\mathbf{v}_1 = (a_1, b_1)$ and $\mathbf{v}_2 = (a_2, b_2)$,

$$\mathbf{v}_1 = \mathbf{v}_2 \quad \text{if and only if} \quad a_1 = a_2 \quad \text{and} \quad b_1 = b_2;$$

the **norm**, or **magnitude**, of $\mathbf{v} = (a, b)$ is the real number (scalar)

$$\|\mathbf{v}\| = \sqrt{a^2 + b^2};$$

for $\|\mathbf{v}\| \neq 0$, the **direction angle** of $\mathbf{v} = (a, b)$ is the angle α such tha

$$\cos \alpha = \frac{a}{\|\mathbf{v}\|} \quad \text{and} \quad \sin \alpha = \frac{b}{\|\mathbf{v}\|} \quad (-180° < \alpha \leq 180°).$$

The **norm** (or **magnitude**) of a vector, $\|\mathbf{v}\|$, is the length of the associate
geometric vector $\vec{v}$ and the **direction angle** of the vector is the angle
where $-180° < \alpha \leq 180°$, measured from the positive x-axis to th
geometric vector corresponding to $\mathbf{v}$.

For all $\mathbf{v}_1, \mathbf{v}_2 \in \mathbf{V}$, where $\mathbf{v}_1 = (a_1, b_1)$, $\mathbf{v}_2 = (a_2, b_2)$, and $c \in R$,

$$\mathbf{v}_1 + \mathbf{v}_2 = (a_1, b_1) + (a_2, b_2) = (a_1 + a_2, b_1 + b_2);$$
$$\mathbf{v}_1 - \mathbf{v}_2 = (a_1, b_1) - (a_2, b_2) = (a_1 - a_2, b_1 - b_2)$$
$$c\mathbf{v} = c(a_1, b_1) = (ca_1, cb_1).$$

If $\mathbf{i} = (1, 0)$ and $\mathbf{j} = (0, 1)$, then each vector $\mathbf{v}$ can be expressed as

$$\mathbf{v} = (a, b) = a\mathbf{i} + b\mathbf{j}.$$

[6.6]   The **inner (dot, scalar) product** of vectors $\mathbf{v}_1$ and $\mathbf{v}_2$ is

$$\mathbf{v}_1 \cdot \mathbf{v}_2 = (a_1, b_1) \cdot (a_2, b_2) = a_1 a_2 + b_1 b_2.$$

If γ is the angle between the geometric vectors corresponding to $\mathbf{v}_1$ an
$\mathbf{v}_2$, then

$$\mathbf{v}_1 \cdot \mathbf{v}_2 = \|\mathbf{v}_1\| \|\mathbf{v}_2\| \cos \gamma.$$

Equivalently,

$$\cos \gamma = \frac{\mathbf{v}_1 \cdot \mathbf{v}_2}{\|\mathbf{v}_1\| \|\mathbf{v}_2\|};$$

$$\|\mathbf{v}_1\| \cos \gamma = \frac{\mathbf{v}_1 \cdot \mathbf{v}_2}{\|\mathbf{v}_2\|}, \quad \text{the scalar projection of } \mathbf{v}_1 \text{ on } \mathbf{v}_2;$$

$$\|\mathbf{v}_2\| \cos \gamma = \frac{\mathbf{v}_1 \cdot \mathbf{v}_2}{\|\mathbf{v}_1\|}, \quad \text{the scalar projection of } \mathbf{v}_2 \text{ on } \mathbf{v}_1.$$

The vectors $\mathbf{v}_1 = (a_1, b_1)$ and $\mathbf{v}_2 = (a_2, b_2)$ are **orthogonal** if an
only if the inner product

$$\mathbf{v}_1 \cdot \mathbf{v}_2 = a_1 a_2 + b_1 b_2 = 0.$$

A list of symbols introduced in this chapter is shown inside the front cover, and important properties are shown inside the back cover.

Review Exercises

A

[6.1] ▪ *Solve each triangle.*

    1. $c = 13.6$, $\alpha = 30.3°$, $\beta = 72.2°$

    2. $c = 1.4$, $\alpha = 135°\ 6'$, $\gamma = 34°\ 54'$

    3. $b = 1.8$, $\beta = 14.7°$, $\alpha = 80.2°$

    4. $a = 12.2$, $\alpha = 61.3°$, $\gamma = 28.2°$

▪ *Determine the number of triangles that satisfy the conditions in each exercise and solve each triangle.*

    5. $b = 3.9$, $a = 5$, $\beta = 21.7°$ **6.** $b = 37$, $a = 51$, $\alpha = 135°\ 30'$

[6.2] ▪ *Solve each triangle.*

    7. $a = 1.9$, $b = 2.3$, $\gamma = 58°\ 12'$ **8.** $b = 2.3$, $c = 5.7$, $\alpha = 23.1°$

    9. $a = 3.6$, $b = 4.2$, $c = 6.1$ **10.** $a = 2.1$, $b = 3.4$, $c = 4.1$

    11. $b = 10.2$, $c = 6.3$, $\alpha = 14.6°$ **12.** $a = 0.8$, $b = 0.5$, $c = 0.7$

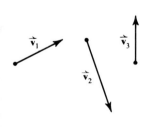

[6.3] ▪ *Use freehand methods and the geometric vectors shown at the left to draw a single vector to represent each of the following.*

    13. $(\vec{v}_1 + \vec{v}_2) + \vec{v}_3$ **14.** $-3\vec{v}_1$

    15. $2\vec{v}_2 + \vec{v}_3$ **16.** $2\vec{v}_1 - 3\vec{v}_3$

[6.4] ▪ **17.** If the direction angle $\alpha = 60°$, and $\|\vec{v}\| = 12$, find $\|\vec{v}_x\|$ and $\|\vec{v}_y\|$.

    18. If $\vec{v}_x$ and $\vec{v}_y$ are directed in a positive direction; $\|\vec{v}_x\| = 4$, with initial point at $(1, 0)$; and $\|\vec{v}_y\| = 3$, with initial point at $(0, 2)$, find $\|\vec{v}\|$ and the direction angle α.

    19. The resultant of two forces acting at an angle of $90°$ with respect to each other is 130 newtons. If one of the forces is 50 newtons, find the other force and the angle it makes with the resultant.

    20. A person can swim at a speed of 3 miles per hour in still water. If he heads across a river at right angles to a current of 6 miles per hour,

find his speed in relation to the land and the direction in which h[
actually moves.

21. An airplane is headed on a bearing of 280° with an air speed o[
800 kilometers per hour. The course has a bearing of 292°. Th[
ground speed is 860 kilometers per hour. Find the wind directio[
and wind speed.

22. A force of 150 pounds is needed to keep a steel ball weighin[
250 pounds from rolling down an inclined plane. What angles doe[
the plane make with the horizontal?

[6.5] ▪ (a) *Sketch the geometric vector corresponding to each vector.*
(b) *Find the norm and the direction angle (to the nearest six minutes) o[
each vector.*

23. $(-4, 3)$ 24. $(5, -5)$

▪ *Given the following norms and direction angles, find the vector* **v**.

25. $\|\mathbf{v}\| = 10$ and $\alpha = -135°$ 26. $\|\mathbf{v}\| = 2$ and $\alpha = 150°$

▪ *Write the sum of each pair of vectors as an ordered pair.*

27. $\mathbf{v}_1 = (4, -5)$ and $\mathbf{v}_2 = (8, 1)$

28. $\mathbf{v}_1 = (-3, 7)$ and $\mathbf{v}_2 = (-7, 3)$

▪ *Let* $\mathbf{v}_1 = (3, -2)$, $\mathbf{v}_2 = (5, 4)$, $\mathbf{v}_3 = (1, -3)$, $c_1 = 2$, *and* $c_2 = -1$[
Express each of the following as an ordered pair.

29. $\mathbf{v}_2 - c_2\mathbf{v}_1$ 30. $c_1(\mathbf{v}_3 + \mathbf{v}_2)$ 31. $\dfrac{\mathbf{v}_3}{c_2}$ 32. $\dfrac{\mathbf{v}_2}{c_1\|\mathbf{v}_1\|}$

▪ *Express each vector as the sum of scalar multiples of* **i** *and* **j**.

33. $(7, -9)$ 34. $(-8, -1)$

▪ *Find the norm and direction angle (to the nearest six minutes) of eac[
vector.*

35. $-10\mathbf{i} + 4\mathbf{j}$ 36. $7\mathbf{i} - 9\mathbf{j}$

[6.6] ▪ *Find the inner product* $\mathbf{v}_1 \cdot \mathbf{v}_2$ *for given vectors.*

37. $\mathbf{v}_1 = (6, 6)$ and $\mathbf{v}_2 = (-5, -4)$

38. $\mathbf{v}_1 = 8\mathbf{i} - \mathbf{j}$ and $\mathbf{v}_2 = 9\mathbf{i} - 3\mathbf{j}$

39. $\|\mathbf{v}_1\| = 8$ and $\alpha_1 = 60°$; $\|\mathbf{v}_2\| = 4$ and $\alpha_2 = 30°$

40. $\|\mathbf{v}_1\| = 10$ and $\alpha_1 = -45°$; $\|\mathbf{v}_2\| = 5$ and $\alpha_2 = -15°$

■ *Determine the angle γ to the nearest 6′, where $0° \leq \gamma \leq 180°$, formed by each pair of vectors.*

41. $\mathbf{v}_1 = (-7, 5)$ and $\mathbf{v}_2 = (1, -5)$

42. $\mathbf{v}_1 = -7\mathbf{i} - \mathbf{j}$ and $\mathbf{v}_2 = -8\mathbf{i} + 9\mathbf{j}$

■ *Find the scalar projections of $\mathbf{v}_1$ on $\mathbf{v}_2$ and of $\mathbf{v}_2$ on $\mathbf{v}_1$.*

43. $\mathbf{v}_1 = (1, 7)$ and $\mathbf{v}_2 = (10, -1)$

44. $\mathbf{v}_1 = 9\mathbf{i} + 2\mathbf{j}$ and $\mathbf{v}_2 = -8\mathbf{i} - 5\mathbf{j}$

■ *Verify that each pair of vectors is orthogonal.*

45. $(-4, -5)$ and $\left(4, -\dfrac{16}{5}\right)$ **46.** $7\mathbf{i} + 6\mathbf{j}$ and $5\mathbf{i} - \dfrac{35}{6}\mathbf{j}$

COMPLEX NUMBERS; POLAR COORDINATES

Recall from your prerequisite courses in algebra that the set of re[al] numbers does not contain an element whose square is a negative re[al] number. Thus, any symbol of the form $\sqrt{-b}$, $b \in \mathsf{R}$, $b > 0$, does n[ot] represent a real number, and equations such as

$$x^2 + 4 = 0 \qquad \text{equivalent to} \qquad x = \pm\sqrt{-4} \qquad ($$

and

$$x^2 - 2x + 5 = 0 \qquad \text{equivalent to} \qquad x = \frac{2 \pm \sqrt{-16}}{2} \qquad ($$

have no real solutions. Recall that such numbers are called **comple[x] numbers.**

We will first consider algebraic representations of complex number[s,] some basic operations on these numbers, and their graphs. Then we w[ill] introduce a trigonometric representation of complex numbers tha[t] facilitates computations involving products, quotients, powers, an[d] roots.

7.1 Sums and Products

A complex number is often represented by the letter z, where

$$z = a + bi, \quad i^2 = -1,$$

and a and b are real numbers.

In this chapter we will restrict the use of the variable z to represent elements in the set of complex numbers C. It is sometimes convenient to refer to a as the **real part** of the complex number $a + bi$ and b as the **imaginary part,** with i the **imaginary unit.**

Computing Sums and Products   If complex numbers are treated as if i were a variable and $a + bi$ were a binomial in the variable i, then the sum or product of two complex numbers can be found by applying the ordinary rules of the algebra of real numbers. Thus,

$$(2 + 3i) + (5 - 2i) = (2 + 5) + (3 - 2)i$$
$$= 7 + i;$$
$$(2 + 3i)(5 - 2i) = 10 + 11i - 6i^2.$$

However, replacing i^2 with -1, we have

$$(2 + 3i)(5 - 2i) = 10 + 11i - 6(-1)$$
$$= 16 + 11i.$$

More formally we have the following definition.

If $z_1 = a_1 + b_1 i$ *and* $z_2 = a_2 + b_2 i$, *then*

$$z_1 + z_2 = (a_1 + a_2) + (b_1 + b_2)i,$$

and

$$z_1 z_2 = (a_1 a_2 - b_1 b_2) + (a_1 b_2 + a_2 b_1)i.$$

Using such an approach for sums and products, we find that

1. C will contain elements $a + 0i$, or a, that can be identified with the real numbers for these operations, and
2. C will contain elements $0 + bi$, or bi, whose squares are negative real numbers.

For example,

$$(2) + (3) = (2 + 0i) + (3 + 0i)$$
$$= 5 + 0i = 5,$$
$$(2)(3) = (2 + 0i)(3 + 0i)$$
$$= 6 + 0i + 0i^2 = 6,$$
$$(3i)^2 = 9i^2 = -9.$$

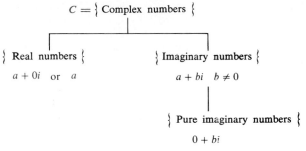

Figure 7.1

Because the complex number $a + bi$, where $b = 0$, behaves like t[
real number a, we will simply write a for $a + 0i$ as convenient and refer
a or $a + 0i$ as a real number.

If $b \neq 0$, the complex number $a + bi$ is called an **imaginary numb**[
and for the special case where $a = 0$ and $b \neq 0$, the complex number
called a **pure imaginary number.** The relationship of the set C and oth[
sets of numbers is shown in Figure 7.1. For example, 3 and -5 a[
elements of C that are real; $2 + 5i$ and $5i$ are elements of C that a[
imaginary numbers. Furthermore, $5i$ is a pure imaginary number.

Periodic Property of i   Powers of the imaginary unit i have an interesting periodic proper[
Observe that

$$i^1 = i,$$

$$i^2 = -1$$

$$i^3 = i^2 \cdot i = -1 \cdot i = -i,$$

$$i^4 = i^2 \cdot i^2 = (-1)(-1) = 1,$$

$$i^5 = i^4 \cdot i = 1 \cdot i = i,$$

$$i^6 = i^4 \cdot i^2 = 1 \cdot (-1) = -1,$$

$$i^7 = i^4 \cdot i^3 = 1 \cdot (-i) = -i,$$

etc.

Thus, any integral power of i can be expressed as one of the numbe[
$i, -1, -i,$ or 1.

Radical Notation   Another notation used to represent a complex number involves t[
square root symbol. In the set C, we make the following definition.

For all $b \in R$, $b > 0$,

$$\sqrt{-b} = \sqrt{-1}\,\sqrt{b} = i\sqrt{b}.$$

Observe that for the special case where $b = 1$,

$$\sqrt{-1} = \sqrt{-1}\sqrt{1} = i \cdot 1 = i.$$

Note that for a and b, positive real numbers,

$$\sqrt{-a}\sqrt{-b} = (i\sqrt{a})(i\sqrt{b}) = i^2\sqrt{a}\sqrt{b} = -\sqrt{ab},$$

a negative real number. Hence, the property of real numbers

$$\sqrt{a}\sqrt{b} = \sqrt{ab}, \quad a, b \geq 0,$$

is not applicable for imaginary numbers. For example,

$$\sqrt{-4}\sqrt{-9} \neq \sqrt{(-4)(-9)} = 6.$$

In this case,

$$\sqrt{-4}\sqrt{-9} = 2i \cdot 3i = 6i^2 = -6.$$

To avoid difficulty in rewriting products, expressions of the form $\sqrt{-b}$ $(b > 0)$ should first be expressed in the form $\sqrt{b}\,i$ or $i\sqrt{b}$.

Examples

a. $\sqrt{-3}(5 + \sqrt{-3})$

$\quad = i\sqrt{3}(5 + i\sqrt{3})$

$\quad = 5i\sqrt{3} + i^2(3)$

$\quad = 5i\sqrt{3} + (-1)(3)$

$\quad = -3 + 5i\sqrt{3}$

b. $(3 - \sqrt{-2})(3 + \sqrt{-2})$

$\quad = (3 - i\sqrt{2})(3 + i\sqrt{2})$

$\quad = 9 - i^2(2)$

$\quad = 9 - (-1)(2)$

$\quad = 11$

EXERCISE SET 7.1

A

■ *Write each expression in the form a + bi.*

Example $(3 + 2i) + (1 - 3i) = (3 + 1) + [2 + (-3)]i$

$$= 4 - i.$$

1. $(4 + 8i) + (-3 + 2i)$

2. $(-6 + 6i) + (4 + 2i)$

3. $(-9 - 5i) + (-1 - 6i)$

4. $(-4 + 3i) + (5 - i)$

5. $(3 + 3i) + (-4 + 3i)$

6. $(-1 + 6i) + (2 - 2i)$

7. $(a + bi) + (-a - bi)$

8. $(a + bi) + (a - bi)$

Examples

a. $2(1 - 5i) = 2(1) + 2(-5)i$ b. $-3(4 + i) = -3(4) - 3(1)i$

$= 2 - 10i$ $= -12 - 3i$

9. $4(3 - 3i)$ **10.** $5(2 + 4i)$ **11.** $-6(6 + 2i)$

12. $-4(8 - i)$ **13.** $\dfrac{1}{2}(4 + 6i)$ **14.** $\dfrac{2}{3}(9 - 6i)$

Example $(-3 + 4i)(1 + 6i) = -3 + 4i - 18i + 24i^2$

$= -3 - 14i + 24(-1) = -27 - 14i.$

15. $(-3 + 8i)(1 + 2i)$ **16.** $(4 + 2i)(-8 - 2i)$

17. $(-6 - i)(-9 + 8i)$ **18.** $(-3 - 9i)(3 + 4i)$

19. $(5 - 6i)(4 + i)$ **20.** $(5 + i)(-1 + i)$

21. $(a + bi)(a - bi)$ **22.** $(a + bi)(a + bi)$

■ *Write each expression as one of the complex numbers, i, -1, $-i$, or 1.*

23. i^5 **24.** i^8 **25.** i^{10} **26.** i^{11}

27. i^{14} **28.** i^{16} **29.** i^{19} **30.** i^{23}

■ *Write each expression in the form bi or a + bi.*

Examples

a. $\sqrt{-16} = \sqrt{16}\sqrt{-1}$ b. $4 + \sqrt{-8} = 4 + \sqrt{4}\sqrt{2}\sqrt{-}$

$= 4i$ $= 4 + 2i\sqrt{2}$

31. $\sqrt{-25}$ **32.** $\sqrt{-49}$ **33.** $\sqrt{-12}$ **34.** $\sqrt{-27}$

35. $1 + \sqrt{-9}$ **36.** $2 - \sqrt{-64}$ **37.** $3 - \sqrt{-20}$ **38.** $5 + \sqrt{-4}$

Examples

a. $\sqrt{-9}(2 - 3\sqrt{-9}) = 3i(2 - 3 \cdot 3i)$

$= 3i(2 - 9i)$

$= 6i - 27i^2$

$= 6i - 27(-1)$

$= 27 + 6i$

b. $(1 + \sqrt{-5})(1 - 2\sqrt{-5}) = (1 + i\sqrt{5})(1 - 2i\sqrt{5})$

$$= 1 + i\sqrt{5} - 2i\sqrt{5} - 2(\sqrt{5})^2 i^2$$

$$= 1 - i\sqrt{5} - 2 \cdot 5(-1) = 11 - i\sqrt{5}$$

39. $\sqrt{-4}(1 - 2\sqrt{-4})$ **40.** $\sqrt{-8}(3 + 4\sqrt{-2})$

41. $\sqrt{-16}(2 + \sqrt{-4})$ **42.** $\sqrt{-9}(3 - \sqrt{-25})$

43. $(\sqrt{-50} + 2)\sqrt{-10}$ **44.** $(\sqrt{-48} - 1)\sqrt{-21}$

45. $(5 + 2\sqrt{-4})(3 - \sqrt{-4})$ **46.** $(1 - 4\sqrt{-9})(7 + 2\sqrt{-9})$

47. $(2 + \sqrt{-3})(3 - \sqrt{-3})$ **48.** $(3 - \sqrt{-7})(3 + \sqrt{-7})$

49. Show that $2i$ and $-2i$ satisfy $x^2 + 4 = 0$.

50. Show that $1 + 2i$ and $1 - 2i$ satisfy $x^2 - 2x + 5 = 0$.

B

51. Show that $\left(-\dfrac{1}{2} + i\dfrac{\sqrt{3}}{2}\right)^3 = 1$ and $\left(-\dfrac{1}{2} - i\dfrac{\sqrt{3}}{2}\right)^3 = 1$.

52. Show that $\left(\dfrac{1}{2} + i\dfrac{\sqrt{3}}{2}\right)^3 = -1$ and $\left(\dfrac{1}{2} - i\dfrac{\sqrt{3}}{2}\right)^3 = -1$.

■ *Find a and b for which each statement is true.*

53. $(a + bi)^2 = -1$.   **54.** $(a + bi)^2 = 1$.

■ *The formula* $e^{ix} = \cos x + i \sin x$, *where* $e = 2.718$ ..., *is called* **Euler's formula.** (*It is usually derived in a course in calculus.*) *Use this formula to show that the following equations are identities.*

55. $e^{\pi i} = -1$ **56.** $e^{-ix} = \cos x - i \sin x$

57. $\sin x = \dfrac{e^{ix} - e^{-ix}}{2i}$ **58.** $\cos x = \dfrac{e^{ix} + e^{-ix}}{2}$

59. $\sin^2 x + \cos^2 x = 1$.   *Hint:* Use Exercises 57 and 58.

7.2 Differences and Quotients

The operations of addition and multiplication of complex numbers have been defined so that many of the properties of real numbers are valid in the set of complex numbers.

Two other useful operations, subtraction and division, are defined i
terms of addition and multiplication. These definitions are similar to th
definitions for a difference and a quotient in the set of real numbers.

For complex numbers z_1 and z_2,

$$\textbf{1. } z_1 - z_2 = z_1 + (-z_2)$$

$$\textbf{2. } \frac{z_1}{z_2} = z_1\left(\frac{1}{z_2}\right), \quad z_2 \neq 0.$$

The property

$$\frac{z_1}{z_2} = \frac{z_1 z_3}{z_2 z_3}, \quad z_2, z_3 \neq 0,$$

which is similar to the fundamental principle of fractions, is useful i
rewriting quotients of complex numbers. We first make the followin
definition.

For all $z = a + bi$, the **conjugate** *of z is*

$$\bar{z} = a - bi.$$

Examples

a. The conjugate of $5 + 4i$ is $5 - 4i$.

b. The conjugate of $2 - 3i$ is $2 + 3i$.

A quotient of complex numbers, z_1/z_2, can be expressed in the for
$a + bi$ by first multiplying both the numerator and the denominato
by the conjugate of the denominator. If the denominator is a pu
imaginary number, then it is necessary only to multiply the numerato
and denominator by i. If the denominator is a real number and th
numerator is of the form $a + bi$, then it is necessary only to write th
quotient as two terms.

Examples

a.
$$\frac{3}{1 + 2i} = \frac{3(1 - 2i)}{(1 + 2i)(1 - 2i)}$$
$$= \frac{3 - 6i}{1 - 4i^2}$$
$$= \frac{3 - 6i}{5} = \frac{3}{5} - \frac{6}{5}i$$

b.
$$\frac{4 - i}{2 - 3i} = \frac{(4 - i)(2 + 3i)}{(2 - 3i)(2 + 3i)}$$
$$= \frac{8 + 10i - 3i^2}{4 - 9i^2}$$
$$= \frac{11 + 10i}{13} = \frac{11}{13} + \frac{10}{13}$$

c. $\dfrac{3-i}{3i} = \dfrac{(3-i)i}{3i \cdot i} = \dfrac{3i - i^2}{3i^2}$ d. $\dfrac{4+5i}{7} = \dfrac{4}{7} + \dfrac{5}{7}i$

$$= \dfrac{3i+1}{-3} = -\dfrac{1}{3} - i$$

In Example a, note that the product of the complex number $1 + 2i$ and its conjugate $1 - 2i$ is the real number 5. In Example b, note that the product of the complex number $2 - 3i$ and its conjugate $2 + 3i$ is the real number 13. In fact, observe that for all $a + bi \in C$,

$$(a + bi)(a - bi) = a^2 - b^2i^2 = a^2 + b^2,$$

a real number.

As we noted in Section 7.1, to avoid difficulty in rewriting products, expressions of the form $\sqrt{-b}$ $(b > 0)$ should first be expressed in the form $\sqrt{b}i$ or $i\sqrt{b}$. Hence, quotients that involve this form should also be rewritten before using Equation (2) on page 246.

Examples

a. $\dfrac{5}{\sqrt{-4}} = \dfrac{5}{2i}$ b. $\dfrac{4}{3 - \sqrt{-2}}$

$$= \dfrac{5 \cdot i}{2i \cdot i} \qquad\qquad\qquad = \dfrac{4(3 + i\sqrt{2})}{(3 - i\sqrt{2})(3 + i\sqrt{2})}$$

$$= \dfrac{5i}{2(-1)} \qquad\qquad\qquad = \dfrac{12 + 4i\sqrt{2}}{9 - i^2(2)}$$

$$= -\dfrac{5}{2}i \qquad\qquad\qquad\;\; = \dfrac{12 + 4i\sqrt{2}}{9 - (-1)(2)}$$

$$\qquad\qquad\qquad\qquad\qquad = \dfrac{12}{11} + \dfrac{4\sqrt{2}}{11}i$$

EXERCISE SET 7.2

A

▪ *Express each difference in the form a + bi.*

Example $(1 - 4i) - (-1 + 10i) = (1 - 4i) + (1 - 10i)$

$$= (1 + 1) + (-4 - 10)i = 2 - 14i$$

1. $(6 + i) - (-7 + 5i)$ **2.** $(-7 - 3i) - (-5 + 5i)$

3. $(3 - i) - (1 - 3i)$ **4.** $(3 + 2i) - (-5 + 8i)$

5. $(5 + 2i) - 3i$ **6.** $4 - (2 - 3i)$

▪ *Write the conjugate of each complex number in the form a + bi.*

Examples   a. $z = -3 + 4i$ b. $z = 5 - 2i$

Solution   a. $\bar{z} = -3 - 4i$ b. $\bar{z} = 5 + 2i$

7. $6 + 3i$ **8.** $5 - 5i$ **9.** $-3 - 3i$

10. $-4 + i$ **11.** $3i$ **12.** 7

▪ *Express each quotient in the form a + bi.*

Examples

a. $\dfrac{8 - 2i}{8i} = \dfrac{(8 - 2i)i}{8i \cdot i}$

$= \dfrac{8i - 2i^2}{8i^2}$

$= \dfrac{8i + 2}{-8} = -\dfrac{1}{4} - i$

b. $\dfrac{-4}{-3 + i} = \dfrac{(-4)(-3 - i)}{(-3 + i)(-3 - i)}$

$= \dfrac{12 + 4i}{9 - i^2}$

$= \dfrac{6}{5} + \dfrac{2}{5}i$

13. $\dfrac{-9 - 6i}{-4i}$ **14.** $\dfrac{-2 + 3i}{3i}$ **15.** $\dfrac{4}{2 - 9i}$ **16.** $\dfrac{2}{-8 + 5i}$

17. $\dfrac{2 + i}{4 + i}$ **18.** $\dfrac{2 - 5i}{-1 + 6i}$ **19.** $\dfrac{-5 + 9i}{6 - 2i}$ **20.** $\dfrac{3 - 2i}{3 + 2i}$

Examples

a. $\dfrac{3}{\sqrt{-2}} = \dfrac{3(i)}{i\sqrt{2}(i)}$

$= \dfrac{3i}{i^2\sqrt{2}}$

$= \dfrac{3i}{(-1)\sqrt{2}}$

$= -\dfrac{3}{\sqrt{2}}i$

b. $\dfrac{5}{2 + \sqrt{-3}}$

$= \dfrac{5(2 - i\sqrt{3})}{(2 + i\sqrt{3})(2 - i\sqrt{3})}$

$= \dfrac{10 - 5i\sqrt{3}}{4 - i^2(3)}$

$= \dfrac{10 - 5i\sqrt{3}}{4 - (-1)(3)}$

$= \dfrac{10}{7} - \dfrac{5\sqrt{3}}{7}i$

21. $\dfrac{1}{\sqrt{-5}}$ **22.** $\dfrac{2}{\sqrt{-7}}$ **23.** $\dfrac{4}{\sqrt{-8}}$

24. $\dfrac{7}{\sqrt{-12}}$ **25.** $\dfrac{3}{4+\sqrt{-5}}$ **26.** $\dfrac{7}{5-\sqrt{-3}}$

27. $\dfrac{6}{\sqrt{-2}+3}$ **28.** $\dfrac{4}{\sqrt{-7}-6}$ **29.** $\dfrac{6}{\sqrt{2}+\sqrt{-5}}$

30. $\dfrac{10}{\sqrt{3}-\sqrt{-6}}$ **31.** $\dfrac{2}{\sqrt{5}+\sqrt{-8}}$ **32.** $\dfrac{3}{\sqrt{6}-\sqrt{-12}}$

7.3 Graphical Representation

Any complex number $a + bi$ can be placed in a one-to-one correspondence with the set of points in a plane by associating a with the
abscissa of a point and b with the ordinate. In this way each point in a
plane can be viewed as the graph of a complex number. A plane on which
complex numbers are thus represented is called a **complex plane.** Any
real number $a + 0i$, or simply a, is associated with a point on the x-axis,
called the **real axis,** and $0 + bi$, or simply bi, is associated with a point on
the y-axis, called the **imaginary axis.**

Examples   Graph each number z, its negative, and its conjugate.

a. $z = 3 - 2i$ b. $z = -2 - 4i$

Solutions

a. The negative of $3 - 2i$ is $-3 + 2i$; the conjugate is $3 + 2i$.

b. The negative of $-2 - 4i$ is $2 + 4i$; the conjugate is $-2 + 4i$.

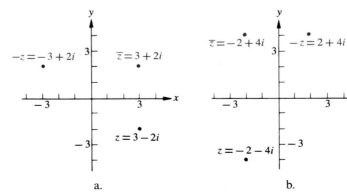

a.

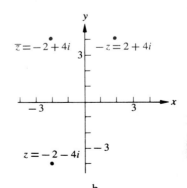

b.

Sometimes the graph of a complex number is shown as a geometric vector with its initial point located at the origin. With this representation, complex numbers can be added graphically by the parallelogram law applicable to geometric vectors.

Examples Represent each sum graphically using geometric vectors and check the results by analytic methods.

a. $(2 + 3i) + (4 - 5i)$ b. $(2 + 2i) + (-4 + 5i)$

Solutions

a.
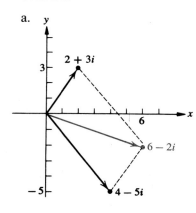

b.
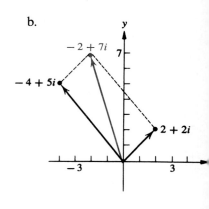

a. Check: $(2 + 3i) + (4 - 5i) = 6 - 2i$

b. Check: $(2 + 2i) + (-4 + 5i) = -2 + 7i$

Complex Numbers as Ordered Pairs   The elements $a + bi$ in the set of complex numbers C and the ordered pairs (a, b) have each been associated with points in a plane. Hence elements in C can also be associated with, and represented by, ordered pairs of real numbers (a, b).

Examples
a. $2 + 3i = (2, 3)$

b. $-5 = -5 + 0i$

$= (-5, 0)$

c. $6i = 0 + 6i = (0, 6)$

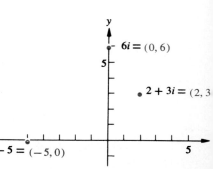

Absolute Value and Argument

Corresponding to the norm $\|\vec{v}\|$ and the direction angle α for each geometric vector $\vec{v}$, there are a real number and a *set* of angles associated with each complex number.

*For all $z \in C$, the **absolute value** or **modulus** of z, where*

$$z = a + bi = (a, b),$$

is

$$\rho = |z| = |a + bi|$$
$$= |(a, b)|$$
$$= \sqrt{a^2 + b^2}.$$

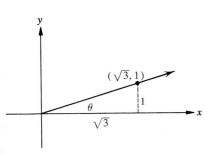

Figure 7.2

The modulus ρ (the Greek letter "rho") is the distance from the origin to the graph of (a, b), as shown in Figure 7.2.

*For all $z \in C$, an **argument** or **amplitude** of z, where*

$$z = a + bi = (a, b),$$

denoted by $\arg(a + bi)$ or $\arg(a, b)$, is an angle θ such that

$$\cos \theta = \frac{a}{\sqrt{a^2 + b^2}} = \frac{a}{\rho} \quad and$$

$$\sin \theta = \frac{b}{\sqrt{a^2 + b^2}} = \frac{b}{\rho} \quad (a^2 + b^2 \neq 0).$$

Thus, $\arg(a + bi)$ is an angle θ with initial side the positive x-axis and terminal side the ray from the origin through the graph of (a, b), as shown in Figure 7.2.

If θ is an argument of $a + bi$, then for $k \in J$, so is $(\theta + k \cdot 360)°$. The angle θ with the least nonnegative measure is sometimes called the **principal argument.** For $a + bi = 0 + 0i$, any angle θ can be used as an argument.

Example Find the absolute value and principal argument of $\sqrt{3} + i$.

Solution $\rho = |\sqrt{3} + i| = \sqrt{(\sqrt{3})^2 + 1^2} = \sqrt{3 + 1} = 2;$

$$\cos \theta = \frac{a}{\rho} = \frac{\sqrt{3}}{2} \quad \text{and} \quad \sin \theta = \frac{b}{\rho} = \frac{1}{2}.$$

(continued)

Because both $\cos\theta$ and $\sin\theta$ are positive, θ is in Quadrant I, and hence $\theta = 30°$.

There is another method that is often used to determine θ. Consider the number $a + bi = \sqrt{3} + i$ from the example above. First plot the point $(a, b) = (\sqrt{3}, 1)$ and form angle θ. From the sketch, θ is in Quadrant I, and $\tan\theta = 1/\sqrt{3}$. Hence $\theta = 30°$.

Applications

Applications in physics and electronics often involve imaginary numbers. For example, the resistance R (in ohms) in a circuit is associated with real numbers in a vector diagram, and reactances (also in ohms) are displaced 90° from the resistances. Inductive reactances X_L are prefixed with i, and capacitive reactances X_C are prefixed with $-i$. The total impedance Z in a circuit is the vector sum of the resistance and reactances.

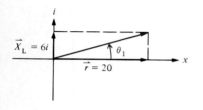

Example   If $R = 20$ and $X_L = 6$,

$$Z_1 = 20 + 6i;$$

if $R = 20$ and $X_C = 6$,

$$Z_2 = 20 - 6i$$

In either case, the magnitude of the impedance—that is, the magnitude of the resulting vector—is

$$\sqrt{(20)^2 + (6)^2} \approx 20.9 \text{ ohms.}$$

In addition, the impedance angles (or amplitudes) θ_1 and θ_2—that is, the angles formed by R and Z_1 and Z_2—are given by

$$\theta_1 = \text{Tan}^{-1}\frac{6}{20} \approx 16.7°, \quad \text{for } Z_1,$$

and

$$\theta_2 = \text{Tan}^{-1}\frac{-6}{20} \approx -16.7°, \quad \text{for } Z_2.$$

EXERCISE SET 7.3

A

■ *Express each complex number as an ordered pair.*

1. $-1 + 6i$ **2.** $5 - 9i$ **3.** $8 + i$

4. $-2 + 7i$ **5.** $6i$ **6.** -5

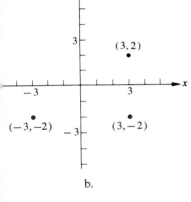

■ *Express each complex number in the form a + bi.*

7. $(-8, 7)$ **8.** $(0, 5)$ **9.** $(-4, 0)$

10. $(6, -3)$ **11.** $(7, -\sqrt{2})$ **12.** $(-\sqrt{3}, 1)$

■ *Graph each complex number z, its negative −z, and its conjugate z̄.*

Examples

a. $-5 - 4i$ b. $(3, 2)$

Solutions

   a. The negative is $5 + 4i$; the conjugate is $-5 + 4i$.

   b. The negative is $(-3, -2)$; the conjugate is $(3, -2)$.

13. $2 - 3i$ **14.** $-6 + 6i$ **15.** 5 **16.** $2i$

17. $(4, 6)$ **18.** $(-5, -3)$ **19.** $(-1, 0)$ **20.** $(0, 3)$

■ *Represent each sum graphically. Check results by algebraic methods.*

Example $(8 + 3i) + (1 - 5i)$

Solution

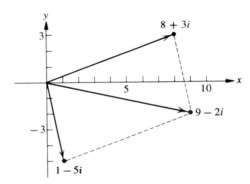

Check: $(8 + 3i) + (1 - 5i) = (8 + 1) + (3 - 5)i = 9 - 2i.$

21. $2 + (-3 + 6i)$ **22.** $(7 - i) + 2i$

23. $(4 - 7i) + (-8 + 2i)$ **24.** $(2 + 4i) + (1 - 5i)$

25. $(-6 + 7i) + (5 - 3i)$ **26.** $(-3 - 5i) + (3 - 2i)$

■ *Find the absolute value and principal argument of each complex number.*

Examples

a. $-7 + 5i$ b. $(3, 9)$ (*continued*)

Solutions

a. $\rho = |-7 + 5i| = \sqrt{(-7)^2 + 5^2} = \sqrt{74}$;

$$\cos \theta = -\frac{7}{\sqrt{74}} \approx -0.8137 \quad \text{and} \quad \sin \theta = \frac{5}{\sqrt{74}} \approx 0.5812.$$

Because $\cos \theta$ is negative and $\sin \theta$ is positive, θ is in Quadrant II; henc
$\theta \approx 144.5°$.

b. $\rho = |(3, 9)| = \sqrt{3^2 + 9^2} = \sqrt{90} = 3\sqrt{10}$;

$$\cos \theta = \frac{3}{3\sqrt{10}} \approx 0.3162 \quad \text{and} \quad \sin \theta = \frac{9}{3\sqrt{10}} \approx 0.9487.$$

Because $\cos \theta$ and $\sin \theta$ are both positive, θ is in Quadrant I; henc
$\theta \approx 71.6°$.

27. $5 + 2i$	**28.** $6 + 8i$	**29.** $-2 - 2i$	**30.** $8 - 3i$
31. 7	**32.** -4	**33.** $-5i$	**34.** $6i$
35. $(3, -9)$	**36.** $(-6, -2)$	**37.** $(4, 0)$	**38.** $(0, -7)$

39. Determine the conditions that a and b must satisfy if the graph of th
complex number $a + bi$ is
   a. on the real axis.   **b.** on the imaginary axis.
   c. above the real axis.   **d.** below the real axis.

40. What condition does $a^2 + b^2$ satisfy if the graph of $a + bi$ is
   a. on a circle with radius of length 5?
   b. inside a circle with radius of length 5?
   c. outside a circle with radius of length 5?

■ *Find the magnitude of the impedance and the impedance angle for eac*
circuit with the given resistance R and reactance(s) for X_L *and/or* X
(to the nearest tenth).

41. $R = 10$ ohms; $X_L = 4$ ohms

42. $R = 6$ ohms; $X_L = 12$ ohms

43. $R = 4$ ohms; $X_C = 12$ ohms

44. $R = 20$ ohms; $X_C = 12$ ohms

45. $R = 6$ ohms; $X_L = 12$ ohms and $X_C = 8$ ohms

46. $R = 12$ ohms; $X_L = 3$ ohms and $X_C = 7$ ohms

B

■ *Graphically, the multiplication or division of a complex number a + bi by the complex number i produces a 90° rotation of the geometric vector corresponding to a + bi, counterclockwise for multiplication and clockwise for division. Graph*

(a) *the given complex number a + bi,* **(b)** *i · (a + bi),* *and* **(c)** $\dfrac{a + bi}{i}$.

Example $2 + 3i$

Solutions

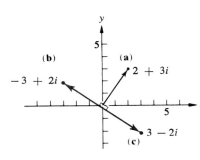

(b) $i \cdot (2 + 3i) = 2i + 3i^2$

$$= -3 + 2i$$

(c) $\dfrac{2 + 3i}{i} = \dfrac{(2 + 3i) \cdot i}{i \cdot i}$

$$= \dfrac{2i + 3i^2}{i^2}$$

$$= \dfrac{2i - 3}{-1} = 3 - 2i$$

47. $5 + 4i$ **48.** $1 - 2i$ **49.** $-6 + i$

50. $-3 - 4i$ **51.** 2 **52.** $-5i$

7.4 Trigonometric Form

In previous sections we have represented complex numbers in the form $a + bi, a + \sqrt{-b^2}$, and (a, b). It is sometimes convenient to represent a complex number in terms of its absolute value $\rho = \sqrt{a^2 + b^2}$ and its argument θ. Since

$$\cos \theta = \frac{a}{\rho} \quad \text{and} \quad \sin \theta = \frac{b}{\rho},$$

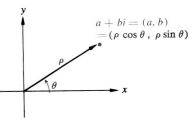

then, as shown in Figure 7.3,

$$a = \rho \cos \theta \quad \text{and} \quad b = \rho \sin \theta,$$

from which we have

$$a + bi = \rho \cos \theta + (\rho \sin \theta)i,$$

$$a + bi = \rho(\cos \theta + i \sin \theta).$$

(The expression $\cos \theta + i \sin \theta$ is sometimes abbreviated as **cis θ**.)

Figure 7.3

The sine and cosine functions are periodic with period of $360°$ or $2\pi^R$. Hence, it is also true that

$$a + bi = \rho[\cos(\theta + k \cdot 360°) + i \sin(\theta + k \cdot 360°)], \quad k \in J,$$

or

$$a + bi = \rho[\cos(\theta + k \cdot 2\pi^R) + i \sin(\theta + k \cdot 2\pi^R)], \quad k \in J$$

where θ is the angle of smallest nonnegative measure. Either equation is called the **trigonometric form** of a complex number.

Example   Represent $-2 + 2i$ in trigonometric form.

Solution   Since

$$\rho = |-2 + 2i| = \sqrt{(-2)^2 + 2^2} = \sqrt{8} = 2\sqrt{2},$$

and

$$\cos \theta = \frac{-2}{2\sqrt{2}} = -\frac{1}{\sqrt{2}} \quad \text{and} \quad \sin \theta = \frac{2}{2\sqrt{2}} = \frac{1}{\sqrt{2}},$$

θ is in Quadrant II and $\theta = 135°$. For $k \in J$,

$$-2 + 2i = 2\sqrt{2}\,[\cos(135° + k \cdot 360°) + i \sin(135° + k \cdot 360°)].$$

For $k = 0$,

$$-2 + 2i = 2\sqrt{2}(\cos 135° + i \sin 135°).$$

Example   Represent $3(\cos 60° + i \sin 60°)$ graphically and express the number in the form $a + bi$.

Solution   Since

$$a = \rho \cos \theta = 3 \cos 60° = 3\left(\frac{1}{2}\right) = \frac{3}{2},$$

and

$$b = \rho \sin \theta = 3 \sin 60° = 3\left(\frac{\sqrt{3}}{2}\right) = \frac{3\sqrt{3}}{2},$$

we have

$$3(\cos 60° + i \sin 60°) = \frac{3}{2} + \frac{3\sqrt{3}}{2}\,i.$$

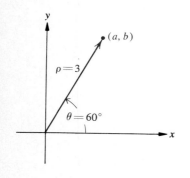

Products and Quotients   Products and quotients of complex numbers expressed in trigonometric form can be found quite readily. If

$$z_1 = \rho_1(\cos \theta_1 + i \sin \theta_1) \qquad \text{and} \qquad z_2 = \rho_2(\cos \theta_2 + i \sin \theta_2),$$

then

$$\begin{aligned} z_1 \cdot z_2 &= \rho_1(\cos \theta_1 + i \sin \theta_1) \cdot \rho_2(\cos \theta_2 + i \sin \theta_2) \\ &= \rho_1 \rho_2 [(\cos \theta_1 \cos \theta_2 - \sin \theta_1 \sin \theta_2) \\ &\quad + i(\cos \theta_1 \sin \theta_2 + \sin \theta_1 \cos \theta_2)]. \end{aligned}$$

Substituting

$$\cos(\theta_1 + \theta_2) \qquad \text{for} \qquad \cos \theta_1 \cos \theta_2 - \sin \theta_1 \sin \theta_2$$

and

$$\sin(\theta_1 + \theta_2) \qquad \text{for} \qquad \cos \theta_1 \sin \theta_2 + \sin \theta_1 \cos \theta_2,$$

we have

$$z_1 \cdot z_2 = \rho_1 \rho_2 [\cos(\theta_1 + \theta_2) + i \sin(\theta_1 + \theta_2)]. \tag{1}$$

It can also be shown in a similar way that

$$\frac{z_1}{z_2} = \frac{\rho_1}{\rho_2}[\cos(\theta_1 - \theta_2) + i \sin(\theta_1 - \theta_2)], \quad z_2 \neq (0, 0). \tag{2}$$

Example   Express $6(\cos 75° + i \sin 75°) \cdot 3(\cos 15° + i \sin 15°)$ in the form $a + bi$.

Solution   From Equation (1),

$$\begin{aligned} 6(\cos 75° &+ i \sin 75°) \cdot 3(\cos 15° + i \sin 15°) \\ &= 6 \cdot 3[\cos(75° + 15°) + i \sin(75° + 15°)] \\ &= 18(\cos 90° + i \sin 90°) \\ &= 18(0 + i) = 18i. \end{aligned}$$

Example   Express $\dfrac{6(\cos 75° + i \sin 75°)}{3(\cos 15° + i \sin 15°)}$ in the form $a + bi$.

Solution   From Equation (2),

$$\frac{6(\cos 75° + i \sin 75°)}{3(\cos 15° + i \sin 15°)} = \frac{6}{3}[\cos(75° - 15°) + i \sin(75° - 15°)]$$

$$= 2(\cos 60° + i \sin 60°)$$

$$= 2\left(\frac{1}{2} + i\frac{\sqrt{3}}{2}\right) = 1 + \sqrt{3}\, i.$$

EXERCISE SET 7.4

A

■ *Represent each complex number in trigonometric form, using the angl[e]
with least nonnegative measure.*

Example $\sqrt{3} - i$

Solution   Since

$$\rho = |\sqrt{3} - i| = \sqrt{3 + 1} = 2,$$

and

$$\cos \theta = \sqrt{3}/2 \qquad \text{and} \qquad \sin \theta = -1/2,$$

angle θ is in Quadrant IV and $\theta = 330°$. Therefore,

$$\sqrt{3} - i = 2[\cos(330° + k \cdot 360°) + i \sin(330° + k \cdot 360°)], \quad k \in J.$$

For $k = 0$, we obtain

$$\sqrt{3} - i = 2(\cos 330° + i \sin 330°).$$

1. $-2 + 2\sqrt{3}\, i$	**2.** $6 + 6i$	**3.** $-5\sqrt{3} - 5i$
4. $3\sqrt{3} + 3i$	**5.** $2 - 2i$	**6.** $-4 - 4i$
7. -8	**8.** $-9i$	

■ *Represent each complex number graphically and write as a + bi.*

Example $6(\cos 315° + i \sin 315°)$

Solution   Since

$$a = \rho \cos \theta = 6\left(\frac{\sqrt{2}}{2}\right) = 3\sqrt{2},$$

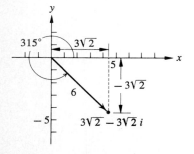

and

$$b = \rho \sin \theta = 6\left(\frac{-\sqrt{2}}{2}\right) = -3\sqrt{2},$$

we have

$$a + bi = 3\sqrt{2} - 3\sqrt{2}\, i.$$

9. $2(\cos 240° + i \sin 240°)$ 10. $2(\cos 90° + i \sin 90°)$

11. $2(\cos 135° + i \sin 135°)$ 12. $3(\cos 30° + i \sin 30°)$

13. $\cos(-45°) + i \sin(-45°)$ 14. $\cos(-210°) + i \sin(-210°)$

15. $\frac{1}{2}(\cos 150° + i \sin 150°)$ 16. $\frac{3}{4}(\cos 330° + i \sin 330°)$

■ *Express each product in the form a + bi.*

Example $2(\cos 140° + i \sin 140°) \cdot 5(\cos 70° + i \sin 70°)$

Solution From Equation (1) in this section,

$$2(\cos 140° + i \sin 140°) \cdot 5(\cos 70° + i \sin 70°)$$
$$= 2 \cdot 5[\cos(140° + 70°) + i \sin(140° + 70°)]$$
$$= 10(\cos 210° + i \sin 210°)$$
$$= 10\left[\frac{-\sqrt{3}}{2} + i\left(-\frac{1}{2}\right)\right]$$
$$= -5\sqrt{3} - 5i.$$

17. $5(\cos 100° + i \sin 100°) \cdot 2(\cos 35° + i \sin 35°)$

18. $3(\cos 130° + i \sin 130°) \cdot \frac{2}{3}(\cos 140° + i \sin 140°)$

19. $4(\cos 85° + i \sin 85°) \cdot \frac{3}{4}(\cos 245° + i \sin 245°)$

20. $4(\cos 20° + i \sin 20°) \cdot \frac{1}{2}(\cos 40° + i \sin 40°)$

21. $2(\cos 40° + i \sin 40°) \cdot (\cos 95° + i \sin 95°)$

22. $2(\cos 5° + i \sin 5°) \cdot 3(\cos 40° + i \sin 40°)$

■ *Express each quotient in the form a + bi.*

Example $\dfrac{6(\cos 85° + i \sin 85°)}{2(\cos 40° + i \sin 40°)}$

(*continued*)

Solution   From Equation (2) in this section,

$$\frac{6(\cos 85° + i \sin 85°)}{2(\cos 40° + i \sin 40°)} = \frac{6}{2}[\cos(85° - 40°) + i \sin(85° - 40°)]$$

$$= 3(\cos 45° + i \sin 45°)$$

$$= \frac{3}{\sqrt{2}} + \frac{3}{\sqrt{2}} i.$$

23. $\dfrac{8(\cos 260° + i \sin 260°)}{2(\cos 80° + i \sin 80°)}$ 24. $\dfrac{6(\cos 215° + i \sin 215°)}{3(\cos 80° + i \sin 80°)}$

25. $\dfrac{2(\cos 255° + i \sin 255°)}{5(\cos 15° + i \sin 15°)}$ 26. $\dfrac{2(\cos 280° + i \sin 280°)}{3(\cos 70° + i \sin 70°)}$

27. $\dfrac{3(\cos 70° + i \sin 70°)}{\cos 130° + i \sin 130°}$ 28. $\dfrac{\cos 19° + i \sin 19°}{4(\cos 259° + i \sin 259°)}$

29. Write $[2(\cos 20° + i \sin 20°)]^3$ in the form $a + bi$.

30. Write $\left(\dfrac{1}{\sqrt{2}} + \dfrac{1}{\sqrt{2}} i\right)^3$ in the form $a + bi$.

31. Show that if

$$a + bi = \rho(\cos \theta + i \sin \theta),$$

then $(a + bi)^2 = \rho^2(\cos 2\theta + i \sin 2\theta).$

32. Use the result of Exercise 31 to show that if

$$a + bi = \rho(\cos \theta + i \sin \theta),$$

then $(a + bi)^3 = \rho^3(\cos 3\theta + i \sin 3\theta).$

33. Use the result of Exercise 32 to show that if

$$a + bi = \rho(\cos \theta + i \sin \theta),$$

then $(a + bi)^4 = \rho^4(\cos 4\theta + i \sin 4\theta).$

34. Using the results of Exercises 31–33, which can you conjectur
about $(a + bi)^n$, $n \in N$, if $a + bi = \rho(\cos \theta + i \sin \theta)$?

35. Show that $\cos \theta + i \sin \theta$ and $\cos(-\theta) + i \sin(-\theta)$ are conjugate
of each other.

36. Show that $\cos \theta + i \sin \theta$ and $\cos(-\theta) + i \sin(-\theta)$ are reciproca
of each other.

7.5 DeMoivre's Theorem; Powers and Roots

Consider the complex number $z = a + bi = \rho(\cos \theta + i \sin \theta)$. From Equation (1) in Section 7.4.

$$[\rho(\cos \theta + i \sin \theta)]^2 = \rho(\cos \theta + i \sin \theta) \cdot \rho(\cos \theta + i \sin \theta)$$
$$= \rho^2[\cos(\theta + \theta) + i \sin(\theta + \theta)]$$
$$= \rho^2(\cos 2\theta + i \sin 2\theta);$$
$$[\rho(\cos \theta + i \sin \theta)]^3 = \rho^2(\cos 2\theta + i \sin 2\theta) \cdot \rho(\cos \theta + i \sin \theta)$$
$$= \rho^3[\cos(2\theta + \theta) + i \sin(2\theta + \theta)]$$
$$= \rho^3(\cos 3\theta + i \sin 3\theta).$$

It can be shown that similar results are valid for each such power for $n \in N$. That is, if $z = \rho(\cos \theta + i \sin \theta)$ and $n \in N$, then

$$z^n = \rho^n(\cos n\theta + i \sin n\theta). \tag{1}$$

This statement is known as **DeMoivre's theorem.** The proof of the statement and extensions of it which are presented in this section involve the process of mathematical induction and will not be shown. However, the informal argument above should make the validity of Equation (1) plausible.

Example   Express $(\sqrt{3} + i)^5$ in the form $a + bi$.

Solution   Since

$$\rho = |\sqrt{3} + i| = \sqrt{3 + 1} = 2$$

and

$$\cos \theta = \frac{\sqrt{3}}{2} \qquad \text{and} \qquad \sin \theta = \frac{1}{2},$$

angle θ is in Quadrant I and $\theta = 30°$. Thus,

$$(\sqrt{3} + i)^5 = [2(\cos 30° + i \sin 30°)]^5$$
$$= 2^5(\cos 5 \cdot 30° + i \sin 5 \cdot 30°)$$
$$= 32(\cos 150° + i \sin 150°)$$
$$= 32\left(-\frac{\sqrt{3}}{2} + \frac{1}{2}i\right)$$
$$= -16\sqrt{3} + 16i.$$

Equation (1) was stated for powers with natural number exponents. this statement is to be valid for integral exponents, consistent meanin; must be assigned to z^0 and z^{-n} for $n \in N$.

Substituting 0 for n in Equation (1), we have

$$z^0 = \rho^0(\cos 0 \cdot \theta + i \sin 0 \cdot \theta) = 1 \cdot (\cos 0 + i \sin 0)$$

$$= 1 \cdot (1 + 0) = 1.$$

Substituting $-n$ for n in Equation (1), we have

$$z^{-n} = \rho^{-n}[\cos(-n\theta) + i \sin(-n\theta)].$$

Since $\cos(-n\theta) = \cos n\theta$ and $\sin(-n\theta) = -\sin n\theta$, we have

$$z^{-n} = \frac{1}{\rho^n} \cdot (\cos n\theta - i \sin n\theta).$$

Multiplying the numerator and the denominator of the right-har member by the conjugate of the numerator yields

$$z^{-n} = \frac{1}{\rho^n} \cdot \frac{(\cos n\theta - i \sin n\theta) \cdot (\cos n\theta + i \sin n\theta)}{\cos n\theta + i \sin n\theta}$$

$$= \frac{1}{\rho^n} \cdot \frac{\cos^2 n\theta + \sin^2 n\theta}{\cos n\theta + i \sin n\theta}$$

$$= \frac{1}{\rho^n} \cdot \frac{1}{\cos n\theta + i \sin n\theta}$$

$$= \frac{1}{\rho^n(\cos n\theta + i \sin n\theta)}.$$

Substituting z^n for $\rho^n(\cos n\theta + i \sin n\theta)$ yields

$$z^{-n} = \frac{1}{z^n}.$$

Thus, if we want Equation (1) to be valid for every exponent $n \in$ z^0 and z^{-n} must be given the following interpretations.

For all $z \neq 0 + 0i$,

I. $z^0 = 1 + 0i = 1.$

II. $z^{-n} = \dfrac{1}{z^n}, \quad n \in N.$

DeMoivre's theorem [Equation (1)] can now be applied to powers with $n \in J$.

Example   Express $(-\sqrt{3} + i)^{-3}$ in the form $a + bi$.

Solution   If $z = -\sqrt{3} + i$,

$$\rho = |-\sqrt{2} + i| = \sqrt{3 + 1} = 2;$$

since

$$\cos \theta = -\frac{\sqrt{3}}{2} \quad \text{and} \quad \sin \theta = \frac{1}{2},$$

θ is in Quadrant II and $\theta = 150°$. Thus,

$$(-\sqrt{3} + i)^{-3} = [2(\cos 150° + i \sin 150°)]^{-3},$$

and from DeMoivre's theorem we have

$$(-\sqrt{3} + i)^{-3} = 2^{-3}[\cos(-3 \cdot 150°) + i \sin(-3 \cdot 150°)]$$

$$= \frac{1}{8}[\cos(-450°) + i \sin(-450°)]$$

$$= \frac{1}{8}[0 - 1 \cdot i] = -\frac{1}{8}i.$$

Another extension of DeMoivre's theorem is possible for rational number exponents $1/n$, where $n \in N$. First, we define an nth root of z in the same way that an nth root of a real number a is defined, when such a number exists.

For $z \in C$, $n \in N$, the number ω is an nth root of z if $\omega^n = z$.

We can now state another extension of DeMoivre's theorem applicable to roots of complex numbers.

If $n \in N$ and $z = \rho(\cos \theta + i \sin \theta)$, $z \neq 0$, then

$$\rho^{1/n}\left[\cos\left(\frac{\theta + k \cdot 360°}{n}\right) + i \sin\left(\frac{\theta + k \cdot 360°}{n}\right)\right], \quad k \in J \quad (2)$$

specifies all nth roots of z.

Using this extension of DeMoivre's theorem, we can now find n distinct complex nth roots for each $z \in C$, where $z \neq 0$. For $k = 0$, one nth root is

$$\rho^{1/n}\left(\cos\frac{\theta}{n} + i\sin\frac{\theta}{n}\right),$$

where θ is the angle with the least positive measure. This root is called the **principal nth root** of z.

Example   Express each of the five fifth roots of $z = \sqrt{2} + \sqrt{2}\,i$ in trigonometric form.

Solution   Since

$$\rho = |\sqrt{2} + \sqrt{2}\,i| = \sqrt{2 + 2} = 2,$$

and

$$\cos\theta = \frac{\sqrt{2}}{2} \quad\text{and}\quad \sin\theta = \frac{\sqrt{2}}{2},$$

the argument θ is in Quadrant I and $\theta = 45°$. Thus, in trigonometric form,

$$z = 2(\cos 45° + i\sin 45°),$$

and by Equation (2),

$$2^{1/5}\left[\cos\left(\frac{45° + k\cdot 360°}{5}\right) + i\sin\left(\frac{45° + k\cdot 360°}{5}\right)\right], \quad k \in J,$$

is a fifth root of z. Taking $k = 0, 1, 2, 3$, and 4 in turn yields the roots

$$2^{1/5}(\cos 9° + i\sin 9°),$$
$$2^{1/5}(\cos 81° + i\sin 81°),$$
$$2^{1/5}(\cos 153° + i\sin 153°),$$
$$2^{1/5}(\cos 225° + i\sin 225°),$$
$$2^{1/5}(\cos 297° + i\sin 297°).$$

Notice that the substitution of any other integer for k will simply produce one of these five complex numbers. For example, if $k = 5$,

$$2^{1/5}[\cos(9 + 360)° + i\sin(9 + 360)°] = 2^{1/5}(\cos 9° + i\sin 9°).$$

EXERCISE SET 7.5

A

■ *Express each power in the form a + bi.*

Example $(\sqrt{3} + i)^{-2}$

Solution   Since

$$\rho = |\sqrt{3} + i| = \sqrt{(\sqrt{3})^2 + (1)^2} = \sqrt{4} = 2;$$

and

$$\cos \theta = \frac{\sqrt{3}}{2} \quad \text{and} \quad \sin \theta = \frac{1}{2},$$

the argument θ is in Quadrant I and $\theta = 30°$. Thus,

$$(\sqrt{3} + i)^{-2} = [2(\cos 30° + i \sin 30°)]^{-2}.$$

From DeMoivre's theorem,

$$(\sqrt{3} + i)^{-2} = 2^{-2}[\cos(-2 \cdot 30°) + i \sin(-2 \cdot 30°)]$$

$$= \frac{1}{4}[\cos(-60°) + i \sin(-60°)]$$

$$= \frac{1}{4}\left(\frac{1}{2} - \frac{\sqrt{3}}{2}i\right) = \frac{1}{8} - \frac{\sqrt{3}}{8}i.$$

1. $[3(\cos 20° + i \sin 20°)]^3$ **2.** $[2(\cos 42° + i \sin 42°)]^5$

3. $(-1 + i)^6$ **4.** $(2\sqrt{3} - 2i)^5$

5. $\left(\frac{1}{4} + \frac{\sqrt{3}}{4}i\right)^3$ **6.** $\left(-\frac{\sqrt{3}}{2} + \frac{1}{2}i\right)^6$

7. $[9(\cos 315° + i \sin 315°)]^{-2}$ **8.** $[8(\cos 120° + i \sin 120°)]^{-3}$

9. $(2 - 2i)^{-3}$ **10.** $(\sqrt{3} - i)^{-3}$

11. $(-1 - \sqrt{3}i)^{-5}$ **12.** $(-2 - 2i)^{-4}$

■ *In Exercises 13–18, find the roots indicated and express each of them in trigonometric form.*

Example   The two square roots of $\dfrac{\sqrt{3}}{2} - \dfrac{1}{2}i.$

(continued)

Solution

$$\rho = \left| \frac{\sqrt{3}}{2} - \frac{1}{2}i \right| = \sqrt{\frac{3}{4} + \frac{1}{4}} = 1.$$

Thus, $\cos \theta = \sqrt{3}/2$ and $\sin \theta = -1/2$, and θ is in Quadrant IV. Hence, $\theta = 330°$ and

$$\frac{\sqrt{3}}{2} - \frac{1}{2}i = \cos(330° + k \cdot 360°) + i \sin(330° + k \cdot 360°).$$

The square roots of $\dfrac{\sqrt{3}}{2} - \dfrac{1}{2}i$, using DeMoivre's theorem, are

$$1^{1/2}\left[\cos\left(\frac{330° + k \cdot 360°}{2}\right) + i \sin\left(\frac{330° + k \cdot 360°}{2}\right) \right],$$

where $k = 0$ and $k = 1$. For $k = 0$, we have

$$\cos 165° + i \sin 165°,$$

and for $k = 1$, we have

$$\cos 345° + i \sin 345°.$$

13. The two square roots of

    a. $\cos 80° + i \sin 80°$ **b.** $\dfrac{1}{2} + \dfrac{\sqrt{3}}{2}i$

14. The two square roots of

    a. $9(\cos 140° + i \sin 140°)$ **b.** $-4i$

15. The three cube roots of

    a. $8(\cos 120° + i \sin 120°)$ **b.** $\dfrac{1}{2} + \dfrac{\sqrt{3}}{2}i$

16. The three cube roots of

    a. $2(\cos 75° + i \sin 75°)$ **b.** $-27i$

17. The four fourth roots of

    a. $\cos 240° + i \sin 240°$ **b.** $\dfrac{1}{2} - \dfrac{\sqrt{3}}{2}i$

18. The four fourth roots of

    a. $81(\cos 60° + i \sin 60°)$ **b.** $-16i$

■ *Solve each equation for all* $x \in C$.

19. $x^3 + 1 = 0$.   *Hint:* $x^3 = -1 = -1 + 0i$. Thus, x is a cube root of $-1 + 0i$.

20. $x^5 + 1 = 0$ **21.** $x^3 - 1 = 0$

22. $x^5 - 1 = 0$ **23.** $x^3 = \dfrac{\sqrt{3}}{4} - \dfrac{1}{4}i$

24. $x^5 = 12 - 5i$

B

25. Graph each of the following on a separate Cartesian coordinate system; make a conjecture about the graphs of all nth roots of 1.
  a. The two square roots of 1   **b.** The three cube roots of 1
  c. The four fourth roots of 1   **d.** The five fifth roots of 1
  e. The six sixth roots of 1   **f.** The eight eighth roots of 1

26. Use the results of Exercise 21 to show that the reciprocal of any cube root of unity is itself a cube root of unity.

27. Use DeMoivre's theorem to find an identity for $\cos 3\theta$ in terms of $\cos \theta$.   *Hint:* $(\cos \theta + i \sin \theta)^3 = \cos 3\theta + i \sin 3\theta$.

28. Use the hint in Exercise 27 to find an identity for $\sin 3\theta$ in terms of $\sin \theta$.

7.6 Polar Coordinates

The basis that we used to graph functions in Chapter 3 was the fact that we associated a point in the plane with an ordered pair of real numbers (x, y) called the Cartesian coordinates of the point. The first component of the ordered pair was the directed distance of the point from the y-axis, and the second component the directed distance from the x-axis.

Graphing Ordered Pairs (ρ, θ) The location of a point A in the plane can also be specified by ρ, the distance (a positive number) from the pole O to A and the measure of an angle θ^* that the ray through point A makes with a ray called the **polar axis.** See Figure 7.4 (page 268). If we extend the replacement set of ρ to include the negative real numbers and the replacement set of θ

---

* The measure of θ can be expressed in degrees or radians.

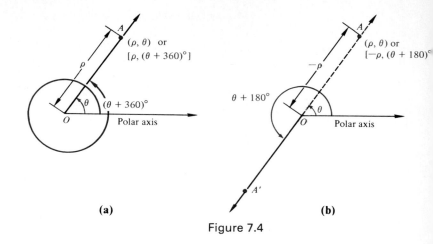

Figure 7.4

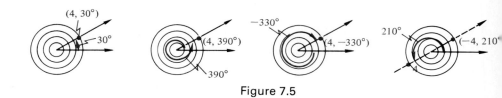

Figure 7.5

to include the set of all angles, we can specify point A by many ordered
pairs of the form (ρ, θ). The two components of each of these ordered
pairs are called **polar coordinates** of A. For example, in Figure 7.4a, if
(ρ, θ) are polar coordinates of A, then so are $[\rho, (\theta + k \cdot 360)^\circ]$, $k \in$
Also, if $-\rho \ (\rho > 0)$ denotes the directed distance from O to A along
the negative extension of the ray OA', as shown in Figure 7.4b, then
$[-\rho, (\theta + 180 + k \cdot 360)^\circ]$ are also acceptable coordinates of A. For
example, in Figure 7.5, the point having polar coordinates $(4, 30^\circ)$ also
has polar coordinates $(4, 390^\circ)$ and $(4, -330^\circ)$ for positive values of ρ
and $(-4, 210^\circ)$ for a negative value of ρ. Of course, there are infinitely
many other possible polar coordinates for this point.

  If point A is at the pole, then $\rho = 0$, and the coordinates of A are
$(0, \theta)$, where θ is an arbitrary angle, not necessarily zero. For example
$(0, 90^\circ)$ and $(0, 60^\circ)$ are both coordinates of the pole.

**Cartesian and Polar
Coordinates**

  If the pole coincides with the origin, and the polar axis coincides
with the positive x-axis, as in Figure 7.6, then for any point on the terminal ray of angle θ with Cartesian coordinates x and y, $\cos \theta = x/$

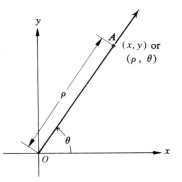

Figure 7.6

and $\sin \theta = y/\rho$. Hence, the Cartesian and polar coordinates can be related as follows.

$$\left. \begin{array}{l} x = \rho \cos \theta \\ y = \rho \sin \theta \end{array} \right\} \quad (1)$$

$$\left. \begin{array}{l} \rho = \pm\sqrt{x^2 + y^2} \\[2mm] \cos \theta = \dfrac{x}{\rho} \\[2mm] \sin \theta = \dfrac{y}{\rho}, \quad \rho \neq 0 \end{array} \right\} \quad (2)$$

Equations (1) can be used to find the rectangular coordinates for a point if the polar coordinates are known, and Equations (2) can be used to find the polar coordinates for a point if the Cartesian coordinates are known. The positive or negative radical expression is chosen depending on the particular problem.

Example   Find the Cartesian coordinates of the point with polar coordinates $(3, 60°)$.

Solution   From Equations (1), it follows that

$$x = 3 \cos 60° = 3 \cdot \frac{1}{2} = \frac{3}{2},$$

$$y = 3 \sin 60° = 3 \cdot \frac{\sqrt{3}}{2} = \frac{3\sqrt{3}}{2}.$$

The point and the corresponding ordered pairs are shown in the figure.

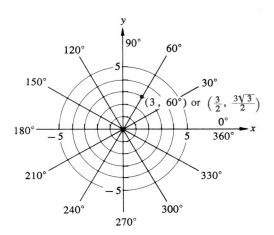

Example Find an approximation for the polar coordinates of the poin
with Cartesian coordinates $(5, -3)$, where $0° < \theta < 360°$ and $\rho > 0$
Graph the point.

Solution From Equations (2) and since $\rho > 0$,

$$\rho = \sqrt{x^2 + y^2} = \sqrt{25 + 9} = \sqrt{34} \approx 5.831;$$

$$\cos \theta = \frac{5}{5.831} \approx 0.8575.$$

Since the point is in Quadrant IV, then $\theta \approx 329.0°$. Therefore, the pai
of polar coordinates is $(\sqrt{34}, 329°)$. The point and the corresponding
ordered pairs are shown in the figure.

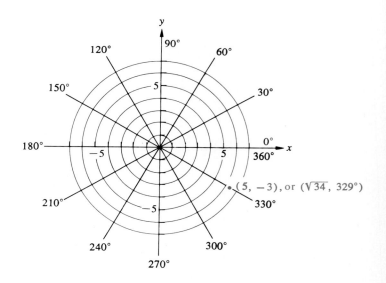

Rewriting Equations   Equations (1) and (2) on page 269 can be used to transform an equation in Cartesian form to polar form, and vice versa.

Example Rewrite $x^2 + y^2 - 4x = 0$ in polar form.

Solution Substituting ρ^2 for $x^2 + y^2$ and $\rho \cos \theta$ for x in

$$x^2 + y^2 - 4x = 0$$

yields

$$\rho^2 - 4\rho \cos \theta = 0,$$

from which

$$\rho(\rho - 4 \cos \theta) = 0.$$

Hence,

$$\rho = 0 \quad \text{or} \quad \rho - 4 \cos \theta = 0.$$

Note that the graph of $\rho = 0$ is the origin and also that the graph of $\rho = 4 \cos \theta$ includes the origin as one of its points. (For example, for $\theta = 90°$, we have $\rho = 0$.) Hence, the equation $\rho = 4 \cos \theta$ corresponds to $x^2 + y^2 - 4x = 0$, and it is not necessary to write $\rho = 0$ as part of the answer.

Example   Rewrite $\rho(1 - 3 \cos \theta) = 2$ in Cartesian form.

Solution   Substituting $\pm\sqrt{x^2 + y^2}$ for ρ and x/ρ, or $x/\pm\sqrt{x^2 + y^2}$, for $\cos \theta$, we have

$$\pm\sqrt{x^2 + y^2}\left[1 - 3\left(\frac{x}{\pm\sqrt{x^2 + y^2}}\right)\right] = 2$$

from which

$$\pm\sqrt{x^2 + y^2} - 3x = 2,$$
$$\pm\sqrt{x^2 + y^2} = 3x + 2.$$

Squaring each member of this equation, we have

$$x^2 + y^2 = 9x^2 + 12x + 4,$$

from which

$$8x^2 - y^2 + 12x + 4 = 0.$$

EXERCISE SET 7.6

A

▪ *Find four additional set of polar coordinates $(-360° < \theta \le 360°)$ for each point whose coordinates are given.*

1. $(3, 380°)$ **2.** $(6, 840°)$ **3.** $(5, -450°)$

4. $(4, -540°)$ **5.** $(-4, 420°)$ **6.** $(-5, -600°)$

▪ *Find the Cartesian coordinates of each point with polar coordinates as given.*

7. $\left(4, \frac{\pi}{6}^R\right)$ **8.** $\left(6, \frac{2\pi}{3}^R\right)$ **9.** $\left(3, -\frac{3\pi}{4}^R\right)$

10. $\left(0, \dfrac{13\pi}{3}^R\right)$ **11.** $\left(-3, \dfrac{5\pi}{6}^R\right)$ **12.** $\left(-4\dfrac{\pi}{4}^R\right)$

■ *Find the pair of polar coordinates (θ in radians) using an angle θ such th* *$0^R < \theta < 2\pi^R$ and $\rho > 0$ for each point with Cartesian coordinates* *given.*

13. $(4, 0)$ **14.** $(0, -3)$ **15.** $(2, 2)$

16. $(-1, -\sqrt{3})$ **17.** $\left(-\dfrac{\sqrt{3}}{2}, \dfrac{1}{2}\right)$ **18.** $\left(\dfrac{1}{2}, -\dfrac{\sqrt{3}}{2}\right)$

■ *Transform each Cartesian equation to an equation in polar form.*

Example $y = 3x$

Solution   Substituting $\rho \sin \theta$ for y and $\rho \cos \theta$ for x yields

$$\rho \sin \theta = 3\rho \cos \theta$$

Multiplying each member by $1/(\rho \cos \theta)$, $\rho \neq 0$, $\theta \neq 90° + k \cdot 180$ yields

$$\frac{\sin \theta}{\cos \theta} = 3,$$

$$\tan \theta = 3.$$

Notice that for values of θ for which the equation is satisfied, $\tan \theta$ is constant, namely, 3.

19. $x = 4$ **20.** $y = -3$

21. $x^2 = 8y$ **22.** $y^2 = 4x$

23. $x^2 + y^2 = 16$ **24.** $4x^2 + y^2 = 4$

25. $x^2 + y^2 - 3y = 0$ **26.** $x^2 + y^2 + 4x = 0$

■ *Transform each polar equation to an equation in Cartesian form.*

Example $\rho = 2 \sec \theta$

Solution   Substituting $1/\cos \theta$ for $\sec \theta$ yields

$$\rho = \frac{2}{\cos \theta},$$

$$\rho \cos \theta = 2.$$

Because $\rho \cos \theta = x$, the Cartesian form of the equation is $x = 2$.

27. $\rho = 6$ **28.** $\rho = 9$

29. $\rho = 4 \sin \theta$ **30.** $\rho = 3 \cos \theta$

31. $\rho = 2 \csc \theta$ **32.** $\rho = -4 \sec \theta$

33. $\rho = \dfrac{4}{1 - \sin \theta}$ **34.** $\rho = \dfrac{4}{1 + 2 \cos \theta}$

7.7 Graphing Equations in Polar Form

In our earlier work we have seen that an equation in two variables, say x and y, serves to pair elements from the replacement sets of the variables and has a graph in the geometric plane in a Cartesian coordinate system. An equation in ρ and θ also has a graph in a polar coordinate system.

Example Graph $\rho = 4 \sin \theta$ in the interval $0° \le \theta < 360°$.

Solution Some ordered pairs with commonly used angles 30°, 45°, etc., are first obtained. The tabulation of the data in the arrangement shown is sometimes helpful. The approximations shown for some values of $\sin \theta$ are sufficient for our purpose. These ordered pairs (ρ, θ) are shown in figure I.

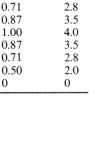

θ	sin θ	ρ or 4 sin θ
0°	0	0
0°	0.50	2.0
5°	0.71	2.8
0°	0.87	3.5
0°	1.00	4.0
0°	0.87	3.5
5°	0.71	2.8
0°	0.50	2.0
0°	0	0

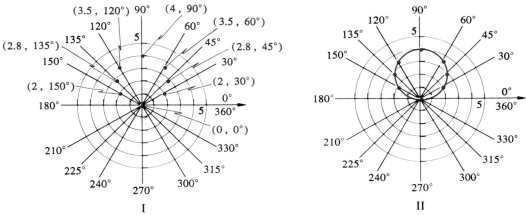

I   II

The complete curve is obtained for $0° \le \theta < 180°$. For values of θ where $180° \le \theta < 360°$, the components of ordered pairs (ρ, θ) are also coordinates of points on the curve shown in figure II.

Example Graph $\rho = \dfrac{3}{\sin \theta + \cos \theta}$.

(continued)

Solution   Ordered pairs are first obtained as shown in the table and figure I. The graph, a straight line, is shown in figure II.

θ	$\sin \theta$	$\cos \theta$	$\sin \theta + \cos \theta$	$\rho = \dfrac{3}{\sin \theta + \cos}$
0°	0	1.00	1.00	3.0
30°	0.50	0.87	1.37	2.2
45°	0.71	0.71	1.42	2.1
60°	0.87	0.50	1.37	2.2
90°	1.00	0	1.00	3.0
120°	0.87	−0.50	0.37	8.1
135°	0.71	−0.71	0	undef.

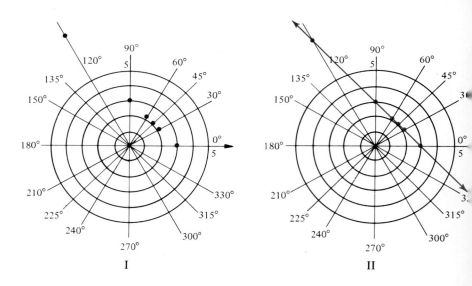

I   II

Frequently, the polar form of an equation is more useful than the rectangular form; furthermore, its graph may be simpler to sketch from the polar representation.

Excluded Values of θ   An analysis of the polar equation for particular features can facilitate graphing. For example, if

$$\rho^2 = 9 \cos 2\theta$$

and we select $\theta = 60°$, then

$$\rho^2 = 9 \cos 120° = \frac{-9}{2},$$

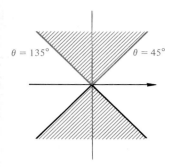

$\theta = 135°$ $\theta = 45°$

Figure 7.7

and $\rho = \pm\sqrt{-9/2}$, an imaginary number. Therefore, 60° must be excluded as a value of θ. In fact, for all θ such that

$$(90° + k \cdot 360°) < 2\theta < (270° + k \cdot 360°),$$

or

$$(45° + k \cdot 180°) < \theta < (135° + k \cdot 180°), \quad k \in J,$$

it follows that $\cos 2\theta < 0$, and hence these values of θ must be excluded (see Figure 7.7).

In order to graph $\rho^2 = 9 \cos 2\theta$, it is therefore necessary to select values of θ that are not excluded. Some ordered pairs (ρ, θ) that are solutions of the equation are shown in Table 7.1, and its graph is shown in Figure 7.8. The graph is called a **lemniscate**.

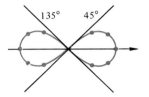

135°   45°

Figure 7.8

TABLE 7.1

θ	2θ	$\cos 2\theta$	ρ^2 or $9 \cos 2\theta$	ρ
0°	0°	1	9	± 3
15°	30°	$\dfrac{\sqrt{3}}{2}$	$\dfrac{9\sqrt{3}}{2}$	± 2.8
30°	60°	$\dfrac{1}{2}$	$\dfrac{9}{2}$	± 2.1
45°	90°	0	0	0
135°	270°	0	0	0
150°	300°	$\dfrac{1}{2}$	$\dfrac{9}{2}$	± 2.1
165°	330°	$\dfrac{\sqrt{3}}{2}$	$\dfrac{9\sqrt{3}}{2}$	± 2.8
180°	360°	1	9	± 3

Tangent Lines at the Pole   If, for some particular value of θ, say α, the value of ρ becomes 0, then the line $\theta = \alpha$ is tangent to the graph at the pole. For example, for the line to be tangent to the graph of $\rho^2 = 9 \cos 2\theta$ at the pole, ρ must be 0. Hence,

$$\cos 2\theta = 0,$$

from which

$$2\theta = \frac{\pi}{2} \quad \text{or} \quad 2\theta = \frac{3\pi}{2},$$

$$\theta = \frac{\pi}{4} \qquad \text{or} \qquad \theta = \frac{3\pi}{4}.$$

Thus, there are two lines tangent to the graph at the pole as shown i
Figure 7.8.

Symmetry   The graph of an equation in polar form is often symmetric wit
respect to a line or a point. In particular, considerations of symmetr
with respect to the polar axis $(\theta = 0°)$, the vertical line through the pol
sometimes called the vertical axis $(\theta = 90°)$, and the pole $(\rho = 0)$ ar
helpful when graphing equations in polar form.

Testing an equation for each of these symmetries consists of makin
certain replacements of ρ and θ. Then, if the new equation is equivaler
to the original, the graph has that particular symmetry. Table 7.2 list
various replacements that are helpful. Note that for each possibl
symmetry, there are two tests. If either test is successful, the graph ha
that particular symmetry.

Note that tests (a), (d), and (f) from Table 7.2 are successful whe
applied to the equation $\rho^2 = 9 \cos 2\theta$ on page 275. Thus, for test (a), if

TABLE 7.2

| Symmetry with respect to | | Replacement for | | Geometric representation |
		ρ	θ	
Polar axis $(\theta = 0°)$	(a)		$-\theta$	
	(b)	$-\rho$	$\pi - \theta$	
Vertical axis $(\theta = 90°)$	(c)		$\pi - \theta$	
	(d)	$-\rho$	$-\theta$	
Pole $(\rho = 0)$	(e)		$\pi + \theta$	
	(f)	$-\rho$		

is replaced by $-\theta$,

$$\rho^2 = 9 \cos 2(-\theta) = 9 \cos(-2\theta)$$
$$= 9 \cos 2\theta.$$

Then for test (d), replacing ρ with $-\rho$ and θ with $-\theta$, we have

$$(-\rho)^2 = 9 \cos 2(-\theta)$$

or

$$\rho^2 = 9 \cos 2\theta.$$

Finally, for test (f), replacing ρ with $-\rho$,

$$(-\rho)^2 = 9 \cos 2\theta$$

or

$$\rho^2 = 9 \cos 2\theta.$$

Hence, as indicated in Figure 7.8, the graph is symmetric with respect to the polar axis, the vertical axis, and the pole.

Example   Test the equation $\rho = 1 + \cos \theta$ for the symmetries in Table 7.2, and then graph the equation.

Test (a):   Replacing θ with $-\theta$:

$$\rho = 1 + \cos(-\theta) = 1 + \cos \theta.$$

The new equation is equivalent to the original; hence, the graph is *symmetric with respect to the polar axis.*

Test (c):   Replacing θ with $\pi - \theta$, we have

$$\rho = 1 + \cos(\pi - \theta) = 1 - \cos \theta.$$

This is not equivalent to the original equation, so we try the second test.

Test (d):   Replacing ρ with $-\rho$, with θ with $-\theta$, we have

$$-\rho = 1 + \cos(-\theta) = 1 + \cos \theta.$$

Since this equation also is not equivalent to the original, *we cannot conclude that the graph is symmetric with respect to the vertical axis.*

(continued)

θ	$\cos \theta$	ρ or $1 + \cos \theta$
0°	1	2
30°	$\dfrac{\sqrt{3}}{2}$	1.87
60°	$\dfrac{1}{2}$	1.5
90°	0	1
120°	$-\dfrac{1}{2}$	$\dfrac{1}{2}$
150°	$-\dfrac{\sqrt{3}}{2}$	0.13
180°	-1	0

Test (e):   Replacing θ with $\pi + \theta$, we have

$$\rho = 1 + \cos(\pi + \theta) = 1 - \cos \theta.$$

which is not equivalent to the original equation. Hence, we try th
second test.

Test (f):   Replacing ρ with $-\rho$, we have

$$-\rho = 1 + \cos \theta.$$

This again is not equivalent to the original equation, so *we cann
conclude that the graph is symmetric with respect to the pole.*

We have shown only that *the graph is symmetric with respect to th
polar axis.* Now, using selected values of θ between 0° and 180°, we fir
graph the equation in this interval. Then, using the fact that the graph
symmetric with respect to the polar axis from test (a), we sketch the enti
graph. The graph is known as a **cardioid.**

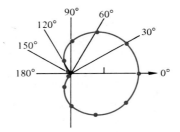

Several additional interesting types of curves that are graphs of pol
equations are shown in Figure 7.9.

LIMACONS

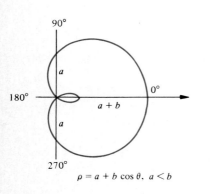

$\rho = a + b \cos \theta, \ a < b$

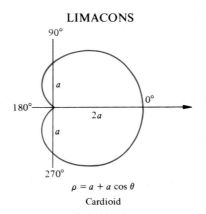

$\rho = a + a \cos \theta$

Cardioid

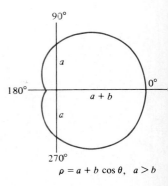

$\rho = a + b \cos \theta, \ a > b$

Figure 7.9

LEMNISCATES

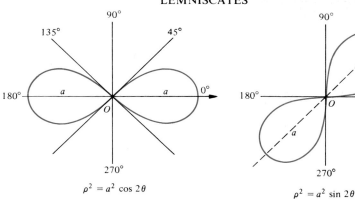

$$\rho^2 = a^2 \cos 2\theta$$

$$\rho^2 = a^2 \sin 2\theta$$

THREE-LEAFED ROSE

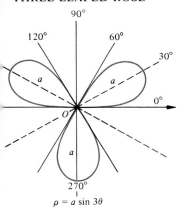

$$\rho = a \sin 3\theta$$

FOUR-LEAFED ROSE

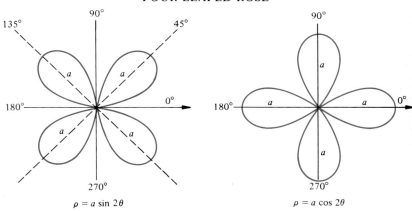

$$\rho = a \sin 2\theta$$

$$\rho = a \cos 2\theta$$

SPIRAL OF ARCHIMEDES

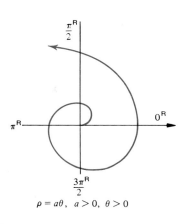

$$\rho = a\theta, \quad a > 0, \ \theta > 0$$

HYPERBOLIC SPIRAL

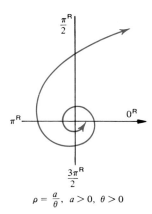

$$\rho = \frac{a}{\theta}, \quad a > 0, \ \theta > 0$$

LOGARITHMIC SPIRAL

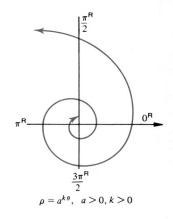

$$\rho = a^{k\theta}, \quad a > 0, k > 0$$

Figure 7.9 (*continued*)

EXERCISE SET 7.7

A

▪ *Graph each equation.*

1. $\rho = 4 \cos \theta$
   (circle)

2. $\rho = 3 \sin \theta$
   (circle)

3. $\rho = \cos 3\theta$
   (three-leaved rose)

4. $\rho = 2 \sin 3\theta$
   (three-leaved rose)

5. $\rho = \sin 2\theta$
   (four-leaved rose)

6. $\rho = 2 \cos 2\theta$
   (four-leaved rose)

7. $\rho = 1 + 2 \cos \theta$
   (limaçon)

8. $\rho = 3 + 3 \cos \theta$
   (cardioid)

9. $\rho = 2 + 2 \sin \theta$
   (cardioid)

10. $\rho = 2 + \cos \theta$
    (limaçon)

11. $\rho^2 = 4 \cos 2\theta$
    (lemniscate)

12. $\rho^2 = 4 \sin 2\theta$
    (lemniscate)

13. $\rho = 4$ *Hint:* Consider the equation $\rho = 0 \cdot \theta + 4$.

14. $\rho = 5$

15. $\theta = 30°$ *Hint:* Consider the equation $\theta = 0 \cdot \rho + 30°$.

16. $\theta = 45°$

B

▪ *Graph each equation.*

17. $\rho = \dfrac{1}{\theta}, \theta > 0$
    (hyperbolic spiral)

18. $\rho = 2\theta, \theta > 0$
    (spiral of Archimedes)

▪ *Graph each of the given pairs of equations and find coordinates for t*
points of intersection.

19. $\rho = \sin \theta$
    $\rho = \cos \theta$

20. $\rho = 1 + \sin \theta$
    $\rho = 1$

21. $\rho = 1 + \sin \theta$
    $\rho = 1 + \cos \theta$

22. $\rho = 2 \sin \theta$
    $\rho = 2 \sin 2\theta$

23. Solve the system in Exercise 19 analytically. Explain why the coo
    dinates of the origin are not in the solution set of the system.

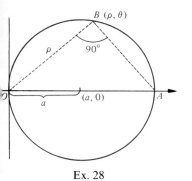

Ex. 28

24. Solve the system in Exercise 20 analytically.

25. Solve the system in Exercise 21 analytically.

26. Solve the system in Exercise 22 analytically.

27. Show that the distance between the points with coordinates (ρ_1, θ_1) and (ρ_2, θ_2) is given by $d = \sqrt{\rho_1^2 + \rho_2^2 - 2\rho_1\rho_2 \cos(\theta_1 - \theta_2)}$.

28. Find an equation in polar form for the circle with center at $(a, 0^R)$ and radius of length a.   *Hint:* Recall that an angle inscribed in a semicircle is a right angle.

Chapter Summary

[7.1]   A **complex number** is defined by

$$z = a + bi, \qquad a, b \text{ real numbers and } i^2 = -1.$$

Sums and products of complex numbers are found by treating $a + bi$ as a binomial in the variable i and replacing i^2 with -1.

A complex number $a + bi$ can be identified with a **real number** if $b = 0$; if $b \neq 0$, the number is an **imaginary number,** and for the special case where $a = 0$, the number is a **pure imaginary number.**

For $b > 0$,

$$\sqrt{-b} = \sqrt{b}\,i = i\sqrt{b}.$$

For $a, b > 0$,

$$\sqrt{-a}\sqrt{-b} = i^2\sqrt{ab} = -\sqrt{ab}.$$

[7.2]   The properties used for the set of complex numbers are similar to the corresponding properties of real numbers. Furthermore,

$$z_1 - z_2 = z_1 + (-z_2);$$

$$\frac{z_1}{z_2} = z_1 \cdot \frac{1}{z_2} \quad (z_2 \neq 0).$$

If $z_2, z_3 \neq 0$, then

$$\frac{z_1}{z_2} = \frac{z_1 z_3}{z_2 z_3}.$$

If $z = a + bi$, then the **conjugate** of z is $\bar{z} = a - bi$. The product of $a + bi$ and its conjugate equals $a^2 + b^2$, a real number.

[7.3]   A plane on which complex numbers are graphed is called a **comple** **plane.** The x-axis is called the **real axis** and the y-axis is called th **imaginary axis.**

Complex numbers can be added graphically by the parallelogram la applicable to geometric vectors.

The **absolute value** or **modulus** of the complex number z is

$$\rho = |z| = |a + bi| = |(a, b)| = \sqrt{a^2 + b^2}.$$

The modulus is the distance from the origin to the graph of (a, b). A **argument** or **amplitude** of the complex number $z = a + bi = (a,$ denoted by **arg$(a + bi)$** or **arg(a, b),** is an angle θ such that

$$\cos \theta = \frac{a}{\rho} \quad \text{and} \quad \sin \theta = \frac{b}{\rho} \quad (a^2 + b^2 \neq 0).$$

If θ is an argument of $a + bi$, then $(\theta + k \cdot 360)^\circ$, $k \in J$, is also a argument of $a + bi$. The angle θ with the least positive measure is calle the **principal argument.**

[7.4]   A nonzero complex number $a + bi$ can be expressed in **trigonometr** **form** as

$$a + bi = \rho[\cos(\theta + k \cdot 360^\circ) + i \sin(\theta + k \cdot 360^\circ)], \quad k \in J.$$

If $z_1 = \rho_1(\cos \theta_1 + i \sin \theta_1)$ and $z_2 = \rho_2(\cos \theta_2 + i \sin \theta_2)$, then

$$z_1 \cdot z_2 = \rho_1 \cdot \rho_2[\cos(\theta_1 + \theta_2) + i \sin(\theta_1 + \theta_2)],$$

$$\frac{z_1}{z_2} = \frac{\rho_1}{\rho_2}[\cos(\theta_1 - \theta_2) + i \sin(\theta_1 - \theta_2)].$$

[7.5]   If $z = \rho(\cos \theta + i \sin \theta)$ and $n \in N$, then

$$z^n = \rho^n(\cos n\theta + i \sin n\theta).$$

This is known as **DeMoivre's theorem.**

If $z \neq 0 + 0i$, then

$$z^0 = 1 + 0i = 1 \quad \text{and} \quad z^{-n} = \frac{1}{z^n}, \quad n \in N.$$

If $n \in N$ and $z = \rho(\cos \theta + i \sin \theta)$, then

$$\rho^{1/n}\left[\cos\left(\frac{\theta + k \cdot 360^\circ}{n}\right) + i \sin\left(\frac{\theta + k \cdot 360^\circ}{n}\right)\right], \quad k \in J$$

is an **nth root of z.** For $k = 0$, the nth root

$$\rho^{1/n}\left(\cos\frac{\theta}{n} + i\sin\frac{\theta}{n}\right)$$

is called the **principal root.**

[7.6]   The components of an ordered pair $(\rho, \theta°)$ or (ρ, θ^R) that locate a point A in the plane are called the **polar coordinates** of A. Each such ordered pair specifies a unique point; however, each point in the plane corresponds to infinitely many ordered pairs of polar coordinates.

Cartesian and polar coordinates are related as follows:

$$x = \rho\cos\theta \qquad\qquad \rho = \pm\sqrt{x^2 + y^2},$$

$$y = \rho\sin\theta$$

$$\cos\theta = \frac{x}{\rho} = \frac{x}{\pm\sqrt{x^2 + y^2}},$$

$$\sin\theta = \frac{y}{\rho} = \frac{y}{\pm\sqrt{x^2 + y^2}} \qquad [(x, y) \neq (0, 0)].$$

[7.7]   Polar equations can be graphed by obtaining a number of ordered pairs (ρ, θ). The notions of excluded values, tangent lines, and symmetry can facilitate the procedures of graphing such equations.

A list of symbols introduced in this chapter is shown inside the front cover, and important properties are shown inside the back cover.

Review Exercises

A

[7.1] ▪ *Write each expression in the form a + bi.*

1. $(5 - 2i) + (3 + i)$ **2.** $(1 + 3i) + (3 - 5i)$

3. $(-4 - i) + (-4 + 9i)$ **4.** $(-3 + i) + (3 - 2i)$

5. $(8 + 6i)(-2 + 5i)$ **6.** $(5 + 3i)(2 - 4i)$

▪ *Write each expression as one of the complex numbers i, -1, $-i$, or 1.*

7. i^{13} **8.** i^{23}

▪ *Write each expression in the form a + bi.*

9. $\sqrt{-5}(3 - \sqrt{-2})$ **10.** $(3 - \sqrt{-5})(3 + \sqrt{-5})$

11. Show that $3i$ and $-3i$ satisfy the equation $x^2 + 9 = 0$.

12. Show that $1 + 3i$ and $1 - 3i$ satisfy $x^2 - 2x + 10 = 0$.

[7.2] ▪ *Express each difference in the form $a + bi$.*

13. $(-7 + 2i) - (8 + 3i)$ **14.** $(6 - 9i) - (3 - 4i)$

▪ *Express each quotient in the form $a + bi$.*

15. $\dfrac{8 + 3i}{2 + 4i}$ **16.** $\dfrac{-7 + i}{6 - 10i}$

▪ *Write each expression in the form bi or $a + bi$.*

17. $\dfrac{-2}{\sqrt{-6}}$ **18.** $\dfrac{4}{2 + \sqrt{-5}}$

[7.3] ▪ *Express each complex number as an ordered pair.*

19. $-4 + i$ **20.** $2\sqrt{3} - 2i$

▪ *Express each complex number in the form $a + bi$.*

21. $(9, -6)$ **22.** $(3, 4)$

▪ *Graph each complex number z, its negative $-z$, and its conjugate $\bar{z}$.*

23. $-3 + 5i$ **24.** $(-2, -6)$

▪ *Represent each sum graphically and check the results by algebra methods.*

25. $(-4 + 3i) + (8 - 2i)$ **26.** $(-1, -8) + (6, 7)$

▪ *Find the absolute value and the principal argument of each comple number.*

27. $4 + 2i$ **28.** $(-2, -5)$

[7.4] ▪ *Represent each complex number in trigonometric form, using the ang with least nonnegative measure.*

29. $1 + i$ **30.** $-3 + 3\sqrt{3}\,i$

■ *Represent each complex number graphically and write it in the form a + bi.*

31. $7(\cos 120° + i \sin 120°)$

32. $2[\cos(-315°) + i \sin(-315°)]$

■ *Express each product in the form a + bi.*

33. $3(\cos 200° + i \sin 200°) \cdot 2(\cos 25° + i \sin 25°)$

34. $\sqrt{3}(\cos 50° + i \sin 50°) \cdot \sqrt{12}[\cos(-20°) + i \sin(-20°)]$

■ *Express each quotient in the form a + bi.*

35. $\dfrac{8(\cos 309° + i \sin 309°)}{9(\cos 129° + i \sin 129°)}$

36. $\dfrac{12[\cos(-315°) + i \sin(-315°)]}{3(\cos 270° + i \sin 270°)}$

[7.5] ■ *Express each power in the form a + bi.*

37. $[3(\cos 15° + i \sin 15°)]^4$ **38.** $(\sqrt{3} - i)^3$

39. $[4(\cos 240° + i \sin 240°)]^{-2}$ **40.** $(4 - 4i)^{-3}$

41. Find the two square roots of $\dfrac{\sqrt{3}}{2} + \dfrac{1}{2}i$.

42. Find the three cube roots of $125(\cos 330° + i \sin 330°)$.

■ *Solve each equation, $x \in C$.*

43. $x^4 + 1 = 0$ **44.** $x^5 = -5 - 5\sqrt{3}\,i$

[7.6] ■ *Find four additional sets of polar coordinates $(-360° < \theta \le 360°)$ for each point.*

45. $(2, 610°)$ **46.** $(-3, 390°)$

■ *Find the Cartesian coordinates for each point.*

47. $\left(3, \dfrac{\pi}{3}^{R}\right)$ **48.** $\left(-2, \dfrac{3\pi}{4}^{R}\right)$

■ *Find the pair of polar coordinates (θ in radians) using an angle θ such the*
   $0^R < \theta < 2\pi^R$ *and* $\rho > 0$ *for each point.*

49. $(2, -2)$ **50.** $(-\sqrt{3}, -1)$

■ *Transform each equation to an equation in polar form.*

51. $x = 3$ **52.** $x^2 + 3y^2 = 5$

■ *Transform each equation to an equation in Cartesian form.*

53. $\rho = 5 \sin \theta$ **54.** $\rho = \dfrac{2}{1 - \cos \theta}$

$[7.7]$ ■ *Graph each equation.*

55. $\rho = 1 + \sin \theta$ **56.** $\rho = 1 - \sin \theta$

57. $\rho = \cos 2\theta$ **58.** $\rho = 2 \cos 3\theta$

ALGEBRA AND GEOMETRY REVIEW

In Appendix A we review the basic ideas of algebra and geometry, using terminology and notation that are needed in the development of various topics in the text.

A.1 Set Notation

Recall that a set is simply a collection of "things." Each "thing" in the set is called a **member** or an **element** of the set. The membership can be described by listing the names of the members in braces, { }, or by stating a rule that identifies the members in the set.

The symbol $\in$ is used to denote membership in a set. For example, we can write

$$3 \in \{3, 4, 5\}$$

to represent the statement "3 is an element of $\{3, 4, 5\}$."

Sets are also designated by means of capital letters, A, B, C, R, etc.

Example   If $A = \{2, 3, 4\}$ and $B = \{4, 5, 6\}$, then

$$3 \in A \qquad \text{but} \qquad 3 \notin B.$$

An unspecified element of a set is usually denoted by a lowercase italic letter such as a, b, c, x, and y or by a lowercase letter from the Greek alphabet such as α, β, γ, and θ. Such a symbol is called a **variable;** a symbol used to denote a specific element is called a **constant.**

There are two important operations on sets.

*The **union** of two sets A and B is the set of all elements that belong to A or to B or to both.*

The symbol $\cup$ is used to denote this operation; $A \cup B$ is read "the union of A and B."

Example If $A = \{2, 3, 4\}$ and $B = \{4, 5, 6\}$, then

$$A \cup B = \{2, 3, 4, 5, 6\}.$$

A second useful operation on sets is the following.

*The **intersection** of two sets A and B is the set of all elements that belong to both A and B.*

The symbol $\cap$ is used to denote this operation; $A \cap B$ is read "the intersection of A and B."

Example If $A = \{2, 3, 4\}$ and $B = \{4, 5, 6\}$, then

$$A \cap B = \{4\}.$$

If $C = \{5, 9\}$ and $D = \{5, 6, 7, 8, 9\}$, then

$$C \cap D = \{5, 9\}.$$

The set that contains no elements is called the **null set** or the **empty set** and is denoted by the symbol $\varnothing$.

Example The intersection of two sets is a set. Therefore, if $C = \{2, 3, 4\}$ and $D = \{5, 6, 7\}$, we can write

$$C \cap D = \{2, 3, 4\} \cap \{5, 6, 7\} = \varnothing.$$

Two sets such as C and D, which do not have any elements in common, are said to be **disjoint.**

A.2 The Set of Real Numbers

Recall from your study of algebra that each real number can be associated with one and only one point on a number line and each point on the line can be associated with one and only one real number. Several

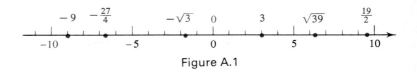

Figure A.1

examples are shown in Figure A.1. The real number corresponding to a point on the line is called the **coordinate** of the point, and the point is called the **graph** of the number. A number whose graph is to the right of the graph of zero is **positive,** and a number whose graph is to the left is **negative.**

The following sets of numbers are contained in the set of **real numbers, R.** The capital letters shown are the ones that are sometimes associated with the respective sets. These are the symbols that we use in this text.

1. The set **N** of **natural numbers,** among whose elements are such numbers as 1, 2, 6, 25, and 624.

2. The set **J** of **integers,** whose elements consist of the natural numbers, their negatives, and zero. Included are such elements as $-24, -9, 0, 7$, and 314.

3. The set **Q** of **rational numbers,** whose elements are all those numbers that can be represented in the form $\dfrac{a}{b}$, or a/b, where a and b are integers and b is not zero. Included are such elements as $-\frac{1}{3}$, $0, \frac{1}{4}$, and 6.

4. The set **H** of **irrational numbers,** whose elements are those numbers that are the coordinates of points on the number line that are not the graphs of rational numbers. An irrational number cannot be represented in the form a/b, where a and b are integers.

Figure A.2 summarizes the relationships among these sets of numbers.

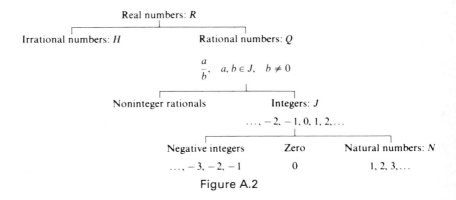

Figure A.2

Radical notation is useful to represent some irrational numbers.

For all $a \in R$ where a is positive or zero, $\sqrt{a}$ is the nonnegative number such that

$$\sqrt{a} \cdot \sqrt{a} = a.$$

Example

a. $\sqrt{9}$ is the positive number such that $\sqrt{9}\sqrt{9} = 9$.

b. For $x > 0$, $\sqrt{x}$ is the positive number such that $\sqrt{x}\sqrt{x} = x$.

Frequently, rational numbers are used as approximations for irratio nal numbers. For example, by using the symbol $\approx$ for "approximatel equal to," we write $\sqrt{2} \approx 1.414$, $\sqrt{3} \approx 1.732$, $\sqrt{5} \approx 2.236$, etc. Table in Appendix C gives rational number approximations for some squar roots that are irrational numbers.

If the elements of a set can be paired with the numerals 1, 2, 3,..., for some natural number n so that each element is paired with exactl one numeral and each numeral is paired with exactly one element, the the set is said to be a **finite set.** A set that is not the empty set and is no finite is called an **infinite set.** Some infinite sets can be designated by usin three dots and some of the members listed to establish a pattern a shown in Figure A.2.

Often we wish to consider the nonnegative member of the pair of rea numbers a and $-a$, where $a \neq 0$. The special notation $|a|$ is used for th purpose. To extend the notation for all $a \in R$, we define $|0| = 0$.

For all $a \in R$,

if a is positive or zero, then $|a| = a$;

if a is negative, then $|a| = -a$.

$|a|$ is called the **absolute value** *of a.*

Examples

a. $|4| = 4$ b. $|-4| = -(-4) = 4$ c. $|0| = 0$

The order of real numbers can be established by the following.

For all $a, b \in R$, b is **less than** *a if for some positive real number c,*

$$b + c = a.$$

For such conditions, a is said to be **greater than** *b.*

Examples

a. 4 is less than 7 because there is a *positive number c* (equal to 3) such that $4 + c = 7$.

b. -9 is less than -5 because there is a *positive number c* (equal to 4) such that $-9 + c = -5$.

The number line shown in Figure A.3 is very helpful in visualizing whether one real number is less than or greater than a second real number. If the graph of b lies to the *left* of the graph of a, then b is less than a.

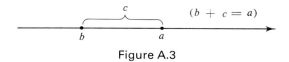

Figure A.3

The following symbols are used in connection with the property of order:

$<$ read "is less than";

$\leq$ read "is less than or equal to";

$>$ read "is greater than";

$\geq$ read "is greater than or equal to."

Observe that $b < a$ and $a > b$ are equivalent statements.

Examples

a. $4 < 7$ and $7 > 4$

b. $-9 < -5$ and $-5 > -9$

c. $x < 7$ is equivalent to $7 > x$

In your study of arithmetic and algebra, you applied the operations of addition and multiplication to elements of the set of real numbers. You also made assumptions, called **axioms,** about these operations and studied the logical consequences of these axioms. For your reference, we include a list of some of these axioms and properties on page 293.

A.3 Equations in Two Variables; Relations and Functions

Recall from your study of algebra that the ordered pair $(2, 9)$ is called a **solution** of the equation $y = x + 7$ because the replacement of x by the first component 2 and the replacement of y by the second component 9 results in a true statement, $9 = 2 + 7$. The ordered pair $(2, 9)$ is said to *satisfy* the equation. The set of all such ordered pairs that satisfy an equation is called the **solution set** of the equation.

The equation $y = x + 7$ does not result in a true statement for all ordered pairs of real numbers. For example, $(3, 5)$ is not a solution because $5 \neq 3 + 7$. Such equations are called **conditional equations.** On the other hand, if an equation such as

$$x + y = y + x$$

results in a true statement for every replacement of x and y, for which both members have meaning, the equation is call an **identity.** The left-hand member and the right-hand member of identities are called **equivalent expressions.**

To show that an equation is not an identity, it is necessary to find only one replacement for the variables that will make the equation false. Such a procedure is referred to as showing a **counterexample.**

Example   We can show by counterexample that $x^2 - 3y = 3y - x^2$ is not an identity by substituting some arbitrary values for the variables. If we use 1 for x and 1 for y, the left-hand member of the equation equals $1^2 - 3 \cdot 1$, or -2, and the right-hand member equals $3 \cdot 1 - 1^2$, or 2. Hence,

$$x^2 - 3y \neq 3y - x^2$$

for the ordered pair $(1, 1)$, and the equation is not an identity.

Example   We can show that the equation

$$(x - y)^2 + 2xy = x^2 + y^2$$

is an identity by simplifying the left-hand member

$$(x - y)^2 + 2xy = x^2 - 2xy + y^2 + 2xy$$
$$= x^2 + y^2.$$

Thus, the left-hand member and the right-hand member are equivalent and the equation is an identity.

Sets of ordered pairs play an important role in mathematics and are given a special name, as follows:

*A **relation** is a set of ordered pairs.*

For a, b, c, $d \in R$:

Equality Axioms

1. $a = a$ Reflexive law
2. If $a = b$, then $b = a$ Symmetric law
3. If $a = b$ and $b = c$, then $a = c$ Transitive law
4. If $a = b$, then a may be substituted   Substitution law
   for b or b for a in any expression.

Order Axioms

1. Exactly one of the following is true: Trichotomy law
   $a < b$, $a = b$, or $a > b$.
2. If $a < b$ and $b < c$, then $a < c$.   Transitive law
3. If $a, b > 0$, then   Closure for positive numbers
   $a + b > 0$ and $a \cdot b > 0$

Axioms for Operations

1. $a + b \in R$ Closure law for addition
2. $a + b = b + a$ Commutative law of addition
3. $(a + b) + c = a + (b + c)$ Associative law of addition
4. $a + 0 = 0 + a = a$ Identity law of addition
5. $a + (-a) = (-a) + a = 0$ Additive inverse law
6. $a \cdot b \in R$ Closure law for multiplication
7. $a \cdot b = b \cdot a$ Commutative law of multiplication
8. $(a \cdot b) \cdot c = a \cdot (b \cdot c)$ Associative law of multiplication
9. $a \cdot 1 = 1 \cdot a = a$ Identity law of multiplication
10. $a \cdot \dfrac{1}{a} = \dfrac{1}{a} \cdot a = 1 \quad (a \neq 0)$ Multiplicative inverse law
11. $a \cdot (b + c) = a \cdot b + a \cdot c$ Distributive law

Some Properties

1. $a \cdot 0 = 0$

2. $-(-a) = a$

3. $-a = -1 \cdot a$

4. If $a = b$, then $a + c = b + c$

5. If $a + c = b + c$, then $a = b$

6. If $a = b$, then $a \cdot c = b \cdot c$

7. If $a \cdot c = b \cdot c$, then $a = b$ $(c \neq 0)$

8. $a - b = a + (-b)$

9. $a \cdot d = b \cdot c$ if $\dfrac{a}{b} = \dfrac{c}{d}$ $(b, d \neq 0)$

10. $\dfrac{a}{b} = \dfrac{a \cdot c}{b \cdot c}$ $(b, c \neq 0)$

11. $\dfrac{a}{c} + \dfrac{b}{c} = \dfrac{a + b}{c}$ $(c \neq 0)$

12. $\dfrac{a}{b} \cdot \dfrac{c}{d} = \dfrac{a \cdot c}{b \cdot d}$ $(b, d \neq 0)$

13. $\dfrac{a}{b} = a \cdot \dfrac{1}{b}$ $(b \neq 0)$

14. $\dfrac{a}{b} \div \dfrac{c}{d} = \dfrac{a}{b} \cdot \dfrac{d}{c}$ $(b, c, d \neq 0)$

15. $\dfrac{a}{b} = \dfrac{-a}{-b} = -\dfrac{a}{-b} = -\dfrac{-a}{b}$;

   $\dfrac{-a}{b} = \dfrac{a}{-b} = -\dfrac{a}{b} = -\dfrac{-a}{-b}$ $(b \neq 0)$

16. $ab = 0$ if and only if $a = 0$ or $b = 0$

The set of first components in a set of ordered pairs is called th
domain of the relation, and the set of second components is called th
range.

Examples

a. $\{(5, 8), (6, 9), (7, 10)\}$ is a relation with domain $\{5, 6, 7\}$ and rang
$\{8, 9, 10\}$.

b. $\{(5, 8), (5, 9), (7, 10)\}$ is a relation with domain $\{5, 7\}$ and rang
$\{8, 9, 10\}$.

The set of ordered pairs exhibited in Example a is a special kind o
relation in which each element in the domain is paired with only on
element in the range. Such relations are given a special name.

A **function** *is a relation in which each element in the domain is
associated with only one element in the range.*

Note that the relation in Example b above is not a function because
an element in the domain, is associated with more than one element i
the range.

The relations above were exhibited by listing the members in the set o
ordered pairs. They can also be described by equations that speci
which ordered pairs are in the relation and which are not; those that ar
in the relation are elements of the solution set. For example,

$$y = x + 3, \quad x \in \{5, 6, 7\}$$

also specifies the function in Example a above:

(Domain elements)

$(5, 8), (6, 9), (7, 10)$

(Range elements)

The elements in the range, 8, 9, and 10, are obtained by replacing x i
the equation $y = x + 3$ with 5, 6, and 7 in turn and solving for y. Th
function consists of three ordered pairs only.

Example   Given the function

$$y = x - 3, \quad x \in \{1, 2, 3\}:$$

1. The domain is $\{1, 2, 3\}$.

2. Replacing x with each element of the domain in turn, you can fir

the associated values of y to be -2, -1, and 0, respectively. Thus, the range is $\{-2, -1, 0\}$.

3. The function is $\{(1, -2), (2, -1), (3, 0)\}$.

If the domain of a function is the set of real numbers R, rather than a finite set, the function consists of an *infinite set* of ordered pairs and obviously not all its elements can be listed. *If the domain is not specified, we will assume that it includes all real numbers for which both members of the defining equation are defined.*

Examples

a. The domain of the function defined by $y = x + 3$ is the set R because $x + 3$, and therefore y, are defined for all real number replacements for x.

b. The domain of the function defined by $y = \dfrac{3}{x - 4}$ is the set R with the restriction that $x \neq 4$, because $x - 4 = 0$ and y is undefined for $x = 4$.

Functions are sometimes designated by names (linear, quadratic, etc.) or by means of a single symbol, commonly the symbol f. If the discussion includes a consideration of more than one function, other letters such as g, h, F, and P are used.

The symbol for the function can be used in conjunction with the variable representing an element in the domain to represent the associated element in the range. For example, $f(x)$ (read "f of x" or "the value of f at x") is the element in the range of f associated or paired with the element x in the domain. Thus, we can discuss functions defined by an equation such as

$$y = x - 2$$
$$f(x) = x - 2,$$
$$g(x) = x - 2,$$

etc.

where the symbols $f(x)$, $g(x)$, etc., are playing exactly the same role as y. The variable x represents an element in the domain, and y, $f(x)$, $g(x)$, etc., represent an element in the range.

The notation $f(x)$ is especially useful because, by replacing x with a specific real number a in the domain, the notation $f(a)$ will then denote the paired, or corresponding, element in the range of the function.

Example If $f(x) = 2x - 5$.

$$f(3) = 2(3) - 5 = 6 - 5 = 1,$$
$$f(0) = 2(0) - 5 = 0 - 5 = -5,$$
$$f(-3) = 2(-3) - 5 = -6 - 5 = -11,$$
$$f(a) = 2(a) - 5 = 2a - 5.$$

A.4 Graphs of Relations and Functions

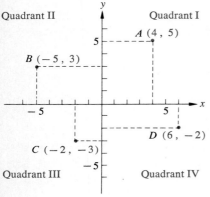

Figure A.4

As you probably recall from your study of algebra, an associatio
exists between ordered pairs of real numbers and the points on
geometric plane, sometimes called the **real plane**. If two number lines a
drawn perpendicular to each other at their origins as in Figure A.4, the
form a rectangular coordinate system. These perpendicular number line
are called **axes** (singular, axis). The horizontal line is usually called th
x-axis, and the vertical line is usually called the **y-axis.** The x-axis an
the y-axis divide the plane into four regions, called **quadrants**, as show:

For a given rectangular coordinate system, any point in the plane ca
be associated with a unique ordered pair of real numbers. For exampl
the point A in Quadrant I, in Figure A.4, can be associated with th
ordered pair (4, 5), where 4 represents the distance of the point to th
right of the y-axis and 5 represents the distance of the point above the :
axis. The points B, C, and D correspond to the ordered pairs (−5, ;
(−2, −3), and (6, −2), respectively.

The *components* of an ordered pair associated with a point in the plan
are called the **coordinates** of the point. The *point* is called the **graph** of th
ordered pair. An ordered pair is sometimes used as a name for th
corresponding point. Thus, we occasionally speak of the point (2, 3) o
the point (4, −3) or, in general, the point (x, y).

Recall that in Section A.3 a function was defined in part as a set o
ordered pairs. Consequently, a function—or at least a part of
function—can now be displayed as a set of points on the plane. Th
phrase "part of a function" is used here since the domain or range o
both may be the infinite set of real numbers, which cannot be shown on
finite line. For example, consider the function

$$y = x - 4.$$

If arbitrary values are assigned to x, say −3, 0, 3, and 6, and th
corresponding values for y are computed, the four ordered pairs

$$(-3, -7), \quad (0, -4), \quad (3, -1), \quad \text{and} \quad (6, 2)$$

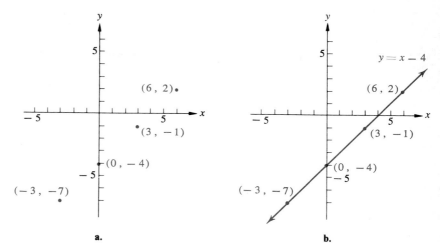

Figure A.5

are obtained, which are solutions to the equation. These points can be located on a coordinate system as shown in Figure A.5a. These points appear to lie on a straight line, and in fact they do. It can be shown that the coordinates of any point on the line (Figure A.5b) constitute a solution of the first-degree equation in two variables, $y = x - 4$ and that, conversely, every solution of $y = x - 4$ corresponds to a point on this line. The line is referred to as the *graph of the function* $y = x - 4$ or, alternatively, as the *graph of the solution set of the equation* or simply as the *graph of the equation*. Obviously, only part of the line can be displayed, so an arrowhead is placed at each end to indicate that the graph, or line, continues in both directions indefinitely.

Furthermore, since the graphs of first-degree equations

$$y = mx + b \qquad \text{or} \qquad ax + by = c$$

are straight lines, it is only necessary to find two ordered pairs in the solution set to sketch the graph. Two pairs that are easy to find are $(0, b)$ and $(a, 0)$, in which the numbers a and b are called the **intercepts**.

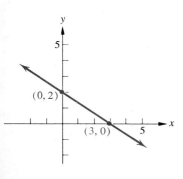

Example   The graph of the function $2x + 3y = 6$ can be obtained by first finding the x and y intercepts.

If $y = 0$, then $x = 3$ and $(3, 0)$ is a solution.

If $x = 0$, then $y = 2$ and $(0, 2)$ is a solution.

The x and y intercepts are 3 and 2, respectively. See the graph.

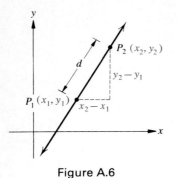

Figure A.6

First-degree equations are also called **linear equations,** and the fun tions defined by these equations are called **linear functions.**

For any two points P_1 and P_2 on a line, the set of points containing and P_2 and all points lying between P_1 and P_2 is called a **line segment.** W will designate the line segment by $\overline{P_1 P_2}$. A line segment has the importa property of length. If we designate the end points of a line segment $\overline{P_1}$ with the pairs of coordinates (x_1, y_1) and (x_2, y_2), as in Figure A.6, t length, or measure, of the line segment, which we will designate by P_1 or some variable such as d, can be determined by an application of t Pythagorean theorem (see Section A.8). From the right triangle show we observe that the lengths of the perpendicular sides are $y_2 - y_1$ a $x_2 - x_1$; the square of the distance between P_1 and P_2 is given by

$$d^2 = (x_2 - x_1)^2 + (y_2 - y_1)^2.$$

By considering only the positive square root of the right-hand membe we have the following result.

Distance Formula

$$d = \sqrt{(x_2 - x_1)^2 + (y_2 - y_1)^2}.$$

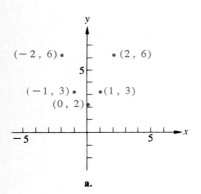

a.

Example The distance between the pair of points $(1, 3)$ and $(5, 6)$ given by

$$d = \sqrt{(5 - 1)^2 + (6 - 3)^2}$$
$$= \sqrt{16 + 9} = \sqrt{25} = 5.$$

As with first-degree equations in two variables, solutions of highe degree equations in two variables are also ordered pairs, which can t found by arbitrarily assigning values to x and finding the associate values of y. For example, in the equation

$$y = x^2 + 2,$$

if $x = -2$, then

$$y = (-2)^2 + 2 = 6,$$

and $(-2, 6)$ is a solution. Similarly, by replacing x with $-1, 0, 1,$ and we obtain $(-1, 3), (0, 2), (1, 3),$ and $(2, 6)$ as additional solutions. Plottir these points on the plane, we have the graph in Figure A.7a. Clearl these points do not lie on a straight line. By plotting additional solutio

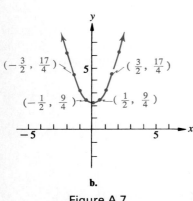

b.

Figure A.7

of $y = x^2 + 2$, such as

$$\left(-\frac{3}{2}, \frac{17}{4}\right), \quad \left(-\frac{1}{2}, \frac{9}{4}\right), \quad \left(\frac{1}{2}, \frac{9}{4}\right), \quad \text{and} \quad \left(\frac{3}{2}, \frac{17}{4}\right),$$

we have the additional points in Figure A.7b. These points can now be connected in sequence from left to right by a smooth curve as shown. We assume that this curve is a good approximation to the graph of $y = x^2 + 2$.

More generally, the graph of any equation of the form

$$y = ax^2 + bx + c \quad (a, b, c, x \in R, a \neq 0),$$

is called a **parabola.** Notice that such an equation defines a function, since with each x, an equation of this form will associate one and only one y. Such functions are called **quadratic functions**.

The graph of

$$x^2 + y^2 = r^2,$$

where r is a real number greater than zero, is a **circle** with center at the origin and radius of length r. However, although this equation defines a relation, it does not define a function.

Whether or not a relation is also a function can be determined by inspecting its graph. For example, consider the graphs of equations $y = ax^2 + bx + c$ and $x^2 + y^2 = r^2$ in Figure A.8. Imagine a vertical line moving across each graph from left to right. Since the line cuts the graph of $y = ax^2 + bx + c$ at *only one point at each position, this equation defines a function;* that is, for each x in this relation, there is only one y. Since the line cuts the graph of $x^2 + y^2 = r^2$ at *more than one point at certain positions, the equation does not define a function,* although it does define a relation. There are ordered pairs in the relation which have the same first components and different second components.

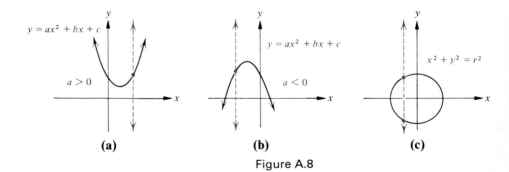

(a)   (b)   (c)

Figure A.8

A.5 Zeros of Functions; Equations in One Variable

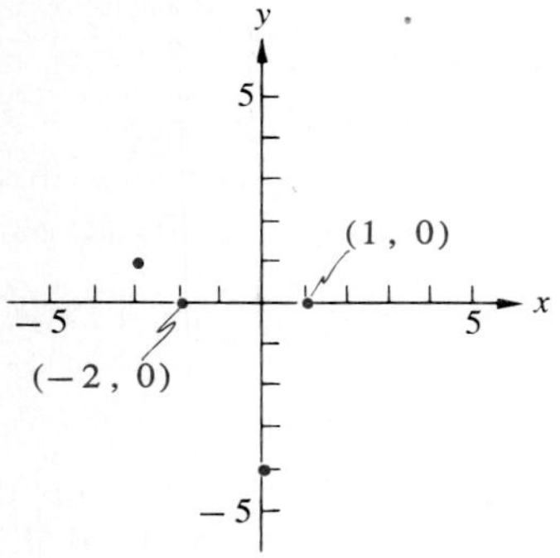

In any function, the element(s) in the domain that are paired with th element 0 in the range are called **zeros of the function.**

Example   The zeros of

$$f = \{(-3, 1), (-2, 0), (-1, -2), (0, -4), (1, 0)\}$$

are -2 and 1 because each of these first components is paired with second component 0.

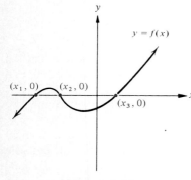

Figure A.9

The graph of each pair containing a zero in the above example lies c the x-axis, as shown in Figure A.9.

For any function

$$y = f(x),$$

the zeros of the function are values of x (x_1, x_2, x_3 in Figure A.10) fc which

$$y = f(x) = 0.$$

These values of x are the solutions of the equation in one variabl $f(x) = 0$, and the set of such numbers is the solution set of the equatio

Note that the ordered pairs $(x_1, 0)$, $(x_2, 0)$, and $(x_3, 0)$ are the cc ordinates of the points of intersection of the graph of $y = f(x)$ and th x-axis.

Thus, there are three different names for a single idea:

1. The zeros of $y = f(x)$.
2. The x intercepts of the graph of $y = f(x)$.
3. Solutions of $f(x) = 0$.

In many simple cases, such as

$$x - 2 = 0,$$

the solution set of an equation in one variable is evident by inspectio However, this is not the case for most first-degree equations. Recall fror your work in algebra that solutions of first-degree equations in on variable can usually be found by generating **equivalent equations** (a set c equations with the same solution set) until you obtain an equation whos solution set is evident by inspection.

Figure A.10

In general, if each member of an equation is multiplied by the same nonzero number, or if the same quantity is added to both members, the result is an equation equivalent to the original equation.

Solutions of quadratic equations in one variable are obtained by various methods. For example, if the left-hand member of a quadratic equation of the form

$$ax^2 + bx + c = 0 \quad (a \neq 0)$$

is factorable, the solutions can be obtained from the fact that the left-hand member will equal zero for values of x for which one or both of the factors equals zero. Such a procedure of solving a quadratic equation is called *solution by factoring.*

Example   We can solve $x^2 + x = 30$, by first writing the equation equivalently as

$$x^2 + x - 30 = 0, \tag{1}$$

in which form the right-hand member is zero. Factoring the left-hand member yields

$$(x + 6)(x - 5) = 0. \tag{2}$$

The left-hand member equals zero for values of x for which

$$x + 6 = 0 \quad \text{or} \quad x - 5 = 0 \quad \text{or both.}$$

By inspection, the solution set of (2) and, therefore, of (1) is $\{-6, 5\}$.

Quadratic equations of the form

$$x^2 = a$$

can be solved by a method called *extraction of roots.* If the equation has a solution, then x must be a square root of a. Since each positive real number a has two square roots, the solution set is $\{\sqrt{a}, -\sqrt{a}\}$. If $a = 0$, the solution set is $\{0\}$.

Example   The method of extraction of roots can be used to find the solution set of the equation $4x^2 = 24$. Multiplying each member by $1/4$ yields

$$x^2 = 6.$$

Hence, $x = \sqrt{6}$ or $x = -\sqrt{6}$ and the solution set is $\{\sqrt{6}, -\sqrt{6}\}$.

From the method of extraction of roots applied to the equation

$$(x - a)^2 = b$$

we obtain

$$x - a = \sqrt{b} \quad \text{and} \quad x - a = -\sqrt{b},$$

from which the solution set $\{a + \sqrt{b}, a - \sqrt{b}\}$ is obtained. If $b = 0$, th
solution set is $\{a\}$. From this it follows that we can find the solution se
of any quadratic equation by first rewriting the equation in the form

$$(x - a)^2 = b.$$

This procedure of solving a quadratic equation is called *completing th
square.*

Applying this procedure to the general quadratic equation

$$ax^2 + bx + c = 0 \quad (a \neq 0)$$

results in the following formula.

Quadratic Formula

$$x = \frac{-b \pm \sqrt{b^2 - 4ac}}{2a}$$

This formula gives us the solutions of any quadratic equation in th
form $ax^2 + bx + c = 0$ expressed in terms of a, b, and c. The solution
set is

$$\left\{\frac{-b + \sqrt{b^2 - 4ac}}{2a}, \frac{-b - \sqrt{b^2 - 4ac}}{2a}\right\}.$$

If $a, b, c \in R$, the elements in the solution set are real numbers if and
only if $b^2 - 4ac \geq 0$.

Example The equation $x^2 - 5x = 6$ can be solved by using the
quadratic formula as follows. First, rewrite the equation as

$$x^2 - 5x - 6 = 0.$$

Then, replacing a with 1, b with -5, and c with -6 in the quadratic formula yields

$$x = \frac{-(-5) \pm \sqrt{(-5)^2 - 4(1)(-6)}}{2(1)},$$

from which

$$x = \frac{5 + \sqrt{25 + 24}}{2} = 6 \quad \text{or} \quad x = \frac{5 - \sqrt{25 + 24}}{2} = -1,$$

and the solution set is $\{6, -1\}$.

A.6 Graphing Techniques

As we noted in Section A.4, functions that are defined by equations can always be graphed by obtaining a sufficient number of ordered pairs that are solutions of the equation. However, some transformations that are determined by the constants in the defining equation can be helpful. We will use the simple function

$$f: y = x^2$$

to illustrate four basic transformations. The graphs in the following examples can be verified by graphing several ordered pairs that are solutions to each equation.

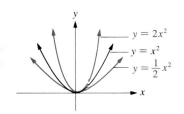

Example These graphs suggest that the graph of an equation of the form

$$y = ax^2$$

is "stretched" vertically if $a > 1$ and "shrunk" vertically if $0 < a < 1$.

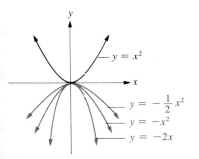

Example These graphs suggest that the graph of an equation of the form

$$y = ax^2 \quad (a > 0)$$

is reflected about the x-axis if $a < 0$.

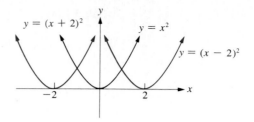

Example   The graphs above suggest that the graph of an equation c
the form

$$y = (x + c)^2$$

is shifted horizontally c units to the left if $c > 0$ and c units to the right i
$c < 0$.

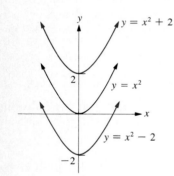

Example   These graphs suggest that the graph of an equation of th
form

$$y = x^2 + d$$

is shifted vertically d units up if $d > 0$ and d units down if $d < 0$.

Example   Graph $y = 2(x - 3)^2 + 1$.

Solution   This equation is of the form

$$y = a(x + c)^2 + d.$$

The graph involves several transformations of the graph of $y = x^2$.
First note that $a = 2$, $c = -3$, and $d = 1$. Then use the appropriat
transformations shown in the previous examples. These transformation
are shown in the series of graphs below.

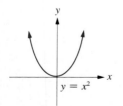

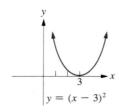

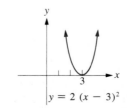

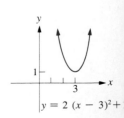

Example   Graph $y = -(x + 2)^2 - 3$.

Solution   The graph involves several transformations with $a = -1$, $c = 2$, and $d = -3$.

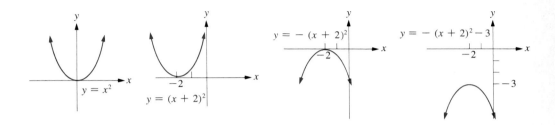

A.7 Inverse of a Function

A function f has been viewed as a relation in which each element in the domain is associated with just one element in the range. If each element in its range is also associated with just one element in its domain, the function f is called a **one-to-one function.** If a function is one-to-one, *any horizontal line* (as well as any vertical line will intersect its graph in *at most one point,* as shown in Figure A.11a. If it is not one-to-one, a horizontal line will intersect the graph at more than one point as shown in Figure A.11b.

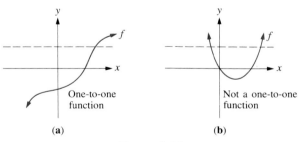

(a)   (b)

Figure A.11

If the components of each ordered pair in a given one-to-one function f are interchanged, the resulting function is called the **inverse** of f and is denoted by f^{-1}. The domain and range of f^{-1} are the range and domain, respectively, of f. Hence, if a one-to-one function is defined by an equation in two variables, the inverse can be obtained by interchanging the variables.

Example   The inverse of the function defined by

$$y = 3x - 5 \qquad ($$

is defined by

$$x = 3y - 5 \qquad (2$$

where the variables in (1) have been interchanged. When y is expressed i terms of x,

$$y = \frac{1}{3}(x + 5). \qquad (.$$

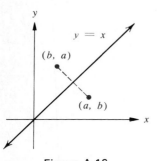

Figure A.12

Equations (2) and (3) are equivalent.

The graphs of inverse relations are always located symmetrically wit respect to the graph of the linear equation $y = x$. To see this, notice th graphs of the ordered pairs (a, b) and (b, a) in Figure A.12. The grap of $y = x$ serves as a reflecting line, or mirror, for these points.

Using the example above, Figure A.13 shows the graphs of

$$y = 3x - 5$$

and its inverse,

$$x = 3y - 5 \qquad \text{or} \qquad y = \frac{1}{3}(x + 5),$$

together with the graph of $y = x$.

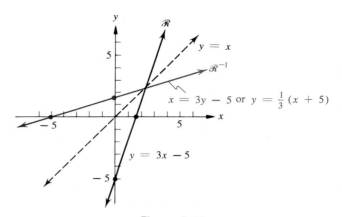

Figure A.13

Functions such as $y = x^2$ whose graph is shown in Figure A.14a are not one-to-one functions and hence do not have inverses. However, if we restrict the domain so that $x \geq 0$, the function will be one-to-one and wil

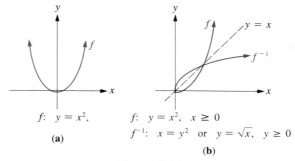

$$f: \quad y = x^2,$$

(a)

$$f: \quad y = x^2, \quad x \geq 0$$
$$f^{-1}: \quad x = y^2 \quad \text{or} \quad y = \sqrt{x}, \quad y \geq 0$$

(b)

Figure A.14

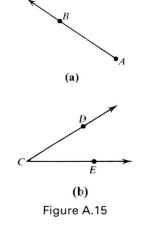

(a)

(b)

Figure A.15

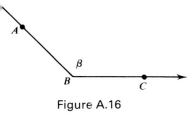

Figure A.16

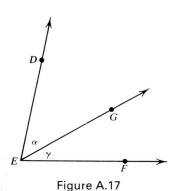

Figure A.17

have an inverse as shown in Figure A.14b, where the graphs of f and f^{-1} are symmetric with respect to the graph of $y = x$.

A.8 Angles and Their Measure

In geometry, a **ray** is a figure such as that shown in Figure A.15a, where the arrow indicates that the ray extends infinitely far in one direction. The point A (points are usually named with capital letters) in A.15a is the **end point** of that ray, and point C is the end point of the two rays in A.15b.

A ray is named by naming its end point and one other point on the ray. In Figure A.15a the ray is named by $\overrightarrow{AB}$ (read "ray AB"), and in A.15b the rays are named by $\overrightarrow{CD}$ and $\overrightarrow{CE}$.

An **angle** is the union of two rays having a common end point called the **vertex** of the angle. The rays are the **sides** of the angle. In Figure A.15b, an angle is formed by $\overrightarrow{CD}$ and $\overrightarrow{CE}$, where $\overrightarrow{CD}$ and $\overrightarrow{CE}$ are the sides, and point C is the vertex.

An angle can be named in several ways. For example, if a point is the vertex of only one angle, as in Figure A.16, then that letter can be used to name the angle; in this case, $\angle B$. Also, an angle can be named using a single letter placed "inside" the angle, such as β (beta), or by using three letters, where the middle letter names the vertex, such as $\angle ABC$ or $\angle CBA$.

If a point is the vertex of more than one angle, as in Figure A.17, each angle can also be named either by using three letters, such as $\angle DEF$, $\angle DEG$, or $\angle GEF$, or by using a single letter "inside" the angle, such as α (alpha) or γ (gamma). However, since three angles have the vertex E, "$\angle E$" cannot be used since it would not clearly name any one of them.

Angles can be measured by various units of measurement, but in geometry they are commonly measured in **degrees.** Such measures can be assigned to an angle by using a circle whose center is at the vertex of the

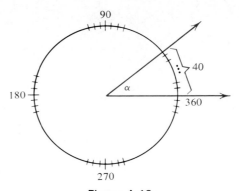

Figure A.18

angle. If a circle is divided into 360 arcs of equal length (Figure A.18 the number of arcs intercepted by a central angle α is the measur of the angle in degree units and is designated by $\alpha°$. (The measure c the angle is independent of the radius of the measuring circle.) Fron Figure A.18, we have that $\alpha = 40°$.

If the measure of an angle is less than 90°, the angle is an **acute angle;** more than 90° and less than 180°, it is an **obtuse angle;** if equal to 90°, it a **right angle.** If two lines, or rays, form a right angle, the lines, or rays, ar **perpendicular,** indicated by the symbol $\perp$. Thus, $\overrightarrow{DE} \perp \overrightarrow{FG}$ is read "ra DE is perpendicular to ray FG."

Examples In the figure, α is an acute angle, β is an obtuse angle, and γ a right angle. Also, $\overrightarrow{AC} \perp \overrightarrow{AB}$.

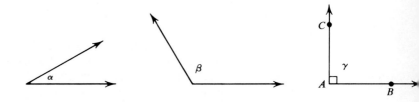

If the sum of the measures of two angles is 90°, the angles ar **complementary;** if the sum is 180°, the angles are **supplementary.**

Examples

a. If $\angle D = 32°$, $\angle E = 148°$, and $\angle F = 58°$, then $\angle D$ and $\angle F$ ar complementary, and $\angle D$ and $\angle E$ are supplementary.

b. In the figure, angles α and β are complementary, and angles γ and (theta) are supplementary.

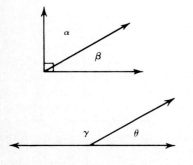

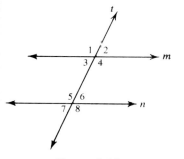

Figure A.19

Frequently, pairs of supplementary angles are introduced through the use of parallel lines. In Figure A.19, in which lines *m* and *n* are intersected by a transversal, line *t*, angles 1, 2, 7, and 8 are called **exterior angles** and angles 3, 4, 5, and 6 are called **interior angles.** Interior angles on opposite sides of the transversal are called **alternate interior angles,** and exterior angles on opposite sides of the transversal are called **alternate exterior angles.** For example, pairs of angles 3 and 6, and 4 and 5, are alternate interior angles, and pairs of angles 1 and 8, and 7 and 2 are alternate exterior angle.

Angles that are on the same side of the transversal and also in the same relative position in relation to the parallel lines are called **corresponding angles.** For example, pairs of angles 1 and 5; 2 and 6; 3 and 7; and 4 and 8 are corresponding angles.

If lines *m* and *n* are parallel, it follows that alternate interior angles are equal and alternate exterior angles are equal, as are corresponding angles. Also, interior angles on the same side of the transversal are supplementary, as are exterior angles on the same side of the transversal.

A.9 Triangles

A **triangle** is a figure formed by three line segments joining three points that are not in the same line. The line segments are the **sides,** and the three points are the **vertices** (plural of **vertex**) of the triangle. The letters assigned to the vertices can be used to name a triangle.

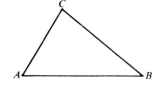

Example In the figure, the points *A*, *B*, and *C* are the vertices of $\triangle ABC$ (read "triangle *ABC*"), which can also be named by $\triangle ACB$, $\triangle BAC$, $\triangle BCA$, $\triangle CAB$, or $\triangle CBA$. The sides are $\overline{AB}$, $\overline{BC}$, and $\overline{CA}$; their lengths (positive real numbers) can be represented by *AB*, *BC*, and *CA*, respectively.

If each angle of a triangle is acute, the triangle is an **acute triangle;** if one angle is obtuse, it is an **obtuse triangle;** if one angle is a right angle, it is a **right triangle.** In a right triangle, the side opposite the right angle is the **hypotenuse,** and the other sides are the **legs.**

Examples In the figure, $\triangle DEF$ is an acute triangle, $\triangle KGH$ is an obtuse triangle, and $\triangle PQR$ is a right triangle. In $\triangle PQR$, $\overline{PQ}$ and $\overline{PR}$ are the legs, and $\overline{RQ}$ is the hypotenuse.

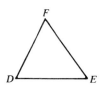
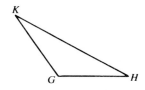
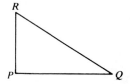

Triangles also have names related to the relative lengths of the sides. I the three sides are of different lengths, the triangle is a **scalene triangle;** two sides have equal lengths, it is an **isosceles triangle;** if all three side have equal lengths, it is an **equilateral triangle.**

Examples In the figure, $\triangle ABC$ is scalene, $\triangle DEF$ is isoscele $(FD = FE)$, and $\triangle HKL$ is equilateral $(HK = KL = LH)$. Note, fur ther, that $\triangle ABC$ is an obtuse triangle, whereas $\triangle DEF$ and $\triangle HK$ are both acute triangles.

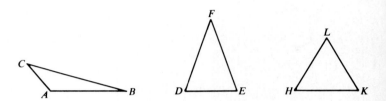

An **altitude** of a triangle is a segment from any vertex of a triangl perpendicular to the opposite side. Sometimes the length of the segmer is called the altitude.

Examples   In the figure, in $\triangle ABC$, h, or $\overline{CD}$, is the altitude to $\overline{AB}$. I $\triangle EFG$, $\overline{GH}$ is the altitude from G, and $\overline{FL}$ is the altitude from F.

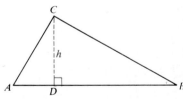

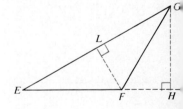

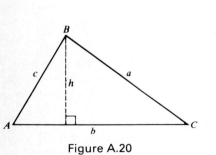

Figure A.20

The following properties hold for any triangle (see Figure A.20).

1. The perimeter is $P = a + b + c$.
2. The area is $\mathscr{A} = \frac{1}{2}bh$.
3. $\angle A + \angle B + \angle C = 180°$.

As in Properties 1 and 2, lowercase letters are often used to indicate th lengths of the sides opposite the angles named by the correspondin capital letters.

Examples
a. The perimeter of $\triangle ABC$ is

$$P = 3 + 5 + 6 = 14.$$

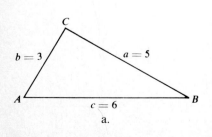

a.

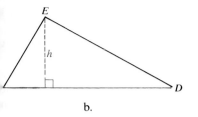

b.

D

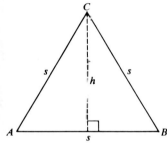

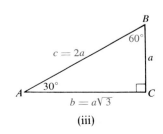

Figure A.21

b. If $h = 6$ and $CD = 10$, the area of $\triangle CDE$ is

$$\mathscr{A} = \frac{1}{2}(10)(6) = 30.$$

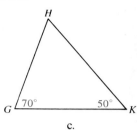

c. If $\angle G = 70°$ and $\angle K = 50°$,

$$70° + 50° + \angle H = 180°,$$

from which we have that $\angle H = 60°$.

In addition to Properties 1, 2, and 3, the following properties also apply to certain special triangles.

4. In an isosceles triangle, the angles opposite the equal sides are equal.

Example   In triangle BCD, if $BD = CD$, then $\angle B = \angle C$.

5. If a triangle is equilateral, then (see Figure A.21)

$$\angle A = \angle B = \angle C = 60°, \qquad h = \frac{s}{2}\sqrt{3}, \qquad \text{and} \qquad \mathscr{A} = \frac{\sqrt{3}}{4}s^2.$$

Example   If $s = 12$, then

$$h = \frac{12}{2}\sqrt{3} = 6\sqrt{3} \qquad \text{and} \qquad \mathscr{A} = \frac{\sqrt{3}}{4}(12)^2 = 36\sqrt{3}.$$

6. In a right triangle ($\angle C = 90°$),

(i) $a^2 + b^2 = c^2$ (the Pythagorean theorem).
(ii) If $a = b$, then $\angle A = \angle B = 45°$, and $c = a\sqrt{2}$.
(iii) If one acute angle is 30°, and the length of the opposite side is a, then the other acute angle is 60°, the length of the hypotenuse is $2a$, and the length of the longer leg is $a\sqrt{3}$.

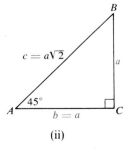

(ii)   (iii)

Examples   Using the measurements of the appropriate triangle:

a. $c^2 = 3^2 + 4^2 = 9 + 16 = 25;\ c = 5$

b. $\angle A = \angle B = 45°;\ c = 5\sqrt{2}$

c. $\angle A = 30°;\ c = 2\cdot 4 = 8;\ b = 4\sqrt{3}$

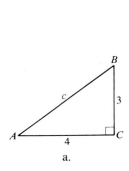

a.

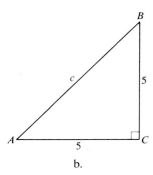

b.

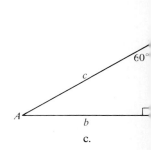

c.

A.10 Congruent and Similar Triangles

If two triangles are the same size and shape, they are said to I
congruent triangles. In Figure A.22, the triangles are congruent. This
indicated by use of the symbol $\cong$ (read "is congruent to"). Thus,
Figure A.22, $\triangle PQR \cong \triangle STV$.

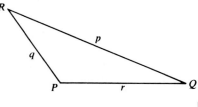

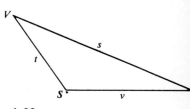

Figure A.22

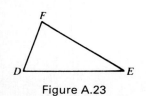

Figure A.23

From this it follows that

$$\angle P = \angle S,\quad p = s,$$
$$\angle Q = \angle T,\quad q = t,$$
$$\angle R = \angle V,\quad r = v.$$

If two triangles are the same shape, but not necessarily the same siz
they are said to be **similar triangles.** In Figure A.23, the triangles a
similar. This is written $\triangle ABC \sim \triangle DEF$, where "$\sim$" is read "is simil
to." In two similar triangles, the corresponding angles are equal—th

is, $\angle A = \angle D$, $\angle B = \angle E$, $\angle C = \angle F$; and the corresponding sides are proportional—that is,

$$\frac{AC}{DF} = \frac{AB}{DE} = \frac{BC}{EF}.$$

Example   Given that $\triangle RST \sim \triangle ADF$.

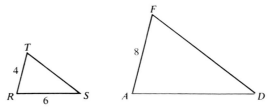

We can find AD as follows. Since the triangles are similar, we have

$$\frac{AD}{RS} = \frac{AF}{RT}, \qquad \frac{AD}{6} = \frac{8}{4},$$

from which

$$AD = 12.$$

The following properties are sufficient to prove that two triangles are similar:

1. two angles of one triangle are equal, respectively, to two angles of the other, or
2. one angle of one triangle is equal to one angle of the other, and the including sides are proportional, or
3. three sides of one triangle are proportional to three sides of the other.

Examples   Using the measurements in the figures:
a. $\angle P = \angle M$ and $\angle Q = \angle N$;

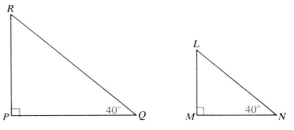

therefore, by Property 1, $\triangle PQR \sim \triangle MNL$.

b. $\angle A = \angle D$, and $\dfrac{AB}{DE} = \dfrac{AC}{DF}$ because $\dfrac{9}{6} = \dfrac{6}{4}$;

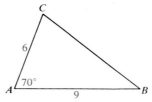
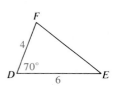

therefore, by Property 2, $\triangle ABC \sim \triangle DEF$.

c. $\dfrac{GK}{SW} = \dfrac{KH}{WT} = \dfrac{GH}{ST}$ because $\dfrac{8}{4} = \dfrac{10}{5} = \dfrac{12}{6}$;

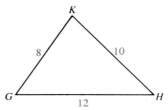

therefore, by Property 3, $\triangle GHK \sim \triangle STW$.

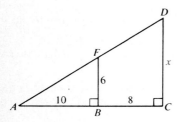

Example To find the length x in the figure, we proceed as follow $\angle A = \angle A$ and $\angle ABF = \angle ACD$; hence, by Property 1, we hav $\triangle ABF \sim \triangle ACD$. Also, $AC = AB + BC = 18$. Because the triangle are similar, the corresponding sides are proportional. Thus,

$$\frac{DC}{FB} = \frac{AC}{AB} \quad \text{and} \quad \frac{x}{6} = \frac{18}{10}, \quad \text{from which}$$

$$x = \frac{6 \cdot 18}{10} = \frac{54}{5}.$$

A list of symbols introduced in Appendix A is shown inside the front cove

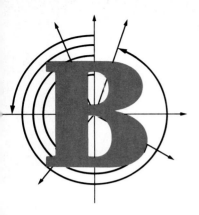

ROUNDING-OFF PROCEDURES

Figure B.1 shows the names of selected place values for the decimal number system: *whole number places* (to the left of the decimal point) are named to thousands; *decimal places* (to the right of the decimal point) are named to ten-thousandths. Familiarity with these place values is sufficient for our work.

Frequently, the result of a computation includes more digits than may be needed for a particular purpose. For example, the result of a calculation that involves the cost of an item might be $32.527. The actual cost, to the nearest cent, is $32.53. We say that 32.527 has been "rounded off to the nearest hundredth" (or "to the nearest cent"). Note that 32.527 is "nearer" to 32.53 than it is to 32.52.

The process shown below indicates the rounding-off process most commonly used. In this procedure, the *round-off digit* is the last digit to be retained; the *test digit* is the digit immediately to the right of the round-off digit. For example, to round off 318.04591 to the nearest *hundredth,* underline the round-off digit 4 in the hundredths place; then

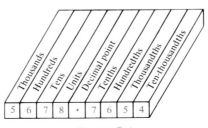

Figure B.1

proceed as follows:

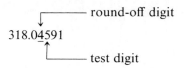

If the test digit is 5 or more, the round-off digit is increased by 1, but if the test digit is less than 5, the round-off digit is not changed. Here the test digit is 5; hence, 1 is added to the round-off digit:

$$5$$
$$318.0\overset{\text{\tiny}}{4}591$$

The round-off digit is in a decimal place, so all digits to the right of the round-off digit are dropped.

$$5$$
$$318.0\overset{}{4}\cancel{591} \longrightarrow 318.05$$

Thus, 318.04591 is written as 318.05 to the nearest hundredth.
   To round off 318.04591 to the nearest *hundred,* we have

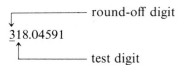

Here the test digit is less than 5, so the round-off digit is not changed. Also, because the round-off digit is to the left of the decimal point, the digits following are replaced with zeros, and all digits to the right of the decimal point are dropped. Thus, to the nearest hundred, we have

$$00$$
$$\underline{3}18.04591 \longrightarrow 3\cancel{18}.\cancel{04591} \longrightarrow 300$$

Examples   In each case the round-off digit is underlined.

   a. Round off to the nearest ten:

$$20$$
$$31\underline{8}.04591 \longrightarrow 3\cancel{18}.\cancel{04591} \longrightarrow 320$$

   b. Round off to the nearest tenth:

$$318.0\underline{4}591 \longrightarrow 318.0\cancel{4591} \longrightarrow 318.0$$

c. Round off to the nearest thousandth:

$$318.04\underline{5}91 \longrightarrow 318.04\overset{6}{\cancel{5}\cancel{9}\cancel{1}} \longrightarrow 318.046$$

Round-off instructions sometimes specify a given number of decimal places.

Examples

a. Round-off to two decimal places:

$$21.5\underline{9}71 \longrightarrow 21.\overset{60}{\cancel{5}\cancel{9}\cancel{7}\cancel{1}} \longrightarrow 21.60$$

b. Round off to three decimal places:

$$21.59\underline{7}1 \longrightarrow 21.597\cancel{1} \longrightarrow 21.597$$

Exact and Approximate Numbers

Any number arrived at by a counting process is an **exact number.** For example, if we say "60 students are enrolled in a class," the number 60 is exact. Exact numbers may also arise by definition. In the statement, "One hour equals 60 minutes," the number 60 is exact. On the other hand, a number that is arrived at by some process of measurement is an **approximate number.** Thus, if the distance between two cities, measured by driving from one city to the other, is given as 60 miles, this distance is an approximation to the actual distance, and the number 60 is an approximate number. The actual distance may be a little more or a little less than 60 miles.

Examples

a. "One dime is equal to 10 cents," the number 10 is *exact;* 10 is defined to be the number of cents in a dime.

b. "This sheet of postage stamps contains 10 stamps," the number 10 is *exact;* 10 is associated with the concept of counting.

c. "The reading shown on a voltmeter is 10 volts," the number 10 is *approximate;* 10 is a *measure* of voltage.

Significant Digits

When working with approximate numbers, the term **significant digits** is used to refer to those digits in a number that have meaning. The following rules can be used to decide which digits of an approximate number are significant.

Significant Digits of Approximate Numbers

1. Nonzero digits (1, 2, 3, 4, 5, 6, 7, 8, 9) are always significant.
2. Zeros that are preceded and followed by nonzero significant digits are always significant.
3. When a number has no nonzero digits on the left of the decimal point, zeros between the decimal point and the first nonzero digit are not significant.
4. Final zeros on the right of a decimal point are significant.
5. Final zeros on a whole number are assumed to be not significant unless further information is available.

In the examples illustrating each rule, the significant digits are underlined.

Examples

a. 62.18 has four significant digits. (Rule 1)

b. 0.042 has two significant digits. (Rules 1 and 3)

c. 23.0017 has six significant digits. (Rule 2)

d. 4005 has four significant digits. (Rule 2)

e. 7.200 has four significant digits. (Rule 4)

f. 0.0160 has three significant digits. (Rules 3 and 4)

g. In the statement "Hawaii is 2400 miles west of Los Angeles," without further information on how the measurement was made, assume that 2400 has only two significant digits. (Rule 5)

h. If a measurement is given as 2400 miles to *the nearest mile,* then the number 2400 has four significant digits. (Rule 5)

Numbers can be rounded off to a specified number of significant digits by using the same procedure that is used to round off a number to a specified place value.

Examples

a. Round off to two significant digits:

$$58.603 \longrightarrow 58.\overset{9}{\cancel{6}}\cancel{0}\cancel{3} \longrightarrow 59$$

b. Round off to four significant digits:

$$58.603 \longrightarrow 58.60\cancel{3} \longrightarrow 58.60$$

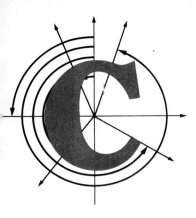

USING TABLES

In previous sections we obtained approximations to powers, square roots, reciprocals, and trigonometric function values using a scientific calculator. Approximations for these values can also be obtained by using tables that have been prepared for this purpose.

Table I in this appendix lists approximations to powers, square roots, and reciprocals of some natural numbers and can be used if a calculator is not available. Tables II and III list trigonometric function values.

Function Values for
Acute Angles

Table II lists rational number approximations for function values of selected angles over the interval $0° \le \alpha \le 90°$. The table is graduated in intervals of six minutes (6′), where one minute is equal to one-sixtieth of a degree, and in intervals of tenths of a degree. Observe that the table reads from top to bottom in the left margin with function values

ART OF TABLE II

θ deg	deg min		$\sin \theta$	$\cos \theta$	$\tan \theta$	$\csc \theta$	$\sec \theta$	$\cot \theta$			
30.0	30	0	0.5000	0.8660	0.5774	2.0000	1.1547	1.7321	60	0	60.0
30.1	30	6	0.5015	0.8652	0.5797	1.9940	1.1559	1.7251	59	54	59.9
30.2	30	12	0.5030	0.8643	0.5820	1.9880	1.1570	1.7182	59	48	59.8
30.3	30	18	0.5045	0.8634	0.5844	1.9821	1.1582	1.7113	59	42	59.7
30.4	30	24	0.5060	0.8625	0.5867	1.9762	1.1594	1.7045	59	36	59.6
30.5	30	30	0.5075	0.8616	0.5890	1.9703	1.1606	1.6977	59	30	59.5
30.6	30	36	0.5090	0.8607	0.5914	1.9645	1.1618	1.6909	59	24	59.4
30.7	30	42	0.5105	0.8599	0.5938	1.9587	1.1630	1.6842	59	18	59.3
30.8	30	48	0.5120	0.8590	0.5961	1.9530	1.1642	1.6775	59	12	59.2
			$\cos \theta$	$\sin \theta$	$\cot \theta$	$\sec \theta$	$\csc \theta$	$\tan \theta$	deg	min	θ deg

identified at the tops of the columns. It reads from bottom to top in the right margin with function values identified at the bottoms of the columns. (See extract of Table II.)

Function values are, in general, irrational numbers, and the four-place decimals shown in the table are mostly rational number approximations.

Examples   Find approximations for

a. sin 30° 12′   b. cot 59.5°

Solutions   From Table II,

a. sin 30° 12′ ≈ 0.5030   b. cot 59.5° ≈ 0.5890

Table III lists function values for angles whose measures are given in radian units in intervals of 0.01ᴿ. (See extracts of Table III.)

PART OF TABLE III

θ radians or Real Number x	θ degrees	sin θ or sin x	cos θ or cos x	tan θ or tan x	csc θ or csc x	sec θ or sec x	cot θ or cot x
0.45	25° 47′	0.4350	0.9004	0.4831	2.299	1.111	2.070
.46	26° 21′	.4439	.8961	.4954	2.253	1.116	2.018
.47	26° 56′	.4529	.8916	.5080	2.208	1.122	1.969
.48	27° 30′	.4618	.8870	.5206	2.166	1.127	1.921
.49	28° 04′	.4706	.8823	.5334	2.125	1.133	1.875
0.50	28° 39′	0.4794	0.8776	0.5463	2.086	1.139	1.830
.51	29° 13′	.4882	.8727	.5594	2.048	1.146	1.788
.52	29° 48′	.4969	.8678	.5726	2.013	1.152	1.747
.53	30° 22′	.5055	.8628	.5859	1.978	1.159	1.707
.54	30° 56′	.5141	.8577	.5994	1.945	1.166	1.668

Examples   Find approximations for

a. cos 0.47ᴿ   b. cot 0.53ᴿ

Solutions   From Table III,

a. cos 0.47ᴿ ≈ 0.8916   b. cot 0.53ᴿ ≈ 1.707

Notice that in both tables the function values sin θ, tan θ, and sec θ increase, while cos θ, cot θ, and csc θ decrease, for increasing measures of θ where 0° ≤ θ ≤ 90°.

Tables II and III consist of function values for angles with the specified measures. To approximate function values for angle measures not listed in the tables, we can read the entry in the table for that angle measure nearest the angle measure with which we are concerned. Thus, from Table III,

$$\sin 0.514^R \approx \sin 0.51^R \approx 0.4882.$$

Function Values for All Angles   Because the trigonometric tables are restricted to acute angles, we have to use a reference angle in conjunction with these tables to find approximations for function values for angles that are not acute. We will use the same procedure that we used to find function values for integral multiples of 30°, 45°, and 60° in Section 2.2.

Examples   Find an approximation for each function value.

a. csc 334° 36′   b. sec 128° 12′.

Solutions   A sketch showing the reference angle is helpful.

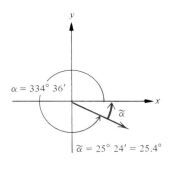

α = 334° 36′

$\tilde{\alpha} = 25° 24′ = 25.4°$

a. $\tilde{\alpha} = 360° - 334° 36′ = 25° 24′.$

Also, since α is in Quadrant IV, csc 334° 36′ is negative. Therefore, using Table II, we obtain

$$\text{csc } 334° 36′ = -\text{csc } \tilde{\alpha}$$
$$= -\text{csc } 25° 24′ \approx -2.3314.$$

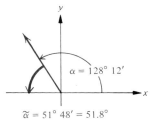

α = 128° 12′

$\tilde{\alpha} = 51° 48′ = 51.8°$

b. $\tilde{\alpha} = 180° - 128° 12′ = 51° 48′.$

Since 128° 12′ is in Quadrant II, sec 128° 12′ is negative. Therefore, using Table II,

$$\text{sec } 128° 12′ = -\text{sec } 51° 48′ = -1.6171.$$

When the measure of an angle is given in radians, and a reference angle is needed, we have to use an approximation for π. We will use 3.14.

Examples   Find an approximation for each function value.

a. 1.95^R   b. cot 5.36^R

Solutions   A sketch showing the reference angle is helpful.

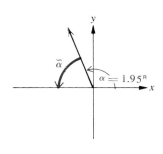

α = 1.95^R

a. Using 3.14 for π,

$$\tilde{\alpha} \approx (3.14 - 1.95)^R = 1.19^R.$$

Since α is in Quadrant II, cos 1.95^R is negative. Therefore, using Table III,

(continued)

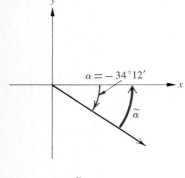

we obtain

$$\cos 1.95^R = -\cos \tilde{\alpha}$$

$$\approx -\cos 1.19^R \approx -0.3717.$$

b. Using 6.28 for 2π,

$$\tilde{\alpha} \approx 6.28^R - 5.36^R = 0.92^R.$$

Since 5.36^R is in Quadrant IV, $\cot 5.36^R$ is negative. Hence, usin
Table III,

$$\cot 5.36^R \approx -\cot 0.92^R \approx -0.7615.$$

The answers in the last example were obtained using $\pi \approx 3.1$
Different results will be obtained with other approximations for π suc
as 3.142 or 3.1416. In fact, if a calculator is used to find an approximatio
in Example a above, we obtain

$$\cos 1.95^R \approx -0.3701808 \approx -0.3702,$$

a significant difference from the value -0.3717 that we obtained usin
the table with $\pi \approx 3.14$. The calculator value is a more accurat
approximation because it is programmed for a more accurate approx
mation to π.

Function values for angles with negative measures can also be foun
using the procedures we used for angles with positive measures.

Examples   Find an approximation for each function value.

a. $\sin(-34° \ 12')$ b. $\cos(-422° \ 30')$

Solutions
a. The reference angle is

$$\tilde{\alpha} = 34° \ 12'.$$

Because $-34° \ 12'$ is in Quadrant IV, we have, using Table II,

$$\sin(-34° \ 12') = -\sin 34° \ 12'$$

$$= -\sin 34.2° \approx -0.5621.$$

b. The reference angle is

$$\tilde{\alpha} = |360° - 422° \ 30'| = 62° \ 30'.$$

Because $-422° \ 30'$ is in Quadrant IV, $\cos(-422° \ 30')$ is positive, and w
have, using Table II,

$$\cos(-422° \ 30') = \cos 62° \ 30' \approx 0.4617.$$

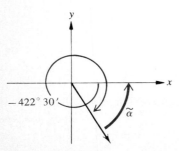

TABLE I   Powers, roots, and reciprocals

N	N^2	$\sqrt{N}$	$\sqrt{10N}$	$1/N$	N	N^2	$\sqrt{N}$	$\sqrt{10N}$	$1/N$
1	1	1.0 00	3.1 62	1.00 00	51	2 601	7.1 41	22.5 83	.019 61
2	4	1.4 14	4.4 72	.500 00	52	2 704	7.2 11	22.8 04	.019 23
3	9	1.7 32	5.4 77	.333 33	53	2 809	7.2 80	23.0 22	.018 87
4	16	2.0 00	6.3 25	.250 00	54	2 916	7.3 48	23.2 38	.018 52
5	25	2.2 36	7.0 71	.200 00	55	3 025	7.4 16	23.4 52	.018 18
6	36	2.4 49	7.7 46	.166 67	56	3 136	7.4 83	23.6 64	.017 86
7	49	2.6 46	8.3 67	.142 86	57	3 249	7.5 50	23.8 75	.017 54
8	64	2.8 28	8.9 44	.125 00	58	3 364	7.6 16	24.0 83	.017 24
9	81	3.0 00	9.4 87	.111 11	59	3 481	7.6 81	24.2 90	.016 95
10	100	3.1 62	10.0 00	.100 00	60	3 600	7.7 46	24.4 95	.016 67
11	121	3.3 17	10.4 88	.090 91	61	3 721	7.8 10	24.6 98	.016 39
12	144	3.4 64	10.9 54	.083 33	62	3 844	7.8 74	24.9 00	.016 13
13	169	3.6 06	11.4 02	.076 92	63	3 969	7.9 37	25.1 00	.015 87
14	196	3.7 42	11.8 32	.071 43	64	4 096	8.0 00	25.2 98	.015 62
15	225	3.8 73	12.2 47	.066 67	65	4 225	8.0 62	25.4 95	.015 38
16	256	4.0 00	12.6 49	.062 50	66	4 356	8.1 24	25.6 90	.015 15
17	289	4.1 23	13.0 38	.058 82	67	4 489	8.1 85	25.8 84	.014 93
18	324	4.2 43	13.4 16	.055 56	68	4 624	8.2 46	26.0 77	.014 71
19	361	4.3 59	13.7 84	.052 63	69	4 761	8.3 07	26.2 68	.014 49
20	400	4.4 72	14.1 42	.050 00	70	4 900	8.3 67	26.4 58	.014 29
21	441	4.5 83	14.4 91	.047 62	71	5 041	8.4 26	26.6 46	.014 08
22	484	4.6 90	14.8 32	.045 45	72	5 184	8.4 85	26.8 33	.013 89
23	529	4.7 96	15.1 66	.043 48	73	5 329	8.5 44	27.0 19	.013 70
24	576	4.8 99	15.4 92	.041 67	74	5 476	8.6 02	27.2 03	.013 51
25	625	5.0 00	15.8 11	.040 00	75	5 625	8.6 60	27.3 86	.013 33
26	676	5.0 99	16.1 25	.038 46	76	5 776	8.7 18	27.5 68	.013 16
27	729	5.1 96	16.4 32	.037 04	77	5 929	8.7 75	27.7 49	.012 99
28	784	5.2 92	16.7 33	.035 71	78	6 084	8.8 32	27.9 28	.012 82
29	841	5.3 85	17.0 29	.034 48	79	6 241	8.8 88	28.1 07	.012 66
30	900	5.4 77	17.3 21	.033 33	80	6 400	8.9 44	28.2 84	.012 50
31	961	5.5 68	17.6 07	.032 26	81	6 561	9.0 00	28.4 60	.012 35
32	1 024	5.6 57	17.8 89	.031 25	82	6 724	9.0 55	28.6 36	.012 20
33	1 089	5.7 45	18.1 66	.030 30	83	6 889	9.1 10	28.8 10	.012 05
34	1 156	5.8 31	18.4 39	.029 41	84	7 056	9.1 65	28.9 83	.011 90
35	1 225	5.9 16	18.7 08	.028 57	85	7 225	9.2 20	29.1 55	.011 76
36	1 296	6.0 00	18.9 74	.027 78	86	7 396	9.2 74	29.3 26	.011 63
37	1 369	6.0 83	19.2 35	.027 03	87	7 569	9.3 27	29.4 96	.011 49
38	1 444	6.1 64	19.4 94	.026 32	88	7 744	9.3 81	29.6 65	.011 36
39	1 521	6.2 45	19.7 48	.025 64	89	7 921	9.4 34	29.8 33	.011 24
40	1 600	6.3 25	20.0 00	.025 00	90	8 100	9.4 87	30.0 00	.011 11
41	1 681	6.4 03	20.2 48	.024 39	91	8 281	9.5 39	30.1 66	.010 99
42	1 764	6.4 81	20.4 94	.023 81	92	8 464	9.5 92	30.3 32	.010 87
43	1 849	6.5 57	20.7 36	.023 26	93	8 649	9.6 44	30.4 96	.010 75
44	1 936	6.6 33	20.9 76	.022 73	94	8 836	9.6 95	30.6 59	.010 64
45	2 025	6.7 08	21.2 13	.022 22	95	9 025	9.7 47	30.8 22	.010 53
46	2 116	6.7 82	21.4 48	.021 74	96	9 216	9.7 98	30.9 84	.010 42
47	2 209	6.8 56	21.6 79	.021 28	97	9 409	9.8 49	31.1 45	.010 31
48	2 304	6.9 28	21.9 09	.020 83	98	9 604	9.8 99	31.3 05	.010 20
49	2 401	7.0 00	22.1 36	.020 41	99	9 801	9.9 50	31.4 64	.010 10
50	2 500	7.0 71	22.3 61	.020 00	100	10 000	10.0 00	31.6 23	.010 00
N	N^2	$\sqrt{N}$	$\sqrt{10N}$	$1/N$	N	N^2	$\sqrt{N}$	$\sqrt{10N}$	$1/N$

TABLE II Values of trigonometric functions (degrees)

θ deg	deg	min	$\sin \theta$	$\cos \theta$	$\tan \theta$	$\csc \theta$	$\sec \theta$	$\cot \theta$			
0.0	0	0	0.0000	1.0000	0.0000	No value	1.0000	No value	90	0	90.0
0.1	0	6	0.0017	1.0000	0.0017	572.96	1.0000	572.96	89	54	89.9
0.2	0	12	0.0035	1.0000	0.0035	286.48	1.0000	286.48	89	48	89.8
0.3	0	18	0.0052	1.0000	0.0052	190.99	1.0000	190.98	89	42	89.7
0.4	0	24	0.0070	1.0000	0.0070	143.24	1.0000	143.24	89	36	89.6
0.5	0	30	0.0087	1.0000	0.0087	114.59	1.0000	114.59	89	30	89.5
0.6	0	36	0.0105	0.9999	0.0105	95.495	1.0001	95.490	89	24	89.4
0.7	0	42	0.0122	0.9999	0.0122	81.853	1.0001	81.847	89	18	89.3
0.8	0	48	0.0140	0.9999	0.0140	71.622	1.0001	71.615	89	12	89.2
0.9	0	54	0.0157	0.9999	0.0157	63.665	1.0001	63.657	89	6	89.1
1.0	1	0	0.0175	0.9998	0.0175	57.299	1.0002	57.290	89	0	89.0
1.1	1	6	0.0192	0.9998	0.0192	52.090	1.0002	52.081	88	54	88.9
1.2	1	12	0.0209	0.9998	0.0209	47.750	1.0002	47.740	88	48	88.8
1.3	1	18	0.0227	0.9997	0.0227	44.077	1.0003	44.066	88	42	88.7
1.4	1	24	0.0244	0.9997	0.0244	40.930	1.0003	40.917	88	36	88.6
1.5	1	30	0.0262	0.9997	0.0262	38.202	1.0003	38.188	88	30	88.5
1.6	1	36	0.0279	0.9996	0.0279	35.815	1.0004	35.801	88	24	88.4
1.7	1	42	0.0297	0.9996	0.0297	33.708	1.0004	33.694	88	18	88.3
1.8	1	48	0.0314	0.9995	0.0314	31.836	1.0005	31.821	88	12	88.2
1.9	1	54	0.0332	0.9995	0.0332	30.161	1.0005	30.145	88	6	88.1
2.0	2	0	0.0349	0.9994	0.0349	28.654	1.0006	28.636	88	0	88.0
2.1	2	6	0.0366	0.9993	0.0367	27.290	1.0007	27.271	87	54	87.9
2.2	2	12	0.0384	0.9993	0.0384	26.050	1.0007	26.031	87	48	87.8
2.3	2	18	0.0401	0.9992	0.0402	24.918	1.0008	24.898	87	42	87.7
2.4	2	24	0.0419	0.9991	0.0419	23.880	1.0009	23.859	87	36	87.6
2.5	2	30	0.0436	0.9990	0.0437	22.926	1.0010	22.904	87	30	87.5
2.6	2	36	0.0454	0.9990	0.0454	22.044	1.0010	22.022	87	24	87.4
2.7	2	42	0.0471	0.9989	0.0472	21.229	1.0011	21.205	87	18	87.3
2.8	2	48	0.0488	0.9988	0.0489	20.471	1.0012	20.446	87	12	87.2
2.9	2	54	0.0506	0.9987	0.0507	19.766	1.0013	19.740	87	6	87.1
3.0	3	0	0.0523	0.9986	0.0524	19.107	1.0014	19.081	87	0	87.0
3.1	3	6	0.0541	0.9985	0.0542	18.492	1.0015	18.464	86	54	86.9
3.2	3	12	0.0558	0.9984	0.0559	17.914	1.0016	17.886	86	48	86.8
3.3	3	18	0.0576	0.9983	0.0577	17.372	1.0017	17.343	86	42	86.7
3.4	3	24	0.0593	0.9982	0.0594	16.862	1.0018	16.832	86	36	86.6
3.5	3	30	0.0610	0.9981	0.0612	16.380	1.0019	16.350	86	30	86.5
3.6	3	36	0.0628	0.9980	0.0629	15.926	1.0020	15.895	86	24	86.4
3.7	3	42	0.0645	0.9979	0.0647	15.496	1.0021	15.464	86	18	86.3
3.8	3	48	0.0663	0.9978	0.0664	15.089	1.0022	15.056	86	12	86.2
3.9	3	54	0.0680	0.9977	0.0682	14.703	1.0023	14.669	86	6	86.1
4.0	4	0	0.0698	0.9976	0.0699	14.336	1.0024	14.301	86	0	86.0
4.1	4	6	0.0715	0.9974	0.0717	13.987	1.0026	13.951	85	54	85.9
4.2	4	12	0.0732	0.9973	0.0734	13.654	1.0027	13.617	85	48	85.8
4.3	4	18	0.0750	0.9972	0.0752	13.337	1.0028	13.300	85	42	85.7
4.4	4	24	0.0767	0.9971	0.0769	13.035	1.0030	12.996	85	36	85.6
4.5	4	30	0.0785	0.9969	0.0787	12.746	1.0031	12.706	85	30	85.5
4.6	4	36	0.0802	0.9968	0.0805	12.469	1.0032	12.429	85	24	85.4
4.7	4	42	0.0819	0.9966	0.0822	12.204	1.0034	12.163	85	18	85.3
4.8	4	48	0.0837	0.9965	0.0840	11.951	1.0035	11.909	85	12	85.2
4.9	4	54	0.0854	0.9963	0.0857	11.707	1.0037	11.665	85	6	85.1
			$\cos \theta$	$\sin \theta$	$\cot \theta$	$\sec \theta$	$\csc \theta$	$\tan \theta$	deg	min	θ deg

TABLE II *(continued)*

θ deg	deg	min	sin θ	cos θ	tan θ	csc θ	sec θ	cot θ			
5.0	5	0	0.0872	0.9962	0.0875	11.474	1.0038	11.430	85	0	85.0
5.1	5	6	0.0889	0.9960	0.0892	11.249	1.0040	11.205	84	54	84.9
5.2	5	12	0.0906	0.9959	0.0910	11.034	1.0041	10.988	84	48	84.8
5.3	5	18	0.0924	0.9957	0.0928	10.826	1.0043	10.780	84	42	84.7
5.4	5	24	0.0941	0.9956	0.0945	10.626	1.0045	10.579	84	36	84.6
5.5	5	30	0.0958	0.9954	0.0963	10.433	1.0046	10.385	84	30	84.5
5.6	5	36	0.0976	0.9952	0.0981	10.248	1.0048	10.199	84	24	84.4
5.7	5	42	0.0993	0.9951	0.0998	10.069	1.0050	10.019	84	18	84.3
5.8	5	48	0.1011	0.9949	0.1016	9.8955	1.0051	9.8448	84	12	84.2
5.9	5	54	0.1028	0.9947	0.1033	9.7283	1.0053	9.6768	84	6	84.1
6.0	6	0	0.1045	0.9945	0.1051	9.5668	1.0055	9.5144	84	0	84.0
6.1	6	6	0.1063	0.9943	0.1069	9.4105	1.0057	9.3573	83	54	83.9
6.2	6	12	0.1080	0.9942	0.1086	9.2593	1.0059	9.2052	83	48	83.8
6.3	6	18	0.1097	0.9940	0.1104	9.1129	1.0061	9.0579	83	42	83.7
6.4	6	24	0.1115	0.9938	0.1122	8.9711	1.0063	8.9152	83	36	83.6
6.5	6	30	0.1132	0.9936	0.1139	8.8337	1.0065	8.7769	83	30	83.5
6.6	6	36	0.1149	0.9934	0.1157	8.7004	1.0067	8.6428	83	24	83.4
6.7	6	42	0.1167	0.9932	0.1175	8.5711	1.0069	8.5126	83	18	83.3
6.8	6	48	0.1184	0.9930	0.1192	8.4457	1.0071	8.3863	83	12	83.2
6.9	6	54	0.1201	0.9928	0.1210	8.3238	1.0073	8.2636	83	6	83.1
7.0	7	0	0.1219	0.9925	0.1228	8.2055	1.0075	8.1444	83	0	83.0
7.1	7	6	0.1236	0.9923	0.1246	8.0905	1.0077	8.0285	82	54	82.9
7.2	7	12	0.1253	0.9921	0.1263	7.9787	1.0079	7.9158	82	48	82.8
7.3	7	18	0.1271	0.9919	0.1281	7.8700	1.0082	7.8062	82	42	82.7
7.4	7	24	0.1288	0.9917	0.1299	7.7642	1.0084	7.6996	82	36	82.6
7.5	7	30	0.1305	0.9914	0.1317	7.6613	1.0086	7.5958	82	30	82.5
7.6	7	36	0.1323	0.9912	0.1334	7.5611	1.0089	7.4947	82	24	82.4
7.7	7	42	0.1340	0.9910	0.1352	7.4635	1.0091	7.3962	82	18	82.3
7.8	7	48	0.1357	0.9907	0.1370	7.3684	1.0093	7.3002	82	12	82.2
7.9	7	54	0.1374	0.9905	0.1388	7.2757	1.0096	7.2066	82	6	82.1
8.0	8	0	0.1392	0.9903	0.1405	7.1853	1.0098	7.1154	82	0	82.0
8.1	8	6	0.1409	0.9900	0.1423	7.0972	1.0101	7.0264	81	54	81.9
8.2	8	12	0.1426	0.9898	0.1441	7.0112	1.0103	6.9395	81	48	81.8
8.3	8	18	0.1444	0.9895	0.1459	6.9273	1.0106	6.8548	81	42	81.7
8.4	8	24	0.1461	0.9893	0.1477	6.8454	1.0108	6.7720	81	36	81.6
8.5	8	30	0.1478	0.9890	0.1495	6.7655	1.0111	6.6912	81	30	81.5
8.6	8	36	0.1495	0.9888	0.1512	6.6874	1.0114	6.6122	81	24	81.4
8.7	8	42	0.1513	0.9885	0.1530	6.6111	1.0116	6.5350	81	18	81.3
8.8	8	48	0.1530	0.9882	0.1548	6.5366	1.0119	6.4596	81	12	81.2
8.9	8	54	0.1547	0.9880	0.1566	6.4637	1.0122	6.3859	81	6	81.1
9.0	9	0	0.1564	0.9877	0.1584	6.3925	1.0125	6.3138	81	0	81.0
9.1	9	6	0.1582	0.9874	0.1602	6.3228	1.0127	6.2432	80	54	80.9
9.2	9	12	0.1599	0.9871	0.1620	6.2547	1.0130	6.1742	80	48	80.8
9.3	9	18	0.1616	0.9869	0.1638	6.1880	1.0133	6.1066	80	42	80.7
9.4	9	24	0.1633	0.9866	0.1655	6.1227	1.0136	6.0405	80	36	80.6
9.5	9	30	0.1650	0.9863	0.1673	6.0589	1.0139	5.9758	80	30	80.5
9.6	9	36	0.1668	0.9860	0.1691	5.9963	1.0142	5.9124	80	24	80.4
9.7	9	42	0.1685	0.9857	0.1709	5.9351	1.0145	5.8502	80	18	80.3
9.8	9	48	0.1702	0.9854	0.1727	5.8751	1.0148	5.7894	80	12	80.2
9.9	9	54	0.1719	0.9851	0.1745	5.8164	1.0151	5.7297	80	6	80.1
			cos θ	sin θ	cot θ	sec θ	csc θ	tan θ	deg	min	θ deg

II

TABLE II (*continued*)

θ deg	deg	min	sin θ	cos θ	tan θ	csc θ	sec θ	cot θ			
10.0	10	0	0.1736	0.9848	0.1763	5.7588	1.0154	5.6713	80	0	80.0
10.1	10	6	0.1754	0.9845	0.1781	5.7023	1.0157	5.6140	79	54	79.9
10.2	10	12	0.1771	0.9842	0.1799	5.6470	1.0161	5.5578	79	48	79.8
10.3	10	18	0.1788	0.9839	0.1817	5.5928	1.0164	5.5027	79	42	79.7
10.4	10	24	0.1805	0.9836	0.1835	5.5396	1.0167	5.4486	79	36	79.6
10.5	10	30	0.1822	0.9833	0.1853	5.4874	1.0170	5.3955	79	30	79.5
10.6	10	36	0.1840	0.9829	0.1871	5.4362	1.0174	5.3435	79	24	79.4
10.7	10	42	0.1857	0.9826	0.1890	5.3860	1.0177	5.2924	79	18	79.3
10.8	10	48	0.1874	0.9823	0.1908	5.3367	1.0180	5.2422	79	12	79.2
10.9	10	54	0.1891	0.9820	0.1926	5.2883	1.0184	5.1929	79	6	79.1
11.0	11	0	0.1908	0.9816	0.1944	5.2408	1.0187	5.1446	79	0	79.0
11.1	11	6	0.1925	0.9813	0.1962	5.1942	1.0191	5.0970	78	54	78.9
11.2	11	12	0.1942	9.9810	0.1980	5.1484	1.0194	5.0504	78	48	78.8
11.3	11	18	0.1959	0.9806	0.1998	5.1034	1.0198	5.0045	78	42	78.7
11.4	11	24	0.1977	0.9803	0.2016	5.0593	1.0201	4.9595	78	36	78.6
11.5	11	30	0.1994	0.9799	0.2035	5.0159	1.0205	4.9152	78	30	78.5
11.6	11	36	0.2011	0.9796	0.2053	4.9732	1.0209	4.8716	78	24	78.4
11.7	11	42	0.2028	0.9792	0.2071	4.9313	1.0212	4.8288	78	18	78.3
11.8	11	48	0.2045	0.9789	0.2089	4.8901	1.0216	4.7867	78	12	78.2
11.9	11	54	0.2062	0.9785	0.2107	4.8496	1.0220	4.7453	78	6	78.1
12.0	12	0	0.2079	0.9781	0.2126	4.8097	1.0223	4.7046	78	0	78.0
12.1	12	6	0.2096	0.9778	0.2144	4.7706	1.0227	4.6646	77	54	77.9
12.2	12	12	0.2113	0.9774	0.2162	4.7321	1.0231	4.6252	77	48	77.8
12.3	12	18	0.2130	0.9770	0.2180	4.6942	1.0235	4.5864	77	42	77.7
12.4	12	24	0.2147	0.9767	0.2199	4.6569	1.0239	4.5483	77	36	77.6
12.5	12	30	0.2164	0.9763	0.2217	4.6202	1.0243	4.5107	77	30	77.5
12.6	12	36	0.2181	0.9759	0.2235	4.5841	1.0247	4.4737	77	24	77.4
12.7	12	42	0.2198	0.9755	0.2254	4.5486	1.0251	4.4374	77	18	77.3
12.8	12	48	0.2215	0.9751	0.2272	4.5137	1.0255	4.4015	77	12	77.2
12.9	12	54	0.2232	0.9748	0.2290	4.4793	1.0259	4.3662	77	6	77.1
13.0	13	0	0.2250	0.9744	0.2309	4.4454	1.0263	4.3315	77	0	77.0
13.1	13	6	0.2267	0.9740	0.2327	4.4121	1.0267	4.2972	76	54	76.9
13.2	13	12	0.2284	0.9736	0.2345	4.3792	1.0271	4.2635	76	48	76.8
13.3	13	18	0.2300	0.9732	0.2364	4.3469	1.0276	4.2303	76	42	76.7
13.4	13	24	0.2317	0.9728	0.2382	4.3150	1.0280	4.1976	76	36	76.6
13.5	13	30	0.2334	0.9724	0.2401	4.2837	1.0284	4.1653	76	30	76.5
13.6	13	36	0.2351	0.9720	0.2419	4.2528	1.0288	4.1335	76	24	76.4
13.7	13	42	0.2368	0.9715	0.2438	4.2223	1.0293	4.1022	76	18	76.3
13.8	13	48	0.2385	0.9711	0.2456	4.1923	1.0297	4.0713	76	12	76.2
13.9	13	54	0.2402	0.9707	0.2475	4.1627	1.0302	4.0408	76	6	76.1
14.0	14	0	0.2419	0.9703	0.2493	4.1336	1.0306	4.0108	76	0	76.0
14.1	14	6	0.2436	0.9699	0.2512	4.1048	1.0311	3.9812	75	54	75.9
14.2	14	12	0.2453	0.9694	0.2530	4.0765	1.0315	3.9520	75	48	75.8
14.3	14	18	0.2470	0.9690	0.2549	4.0486	1.0320	3.9232	75	42	75.7
14.4	14	24	0.2487	0.9686	0.2568	4.0211	1.0324	3.8947	75	36	75.6
14.5	14	30	0.2504	0.9681	0.2586	3.9939	1.0329	3.8667	75	30	75.5
14.6	14	36	0.2521	0.9677	0.2605	3.9672	1.0334	3.8391	75	24	75.4
14.7	14	42	0.2538	0.9673	0.2623	3.9408	1.0338	3.8118	75	18	75.3
14.8	14	48	0.2554	0.9668	0.2642	3.9147	1.0343	3.7849	75	12	75.2
14.9	14	54	0.2571	0.9664	0.2661	3.8890	1.0348	3.7583	75	6	75.1
			cos θ	sin θ	cot θ	sec θ	csc θ	tan θ	deg	min	θ de

TABLE II (*continued*)

θ deg	deg	min	sin θ	cos θ	tan θ	csc θ	sec θ	cot θ			
15.0	15	0	0.2588	0.9659	0.2679	3.8637	1.0353	3.7321	75	0	75.0
15.1	15	6	0.2605	0.9655	0.2698	3.8387	1.0358	3.7062	74	54	74.9
15.2	15	12	0.2622	0.9650	0.2717	3.8140	1.0363	3.6806	74	48	74.8
15.3	15	18	0.2639	0.9646	0.2736	3.7897	1.0367	3.6554	74	42	74.7
15.4	15	24	0.2656	0.9641	0.2754	3.7657	1.0372	3.6305	74	36	74.6
15.5	15	30	0.2672	0.9636	0.2773	3.7420	1.0377	3.6059	74	30	74.5
15.6	15	36	0.2689	0.9632	0.2792	3.7186	1.0382	3.5816	74	24	74.4
15.7	15	42	0.2706	0.9627	0.2811	3.6955	1.0388	3.5576	74	18	74.3
15.8	15	48	0.2723	0.9622	0.2830	3.6727	1.0393	3.5339	74	12	74.2
15.9	15	54	0.2740	0.9617	0.2849	3.6502	1.0398	3.5105	74	6	74.1
16.0	16	0	0.2756	0.9613	0.2867	3.6280	1.0403	3.4874	74	0	74.0
16.1	16	6	0.2773	0.9608	0.2886	3.6060	1.0408	3.4646	73	54	73.9
16.2	16	12	0.2790	0.9603	0.2905	3.5843	1.0413	3.4420	73	48	73.8
16.3	16	18	0.2807	0.9598	0.2924	3.5629	1.0419	3.4197	73	42	73.7
16.4	16	24	0.2823	0.9593	0.2943	3.5418	1.0424	3.3977	73	36	73.6
16.5	16	30	0.2840	0.9588	0.2962	3.5209	1.0429	3.3759	73	30	73.5
16.6	16	36	0.2857	0.9583	0.2981	3.5003	1.0435	3.3544	73	24	74.4
16.7	16	42	0.2874	0.9578	0.3000	3.4800	1.0440	3.3332	73	18	73.3
16.8	16	48	0.2890	0.9573	0.3019	3.4598	1.0446	3.3122	73	12	73.2
16.9	16	54	0.2907	0.9568	0.3038	3.4399	1.0451	3.2914	73	6	73.1
17.0	17	0	0.2924	0.9563	0.3057	3.4203	1.0457	3.2709	73	0	73.0
17.1	17	6	0.2940	0.9558	0.3076	3.4009	1.0463	3.2506	72	54	72.9
17.2	17	12	0.2957	0.9553	0.3096	3.3817	1.0468	3.2305	72	48	72.8
17.3	17	18	0.2974	0.9548	0.3115	3.3628	1.0474	3.2106	72	42	72.7
17.4	17	24	0.2990	0.9542	0.3134	3.3440	1.0480	3.1910	72	36	72.6
17.5	17	30	0.3007	0.9537	0.3153	3.3255	1.0485	3.1716	72	30	72.5
17.6	17	36	0.3024	0.9532	0.3172	3.3072	1.0491	3.1524	72	24	72.4
17.7	17	42	0.3040	0.9527	0.3191	3.2891	1.0497	3.1334	72	18	72.3
17.8	17	48	0.3057	0.9521	0.3211	3.2712	1.0503	3.1146	72	12	72.2
17.9	17	54	0.3074	0.9516	0.3230	3.2536	1.0509	3.0961	72	6	72.1
18.0	18	0	0.3090	0.9511	0.3249	3.2361	1.0515	3.0777	72	0	72.0
18.1	18	6	0.3107	0.9505	0.3268	3.2188	1.0521	3.0595	71	54	71.9
18.2	18	12	0.3123	0.9500	0.3288	3.2017	1.0527	3.0415	71	48	71.8
18.3	18	18	0.3140	0.9494	0.3307	3.1848	1.0533	3.0237	71	42	71.7
18.4	18	24	0.3156	0.9489	0.3327	3.1681	1.0539	3.0061	71	36	71.6
18.5	18	30	0.3173	0.9483	0.3346	3.1515	1.0545	2.9887	71	30	71.5
18.6	18	36	0.3190	0.9478	0.3365	3.1352	1.0551	2.9714	71	24	71.4
18.7	18	42	0.3206	0.9472	0.3385	3.1190	1.0557	2.9544	71	18	71.3
18.8	18	48	0.3223	0.9466	0.3404	3.1030	1.0564	2.9375	71	12	71.2
18.9	18	54	0.3239	0.9461	0.3424	3.0872	1.0570	2.9208	71	6	71.1
19.0	19	0	0.3256	0.9455	0.3443	2.0716	1.0576	2.9042	71	0	71.0
19.1	19	6	0.3272	0.9449	0.3463	3.0561	1.0583	2.8878	70	54	70.9
19.2	19	12	0.3289	0.9444	0.3482	3.0407	1.0589	2.8716	70	48	70.8
19.3	19	18	0.3305	0.9438	0.3502	3.0256	1.0595	2.8556	70	42	70.7
19.4	19	24	0.3322	0.9432	0.3522	3.0106	1.0602	2.8397	70	36	70.6
19.5	19	30	0.3338	0.9426	0.3541	2.9957	1.0608	2.8239	70	30	70.5
19.6	19	36	0.3355	0.9421	0.3561	2.9811	1.0615	2.8083	70	24	70.4
19.7	19	42	0.3371	0.9415	0.3581	2.9665	1.0622	2.7929	70	18	70.3
19.8	19	48	0.3387	0.9409	0.3600	2.9521	1.0628	2.7776	70	12	70.2
19.9	19	54	0.3404	0.9403	0.3620	2.9379	1.0635	2.7625	70	6	70.1
			cos θ	sin θ	cot θ	sec θ	csc θ	tan θ	deg	min	θ deg

TABLE II (*continued*)

θ deg	deg min		sin θ	cos θ	tan θ	csc θ	sec θ	cot θ			
20.0	20	0	0.3420	0.9397	0.3640	2.9238	1.0642	2.7475	70	0	70.0
20.1	20	6	0.3437	0.9391	0.3659	2.9099	1.0649	2.7326	69	54	69.9
20.2	20	12	0.3453	0.9385	0.3679	2.8960	1.0655	2.7179	69	48	69.8
20.3	20	18	0.3469	0.9379	0.3699	2.8824	1.0662	2.7034	69	42	69.7
20.4	20	24	0.3486	0.9373	0.3719	2.8688	1.0669	2.6889	69	36	69.6
20.5	20	30	0.3502	0.9367	0.3739	2.8555	1.0676	2.6746	69	30	69.5
20.6	20	36	0.3518	0.9361	0.3759	2.8422	1.0683	2.6605	69	24	69.4
20.7	20	42	0.3535	0.9354	0.3779	2.8291	1.0690	2.6464	69	18	69.3
20.8	20	48	0.3551	0.9348	0.3799	2.8161	1.0697	2.6325	69	12	69.2
20.9	20	54	0.3567	0.9342	0.3819	2.8032	1.0704	2.6187	69	6	69.1
21.0	21	0	0.3584	0.9336	0.3839	2.7904	1.0711	2.6051	69	0	69.0
21.1	21	6	0.3600	0.9330	0.3859	2.7778	1.0719	2.5916	68	54	68.9
21.2	21	12	0.3616	0.9323	0.3879	2.7653	1.0726	2.5782	68	48	68.8
21.3	21	18	0.3633	0.9317	0.3899	2.7529	1.0733	2.5649	68	42	68.7
21.4	21	24	0.3649	0.9311	0.3919	2.7407	1.0740	2.5517	68	36	68.6
21.5	21	30	0.3665	0.9304	0.3939	2.7285	1.0748	2.5386	68	30	68.5
21.6	21	36	0.3681	0.9298	0.3959	2.7165	1.0755	2.5257	68	24	68.4
21.7	21	42	0.3697	0.9291	0.3979	2.7046	1.0763	2.5129	68	18	68.3
21.8	21	48	0.3714	0.9285	0.4000	2.6927	1.0770	2.5002	68	12	68.2
21.9	21	54	0.3730	0.9278	0.4020	2.6811	1.0778	2.4876	68	6	68.1
22.0	22	0	0.3746	0.9272	0.4040	2.6695	1.0785	2.4751	68	0	68.0
22.1	22	6	0.3762	0.9265	0.4061	2.6580	1.0793	2.4627	67	54	67.9
22.2	22	12	0.3778	0.9259	0.4081	2.6466	1.0801	2.4504	67	48	67.8
22.3	22	18	0.3795	0.9252	0.4101	2.6354	1.0808	2.4383	67	42	67.7
22.4	22	24	0.3811	0.9245	0.4122	2.6242	1.0816	2.4262	67	36	67.6
22.5	22	30	0.3827	0.9239	0.4142	2.6131	1.0824	2.4142	67	30	67.5
22.6	22	36	0.3843	0.9232	0.4163	2.6022	1.0832	2.4023	67	24	67.4
22.7	22	42	0.3859	0.9225	0.4183	2.5913	1.0840	2.3906	67	18	67.3
22.8	22	48	0.3875	0.9219	0.4204	2.5805	1.0848	2.3789	67	12	67.2
22.9	22	54	0.3891	0.9212	0.4224	2.5699	1.0856	2.3673	67	6	67.1
23.0	23	0	0.3907	0.9205	0.4245	2.5593	1.0864	2.3559	67	0	67.0
23.1	23	6	0.3923	0.9198	0.4265	2.5488	1.0872	2.3445	66	54	66.9
23.3	23	12	0.3939	0.9191	0.4286	2.5384	1.0880	2.3332	66	48	66.8
23.3	23	18	0.3955	0.9184	0.4307	2.5282	1.0888	2.3220	66	42	66.7
23.4	23	24	0.3971	0.9178	0.4327	2.5180	1.0896	2.3109	66	36	66.6
23.5	23	30	0.3987	0.9171	0.4348	2.5078	1.0904	2.2998	66	30	66.5
23.6	23	36	0.4003	0.9164	0.4369	2.4978	1.0913	2.2889	66	24	66.4
23.7	23	42	0.4019	0.9157	0.4390	2.4879	1.0921	2.2781	66	18	66.3
23.8	23	48	0.4035	0.9150	0.4411	2.4780	1.0929	2.2673	66	12	66.2
23.9	23	54	0.4051	0.9143	0.4431	2.4683	1.0938	2.2566	66	6	66.1
24.0	24	0	0.4067	0.9135	0.4452	2.4586	1.0946	2.2460	66	0	66.0
24.1	24	6	0.4083	0.9128	0.4473	2.4490	1.0955	2.2355	65	54	65.9
24.2	24	12	0.4099	0.9121	0.4494	2.4395	1.0963	2.2251	65	48	65.8
24.3	24	18	0.4115	0.9114	0.4515	2.4301	1.0972	2.2148	65	42	65.7
24.4	24	24	0.4131	0.9107	0.4536	2.4207	1.0981	2.2045	65	36	65.6
24.5	24	30	0.4147	0.9100	0.4557	2.4114	1.0989	2.1943	65	30	65.5
24.6	24	36	0.4163	0.9092	0.4578	2.4022	1.0998	2.1842	65	24	65.4
24.7	24	42	0.4179	0.9085	0.4599	2.3931	1.1007	2.1742	65	18	65.3
24.8	24	48	0.4195	0.9078	0.4621	2.3841	1.1016	2.1642	65	12	65.2
24.9	24	54	0.4210	0.9070	0.4642	2.3751	1.1025	2.1543	65	6	65.1
			cos θ	sin θ	cot θ	sec θ	csc θ	tan θ	deg	min	θ deg

TABLE II   (*continued*)

θ deg	deg	min	sin θ	cos θ	tan θ	csc θ	sec θ	cot θ			
25.0	25	0	0.4226	0.9063	0.4663	2.3662	1.1034	2.1445	65	0	65.0
25.1	25	6	0.4242	0.9056	0.4684	2.3574	1.1043	2.1348	64	54	64.9
25.2	25	12	0.4258	0.9048	0.4706	2.3486	1.1052	2.1251	64	48	64.8
25.3	25	18	0.4274	0.9041	0.4727	2.3400	1.1061	2.1155	64	42	64.7
25.4	25	24	0.4289	0.9033	0.4748	2.3314	1.1070	2.1060	64	36	64.6
25.5	25	30	0.4305	0.9026	0.4770	2.3228	1.1079	2.0965	64	30	64.5
25.6	25	36	0.4321	0.9018	0.4791	2.3144	1.1089	2.0872	64	24	64.4
25.7	25	42	0.4337	0.9011	0.4813	2.3060	1.1098	2.0778	64	18	64.3
25.8	25	48	0.4352	0.9003	0.4834	2.2976	1.1107	2.0686	64	12	64.2
25.9	25	54	0.4368	0.8996	0.4856	2.2894	1.1117	2.0594	64	6	64.1
26.0	26	0	0.4384	0.8988	0.4877	2.2812	1.1126	2.0503	64	0	64.0
26.1	26	6	0.4399	0.8980	0.4899	2.2730	1.1136	2.0413	63	54	63.9
26.2	26	12	0.4415	0.8973	0.4921	2.2650	1.1145	2.0323	63	48	63.8
26.3	26	18	0.4431	0.8965	0.4942	2.2570	1.1155	2.0233	63	42	63.7
26.4	26	24	0.4446	0.8957	0.4964	2.2490	1.1164	2.0145	63	36	63.6
26.5	26	30	0.4462	0.8949	0.4986	2.2412	1.1174	2.0057	63	30	63.5
26.6	26	36	0.4478	0.8942	0.5008	2.2333	1.1184	1.9970	63	24	63.4
26.7	26	42	0.4493	0.8934	0.5029	2.2256	1.1194	1.9883	63	18	63.3
26.8	26	48	0.4509	0.8926	0.5051	2.2179	1.1203	1.9797	63	12	63.2
26.9	26	54	0.4524	0.8918	0.5073	2.2103	1.1213	1.9711	63	6	63.1
27.0	27	0	0.4540	0.8910	0.5095	2.2027	1.1223	1.9626	63	0	63.0
27.1	27	6	0.4555	0.8902	0.5117	2.1952	1.1233	1.9542	62	54	62.9
27.2	27	12	0.4571	0.8894	0.5139	2.1877	1.1243	1.9458	62	48	62.8
27.3	27	18	0.4586	0.8886	0.5161	2.1803	1.1253	1.9375	62	42	62.7
27.4	27	24	0.4602	0.8878	0.5184	2.1730	1.1264	1.9292	62	36	62.6
27.5	27	30	0.4617	0.8870	0.5206	2.1657	1.1274	1.9210	62	30	62.5
27.6	27	36	0.4633	0.8862	0.5228	2.1584	1.1284	1.9128	62	24	62.4
27.7	27	42	0.4648	0.8854	0.5250	2.1513	1.1294	1.9047	62	18	62.3
27.8	27	48	0.4664	0.8846	0.5272	2.1441	1.1305	1.8967	62	12	62.2
27.9	27	54	0.4679	0.8838	0.5295	2.1371	1.1315	1.8887	62	6	62.1
28.0	28	0	0.4695	0.8829	0.5317	2.1301	1.1326	1.8807	62	0	62.0
28.1	28	6	0.4710	0.8821	0.5339	2.1231	1.1336	1.8728	61	54	61.9
28.2	28	12	0.4726	0.8813	0.5362	2.1162	1.1347	1.8650	61	48	61.8
28.3	28	18	0.4741	0.8805	0.5384	2.1093	1.1357	1.8572	61	42	61.7
28.4	28	24	0.4756	0.8796	0.5407	2.1025	1.1368	1.8495	61	36	61.6
28.5	28	30	0.4772	0.8788	0.5430	2.0957	1.1379	1.8418	61	30	61.5
28.6	28	36	0.4787	0.8780	0.5452	2.0890	1.1390	1.8341	61	24	61.4
28.7	28	42	0.4802	0.8771	0.5475	2.0824	1.1401	1.8265	61	18	61.3
28.8	28	48	0.4818	0.8763	0.5498	2.0758	1.1412	1.8190	61	12	61.2
28.9	28	54	0.4833	0.8755	0.5520	2.0692	1.1423	1.8115	61	6	61.1
29.0	29	0	0.4848	0.8746	0.5543	2.0627	1.1434	1.8040	61	0	61.0
29.1	29	6	0.4863	0.8738	0.5566	2.0562	1.1445	1.7966	60	54	60.9
29.2	29	12	0.4879	0.8729	0.5589	2.0598	1.1456	1.7893	60	48	60.8
29.3	29	18	0.4894	0.8721	0.5612	2.0434	1.1467	1.7820	60	42	60.7
29.4	29	24	0.4909	0.8712	0.5635	2.0371	1.1478	1.7747	60	36	60.6
29.5	29	30	0.4924	0.8704	0.5658	2.0308	1.1490	1.7675	60	30	60.5
29.6	29	36	0.4939	0.8695	0.5681	2.0245	1.1501	1.7603	60	24	60.4
29.7	29	42	0.4955	0.8686	0.5704	2.0183	1.1512	1.7532	60	18	60.3
29.8	29	48	0.4970	0.8678	0.5727	2.0122	1.1524	1.7461	60	12	60.2
29.9	29	54	0.4985	0.8669	0.5750	2.0061	1.1535	1.7391	60	6	60.1
			cos θ	sin θ	cot θ	sec θ	csc θ	tan θ	deg	min	θ deg

TABLE II (*continued*)

θ deg	deg	min	sin θ	cos θ	tan θ	csc θ	sec θ	cot θ			
30.0	30	0	0.5000	0.8660	0.5774	2.0000	1.1547	1.7321	60	0	60.0
30.1	30	6	0.5015	0.8652	0.5797	1.9940	1.1559	1.7251	59	54	59.9
30.2	30	12	0.5030	0.8643	0.5820	1.9880	1.1570	1.7182	59	48	59.8
30.3	30	18	0.5045	0.8634	0.5844	1.9821	1.1582	1.7113	59	42	59.7
30.4	30	24	0.5060	0.8625	0.5867	1.9762	1.1594	1.7045	59	36	59.6
30.5	30	30	0.5075	0.8616	0.5890	1.9703	1.1606	1.6977	59	30	59.5
30.6	30	36	0.5090	0.8607	0.5914	1.9645	1.1618	1.6909	59	24	59.4
30.7	30	42	0.5105	0.8599	0.5938	1.9587	1.1630	1.6842	59	18	59.3
30.8	30	48	0.5120	0.8590	0.5961	1.9530	1.1642	1.6775	59	12	59.2
30.9	30	54	0.5135	0.8581	0.5985	1.9473	1.1654	1.6709	59	6	59.1
31.0	31	0	0.5150	0.8572	0.6009	1.9416	1.1666	1.6643	59	0	59.0
31.1	31	6	0.5165	0.8563	0.6032	1.9360	1.1679	1.6577	58	54	58.9
31.2	31	12	0.5180	0.8554	0.6056	1.9304	1.1691	1.6512	58	48	58.8
31.3	31	18	0.5195	0.8545	0.6080	1.9249	1.1703	1.6447	58	42	58.7
31.4	31	24	0.5210	0.8536	0.6104	1.9194	1.1716	1.6383	58	36	58.6
31.5	31	30	0.5225	0.8526	0.6128	1.9139	1.1728	1.6319	58	30	58.5
31.6	31	36	0.5240	0.8517	0.6152	1.9084	1.1741	1.6255	58	24	58.4
31.7	31	42	0.5255	0.8508	0.6176	1.9031	1.1753	1.6191	58	18	58.3
31.8	31	48	0.5270	0.8499	0.6200	1.8977	1.1766	1.6128	58	12	58.2
31.9	31	54	0.5284	0.8490	0.6224	1.8924	1.1779	1.6066	58	6	58.1
32.0	32	0	0.5299	0.8480	0.6249	1.8871	1.1792	1.6003	58	0	58.0
32.1	32	6	0.5314	0.8471	0.6273	1.8818	1.1805	1.5941	57	54	57.9
32.2	32	12	0.5329	0.8462	0.6297	1.8766	1.1818	1.5880	57	48	57.8
32.3	32	18	0.5344	0.8453	0.6322	1.8714	1.1831	1.5818	57	42	57.7
32.4	32	24	0.5358	0.8443	0.6346	1.8663	1.1844	1.5757	57	36	57.6
32.5	32	30	0.5373	0.8434	0.6371	1.8612	1.1857	1.5697	57	30	57.5
32.6	32	36	0.5388	0.8425	0.6395	1.8561	1.1870	1.5637	57	24	57.4
32.7	32	42	0.5402	0.8415	0.6420	1.8510	1.1883	1.5577	57	18	57.3
32.8	32	48	0.5417	0.8406	0.6445	1.8460	1.1897	1.5517	57	12	57.2
32.9	32	54	0.5432	0.8396	0.6469	1.8410	1.1910	1.5458	57	6	57.1
33.0	33	0	0.5446	0.8387	0.6494	1.8361	1.1924	1.5399	57	0	57.0
33.1	33	6	0.5461	0.8377	0.6519	1.8312	1.1937	1.5340	56	54	56.9
33.2	33	12	0.5476	0.8368	0.6544	1.8263	1.1951	1.5282	56	48	56.8
33.3	33	18	0.5490	0.8358	0.6569	1.8214	1.1964	1.5224	56	42	56.7
33.4	33	24	0.5505	0.8348	0.6594	1.8166	1.1978	1.5166	56	36	56.6
33.5	33	30	0.5519	0.8339	0.6619	1.8118	1.1992	1.5108	56	30	56.5
33.6	33	36	0.5534	0.8329	0.6644	1.8070	1.2006	1.5051	56	24	56.4
33.7	33	42	0.5548	0.8320	0.6669	1.8023	1.2020	1.4994	56	18	56.3
33.8	33	48	0.5563	0.8310	0.6694	1.7976	1.2034	1.4938	56	12	56.2
33.9	33	54	0.5577	0.8300	0.6720	1.7929	1.2048	1.4882	56	6	56.1
34.0	34	0	0.5592	0.8290	0.6745	1.7883	1.2062	1.4826	56	0	56.0
34.1	34	6	0.5606	0.8281	0.6771	1.7837	1.2076	1.4770	55	54	55.9
34.2	34	12	0.5621	0.8271	0.6796	1.7791	1.2091	1.4715	55	48	55.8
34.3	34	18	0.5635	0.8261	0.6822	1.7745	1.2105	1.4659	55	42	55.7
34.4	34	24	0.5650	0.8251	0.6847	1.7700	1.2120	1.4605	55	36	55.6
34.5	34	30	0.5664	0.8241	0.6873	1.7655	1.2134	1.4550	55	30	55.5
34.6	34	36	0.5678	0.8231	0.6899	1.7610	1.2149	1.4496	55	24	55.4
34.7	34	42	0.5693	0.8221	0.6924	1.7566	1.2163	1.4442	55	18	55.3
34.8	34	48	0.5707	0.8211	0.6950	1.7522	1.2178	1.4388	55	12	55.2
34.9	34	54	0.5721	0.8202	0.6976	1.7478	1.2193	1.4335	55	6	55.1
			cos θ	sin θ	cot θ	sec θ	csc θ	tan θ	deg	min	θ deg

TABLE II (*continued*)

θ deg	deg	min	sin θ	cos θ	tan θ	csc θ	sec θ	cot θ			
35.0	35	0	0.5736	0.8192	0.7002	1.7434	1.2208	1.4281	55	0	55.0
35.1	35	6	0.5750	0.8181	0.7028	1.7391	1.2223	1.4229	54	54	54.9
35.2	35	12	0.5764	0.8171	0.7054	1.7348	1.2238	1.4176	54	48	54.8
35.3	35	18	0.5779	0.8161	0.7080	1.7305	1.2253	1.4124	54	42	54.7
35.4	35	24	0.5793	0.8151	0.7107	1.7263	1.2268	1.4071	54	36	54.6
35.5	35	30	0.5807	0.8141	0.7133	1.7221	1.2283	1.4019	54	30	54.5
35.6	35	36	0.5821	0.8131	0.7159	1.7179	1.2299	1.3968	54	24	54.4
35.7	35	42	0.5835	0.8121	0.7186	1.7137	1.2314	1.3916	54	18	54.3
35.8	35	48	0.5850	0.8111	0.7212	1.7095	1.2329	1.3865	54	12	54.2
35.9	35	54	0.5864	0.8100	0.7239	1.7054	1.2345	1.3814	54	6	54.1
36.0	36	0	0.5878	0.8090	0.7265	1.7013	1.2361	1.3764	54	0	54.0
36.1	36	6	0.5892	0.8080	0.7292	1.6972	1.2376	1.3713	53	54	53.9
36.2	36	12	0.5906	0.8070	0.7319	1.6932	1.2392	1.3663	53	48	53.8
36.3	36	18	0.5920	0.8059	0.7346	1.6892	1.2408	1.3613	53	42	53.7
36.4	36	24	0.5934	0.8049	0.7373	1.6852	1.2424	1.3564	53	36	53.6
36.5	36	30	0.5948	0.8039	0.7400	1.6812	1.2440	1.3514	53	30	53.5
36.6	36	36	0.5962	0.8028	0.7427	1.6772	1.2456	1.3465	53	24	53.4
36.7	36	42	0.5976	0.8018	0.7454	1.6733	1.2472	1.3416	53	18	53.3
36.8	36	48	0.5990	0.8007	0.7481	1.6694	1.2489	1.3367	53	12	53.2
36.9	36	54	0.6004	0.7997	0.7508	1.6655	1.2505	1.3319	53	6	53.1
37.0	37	0	0.6018	0.7986	0.7536	1.6616	1.2521	1.3270	53	0	53.0
37.1	37	6	0.6032	0.7976	0.7563	1.6578	1.2538	1.3222	52	54	52.9
37.2	37	12	0.6046	0.7965	0.7590	1.6540	1.2554	1.3175	52	48	52.8
37.3	37	18	0.6060	0.7955	0.7618	1.6502	1.2571	1.3127	52	42	52.7
37.4	37	24	0.6074	0.7944	0.7646	1.6464	1.2588	1.3079	52	36	52.6
37.5	37	30	0.6088	0.7934	0.7673	1.6427	1.2605	1.3032	52	30	52.5
37.6	37	36	0.6101	0.7923	0.7701	1.6390	1.2622	1.2985	52	24	52.4
37.7	37	42	0.6115	0.7912	0.7729	1.6353	1.2639	1.2938	52	18	52.3
37.8	37	48	0.6129	0.7902	0.7757	1.6316	1.2656	1.2892	52	12	52.2
37.9	37	54	0.6143	0.7891	0.7785	1.6279	1.2673	1.2846	52	6	52.1
38.0	38	0	0.6157	0.7880	0.7813	1.6243	1.2690	1.2799	52	0	52.0
38.1	38	6	0.6170	0.7869	0.7841	1.6207	1.2708	1.2753	51	54	51.9
38.2	38	12	0.6184	0.7859	0.7869	1.6171	1.2725	1.2708	51	48	51.8
38.3	38	18	0.6198	0.7848	0.7898	1.6135	1.2742	1.2662	51	42	51.7
38.4	38	24	0.6211	0.7837	0.7926	1.6099	1.2760	1.2617	51	36	51.6
38.5	38	30	0.6225	0.7826	0.7954	1.6064	1.2778	1.2572	51	30	51.5
38.6	38	36	0.6239	0.7815	0.7983	1.6029	1.2796	1.2527	51	24	51.4
38.7	38	42	0.6252	0.7804	0.8012	1.5994	1.2813	1.2482	51	18	51.3
38.8	38	48	0.6266	0.7793	0.8040	1.5959	1.2831	1.2437	51	12	51.2
38.9	38	54	0.6280	0.7782	0.8069	1.5925	1.2849	1.2393	51	6	51.1
39.0	39	0	0.6293	0.7771	0.8098	1.5890	1.2868	1.2349	51	0	51.0
39.1	39	6	0.6307	0.7760	0.8127	1.5856	1.2886	1.2305	50	54	50.9
39.2	39	12	0.6320	0.7749	0.8156	1.5822	1.2904	1.2261	50	48	50.8
39.3	39	18	0.6334	0.7738	0.8185	1.5788	1.2923	1.2218	50	42	50.7
39.4	39	24	0.6347	0.7727	0.8214	1.5755	1.2941	1.2174	50	36	50.6
39.5	39	30	0.6361	0.7716	0.8243	1.5721	1.2960	1.2131	50	30	50.5
39.6	39	36	0.6374	0.7705	0.8273	1.5688	1.2978	1.2088	50	24	50.4
39.7	39	42	0.6388	0.7694	0.8302	1.5655	1.2997	1.2045	50	18	50.3
39.8	39	48	0.6401	0.7683	0.8332	1.5622	1.3016	1.2002	50	12	50.2
39.9	39	54	0.6414	0.7672	0.8361	1.5590	1.3035	1.1960	50	6	50.1
			cos θ	sin θ	cot θ	sec θ	csc θ	tan θ	deg	min	θ deg

II

TABLE II (continued)

θ deg	deg	min	sin θ	cos θ	tan θ	csc θ	sec θ	cot θ			
40.0	40	0	0.6428	0.7660	0.8391	1.5557	1.3054	1.1918	50	0	50.
40.1	40	6	0.6441	0.7649	0.8421	1.5525	1.3073	1.1875	49	54	49.
40.2	40	12	0.6455	0.7638	0.8451	1.5493	1.3092	1.1833	49	48	49.
40.3	40	18	0.6468	0.7627	0.8481	1.5461	1.3112	1.1792	49	42	49.
40.4	40	24	0.6481	0.7615	0.8511	1.5429	1.3131	1.1750	49	36	49.
40.5	40	30	0.6494	0.7604	0.8541	1.5398	1.3151	1.1708	49	30	49.
40.6	40	36	0.6508	0.7593	0.8571	1.5366	1.3171	1.1667	49	24	49.
40.7	40	42	0.6521	0.7581	0.8601	1.5335	1.3190	1.1626	49	18	49.
40.8	40	48	0.6534	0.7570	0.8632	1.5304	1.3210	1.1585	49	12	49.
40.9	40	54	0.6547	0.7559	0.8662	1.5273	1.3230	1.1544	49	6	49.
41.0	41	0	0.6561	0.7547	0.8693	1.5243	1.3250	1.1504	49	0	49.
41.1	41	6	0.6574	0.7536	0.8724	1.5212	1.3270	1.1463	48	54	48.
41.2	41	12	0.6587	0.7524	0.8754	1.5182	1.3291	1.1423	48	48	48.
41.3	41	18	0.6600	0.7513	0.8785	1.5151	1.3311	1.1383	48	42	48.
41.4	41	24	0.6613	0.7501	0.8816	1.5121	1.3331	1.1343	48	36	48.
41.5	41	30	0.6626	0.7490	0.8847	1.5092	1.3352	1.1303	48	30	48.
41.6	41	36	0.6639	0.7478	0.8878	1.5062	1.3373	1.1263	48	24	48.
41.7	41	42	0.6652	0.7466	0.8910	1.5032	1.3393	1.1224	48	18	48.
41.8	41	48	0.6665	0.7455	0.8941	1.5003	1.3414	1.1184	48	12	48.
41.9	41	54	0.6678	0.7443	0.8972	1.4974	1.3435	1.1145	48	6	48.
42.0	42	0	0.6691	0.7431	0.9004	1.4945	1.3456	1.1106	48	0	48.
42.1	42	6	0.6704	0.7420	0.9036	1.4916	1.3478	1.1067	47	54	47.
42.2	42	12	0.6717	0.7408	0.9067	1.4887	1.3499	1.1028	47	48	47.
42.3	42	18	0.6730	0.7396	0.9099	1.4859	1.3520	1.0990	47	42	47.
42.4	42	24	0.6743	0.7385	0.9131	1.4830	1.3542	1.0951	47	36	47.
42.5	42	30	0.6756	0.7373	0.9163	1.4802	1.3563	1.0913	47	30	47.
42.6	42	36	0.6769	0.7361	0.9195	1.4774	1.3585	1.0875	47	24	47.
42.7	42	42	0.6782	0.7349	0.9228	1.4746	1.3607	1.0837	47	18	47.
42.8	42	48	0.6794	0.7337	0.9260	1.4718	1.3629	1.0799	47	12	47.
42.9	42	54	0.6807	0.7325	0.9293	1.4690	1.3651	1.0761	47	6	47.
43.0	43	0	0.6820	0.7314	0.9325	1.4663	1.3673	1.0724	47	0	47.
43.1	43	6	0.6833	0.7302	0.9358	1.4635	1.3696	1.0686	46	54	46.
43.2	43	12	0.6845	0.7290	0.9391	1.4608	1.3718	1.0649	46	48	46.
43.3	43	18	0.6858	0.7278	0.9424	1.4581	1.3741	1.0612	46	42	46.
43.4	43	24	0.6871	0.7266	0.9457	1.4554	1.3763	1.0575	46	36	46.
43.5	43	30	0.6884	0.7254	0.9490	1.4527	1.3786	1.0538	46	30	46.
43.6	43	36	0.6896	0.7242	0.9523	1.4501	1.3809	1.0501	46	24	46.
43.7	43	42	0.6909	0.7230	0.9556	1.4474	1.3832	1.0464	46	18	46.
43.8	43	48	0.6921	0.7218	0.9590	1.4448	1.3855	1.0428	46	12	46.
43.9	43	54	0.6934	0.7206	0.9623	1.4422	1.3878	1.0392	46	6	46.
44.0	44	0	0.6947	0.7193	0.9657	1.4396	1.3902	1.0355	46	0	46.
44.1	44	6	0.6959	0.7181	0.9691	1.4370	1.3925	1.0319	45	54	45.
44.2	44	12	0.6972	0.7169	0.9725	1.4344	1.3949	1.0283	45	48	45.
44.3	44	18	0.6984	0.7157	0.9759	1.4318	1.3972	1.0247	45	42	45.
44.4	44	24	0.6997	0.7145	0.9793	1.4293	1.3996	1.0212	45	36	45.
44.5	44	30	0.7009	0.7133	0.9827	1.4267	1.4020	1.0176	45	30	45.
44.6	44	36	0.7022	0.7120	0.9861	1.4242	1.4044	1.0141	45	24	45.
44.7	44	42	0.7034	0.7108	0.9896	1.4217	1.4069	1.0105	45	18	45.
44.8	44	48	0.7046	0.7096	0.9930	1.4192	1.4093	1.0070	45	12	45.
44.9	44	54	0.7059	0.7083	0.9965	1.4167	1.4118	1.0035	45	6	45.
45.0	45	0	0.7071	0.7071	1.0000	1.4142	1.4142	1.0000	45	0	45.
			cos θ	sin θ	cot θ	sec θ	csc θ	tan θ	deg	min	θ deg

TABLE III Values of trigonometric functions (radians)

θ radians or Real Number x	θ degrees	sin θ or sin x	cos θ or cos x	tan θ or tan x	csc θ or csc x	sec θ or sec x	cot θ or cot x
0.00	0° 00′	0.0000	1.000	0.0000	No value	1.000	No value
.01	0° 34′	.0100	1.000	.0100	100.0	1.000	100.0
.02	1° 09′	.0200	0.9998	.0200	50.00	1.000	49.99
.03	1° 43′	.0300	0.9996	.0300	33.34	1.000	33.32
.04	2° 18′	.0400	0.9992	.0400	25.01	1.001	24.99
0.05	2° 52′	0.0500	0.9988	0.0500	20.01	1.001	19.98
.06	3° 26′	.0600	.9982	.0601	16.68	1.002	16.65
.07	4° 01′	.0699	.9976	.0701	14.30	1.002	14.26
.08	4° 35′	.0799	.9968	.0802	12.51	1.003	12.47
.09	5° 09′	.0899	.9960	.0902	11.13	1.004	11.08
0.10	5° 44′	0.0998	0.9950	0.1003	10.02	1.005	9.967
.11	6° 18′	.1098	.9940	.1104	9.109	1.006	9.054
.12	6° 53′	.1197	.9928	.1206	8.353	1.007	8.293
.13	7° 27′	.1296	.9916	.1307	7.714	1.009	7.649
.14	8° 01′	.1395	.9902	.1409	7.166	1.010	7.096
0.15	8° 36′	0.1494	0.9888	0.1511	6.692	1.011	6.617
.16	9° 10′	.1593	.9872	.1614	6.277	1.013	6.197
.17	9° 44′	.1692	.9856	.1717	5.911	1.015	5.826
.18	10° 19′	.1790	.9838	.1820	5.586	1.016	5.495
.19	10° 53′	.1889	.9820	.1923	5.295	1.018	5.200
0.20	11° 28′	0.1987	0.9801	0.2027	5.033	1.020	4.933
.21	12° 02′	.2085	.9780	.2131	4.797	1.022	4.692
.22	12° 36′	.2182	.9759	.2236	4.582	1.025	4.472
.23	13° 11′	.2280	.9737	.2341	4.386	1.027	4.271
.24	13° 45′	.2377	.9713	.2447	4.207	1.030	4.086
0.25	14° 19′	0.2474	0.9689	0.2553	4.042	1.032	3.916
.26	14° 54′	.2571	.9664	.2660	3.890	1.035	3.759
.27	15° 28′	.2667	.9638	.2768	3.749	1.038	3.613
.28	16° 03′	.2764	.9611	.2876	3.619	1.041	3.478
.29	16° 37′	.2860	.9582	.2984	3.497	1.044	3.351
0.30	17° 11′	0.2955	0.9553	0.3093	3.384	1.047	3.233
.31	17° 46′	.3051	.9523	.3203	3.278	1.050	3.122
.32	18° 20′	.3146	.9492	.3314	3.179	1.053	3.018
.33	18° 54′	.3240	.9460	.3425	3.086	1.057	2.920
.34	19° 29′	.3335	.9428	.3537	2.999	1.061	1.827
0.35	20° 03′	0.3429	0.9394	0.3650	2.916	1.065	2.740
.36	20° 38′	.3523	.9359	.3764	2.839	1.068	2.657
.37	21° 12′	.3616	.9323	.3879	2.765	1.073	2.578
.38	21° 46′	.3709	.9287	.3994	2.696	1.077	2.504
.39	22° 21′	.3802	.9249	.4111	2.630	1.081	2.433
0.40	22° 55′	0.3894	0.9211	0.4228	2.568	1.086	2.365
.41	23° 29′	.3986	.9171	.4346	2.509	1.090	2.301
.42	24° 04′	.4078	.9131	.4466	2.452	1.095	2.239
.43	24° 38′	.4169	.9090	.4586	2.399	1.100	2.180
.44	25° 13′	.4259	.9048	.4708	2.348	1.105	2.124
0.45	25° 47′	0.4350	0.9004	0.4831	2.299	1.111	2.070

TABLE III (*continued*)

θ radians or Real Number x	θ degrees	sin θ or sin x	cos θ or cos x	tan θ or tan x	csc θ or csc x	sec θ or sec x	cot θ or cot x
0.45	25° 47′	0.4350	0.9004	0.4831	2.299	1.111	2.070
.46	26° 21′	.4439	.8961	.4954	2·253	1.116	2.018
.47	26° 56′	.4529	.8916	.5080	2.208	1.122	1.969
.48	27° 30′	.4618	.8870	.5206	2.166	1.127	1.921
.49	28° 04′	.4706	.8823	.5334	2.125	1.133	1.875
0.50	28° 39′	0.4794	0.8776	0.5463	2.086	1.139	1.830
.51	29° 13′	.4882	.8727	.5594	2.048	1.146	1.788
.52	29° 48′	.4969	.8678	.5726	2.013	1.152	1.747
.53	30° 22′	.5055	.8628	.5859	1.978	1.159	1.707
.54	30° 56′	.5141	.8577	.5994	1.945	1.166	1.668
0.55	31° 31′	0.5227	0.8525	0.6131	1.913	1.173	1.631
.56	32° 05′	.5312	.8473	.6269	1.883	1.180	1.595
.57	32° 40′	.5396	.8419	.6410	1.853	1.188	1.560
.58	33° 14′	.5480	.8365	.6552	1.825	1.196	1.526
.59	33° 48′	.5564	.8309	.6696	1.797	1.203	1.494
0.60	34° 23′	0.5646	0.8253	0.6841	1.771	1.212	1.462
.61	34° 57′	.5729	.8196	.6989	1.746	1.220	1.431
.62	35° 31′	.5810	.8139	.7139	1.721	1.229	1.401
.63	36° 06′	.5891	.8080	.7291	1.697	1.238	1.372
.64	36° 40′	.5972	.8021	.7445	1.674	1.247	1.343
0.65	37° 15′	0.6052	0.7961	0.7602	1.652	1.256	1.315
.66	37° 49′	.6131	.7900	.7761	1.631	1.266	1.288
.67	38° 23′	.6210	.7838	.7923	1.610	1.276	1.262
.68	38° 58′	.6288	.7776	.8087	1.590	1.286	1.237
.69	39° 32′	.6365	.7712	.8253	1.571	1.297	1.212
0.70	40° 06′	0.6442	0.7648	0.8423	1.552	1.307	1.187
.71	40° 41′	.6518	.7584	.8595	1.534	1.319	1.163
.72	41° 15′	·6594	.7518	.8771	1.517	1.330	1.140
.73	41° 50′	.6669	.7452	.8949	1.500	1.342	1.117
.74	42° 24′	.6743	.7385	.9131	1.483	1.354	1.095
0.75	42° 58′	0.6816	0.7317	0.9316	1.467	1.367	1.073
.76	43° 33′	.6889	.7248	.9505	1.452	1.380	1.052
.77	44° 07′	.6961	.7179	.9697	1.436	1.393	1.031
.78	44° 41′	.7033	.7109	.9893	1.422	1.407	1.011
.79	45° 16′	.7104	.7038	1.009	1.408	1.421	0.9908
0.80	45° 50′	0.7174	0.6967	1.030	1.394	1.435	0.9712
.81	46° 25′	.7243	.6895	1.050	1.381	1.450	.9520
.82	46° 59′	.7311	.6822	1.072	1.368	1.466	.9331
.83	47° 33′	.7379	.6749	1.093	1.355	1.482	.9146
.84	48° 08′	.7446	.6675	1.116	1.343	1.498	.8964
0.85	48° 42′	0.7513	0.6600	1.138	1.331	1.515	0.8785
.86	49° 16′	.7578	.6524	1.162	1.320	1.533	.8609
.87	49° 51′	.7643	.6448	1.185	1.308	1.551	.8437
.88	50° 25′	.7707	.6372	1.210	1.297	1.569	.8267
.89	51° 00′	.7771	.6294	1.235	1.287	1.589	.8100
0.90	51° 34′	0.7833	0.6216	1.260	1.277	1.609	0.7936
.91	52° 08′	.7895	.6137	1.286	1.267	1.629	.7774
.92	52° 43′	.7956	.6058	1.313	1.257	1.651	.7615
.93	53° 17′	.8016	.5978	1.341	1.247	1.673	.7458
.94	53° 51′	.8076	.5898	1.369	1.238	1.696	.7303
0.95	54° 26′	0.8134	0.5817	1.398	1.229	1.719	0.7151

TABLE III *(continued)*

θ radians or Real Number x	θ degrees	sin θ or sin x	cos θ or cos x	tan θ or tan x	csc θ or csc x	sec θ or sec x	cot θ or cot x
0.95	54° 26′	0.8134	0.5817	1.398	1.229	1.719	0.7151
.96	55° 00′	.8192	.5735	1.428	1.221	1.744	.7001
.97	55° 35′	.8249	.5653	1.459	1.212	1.769	.6853
.98	56° 09′	.8305	.5570	1.491	1.204	1.795	.6707
.99	56° 43′	.8360	·5487	1.524	1.196	1.823	.6563
1.00	57° 18′	0.8415	0.5403	1.557	1.188	1.851	0.6421
1.01	57° 52′	.8468	.5319	1.592	1.181	1.880	.6281
1.02	58° 27′	.8521	.5234	1.628	1.174	1.911	.6142
1.03	59° 01′	.8573	.5148	1.665	1.166	1.942	.6005
1.04	59° 35′	.8624	.5062	1.704	1.160	1.975	.5870
1.05	60° 10′	0.8674	0.4976	1.743	1.153	2.010	0.5736
1.06	60° 44′	.8724	.4889	1.784	1.146	2.046	.5604
1.07	61° 18′	.8772	.4801	1.827	1.140	2.083	.5473
1.08	61° 53′	.8820	.4713	1.871	1.134	2.122	.5344
1.09	62° 27′	.8866	.4625	1.917	1.128	2.162	.5216
1.10	63° 02′	0.8912	0.4536	1.965	1.122	2.205	0.5090
1.11	63° 36′	.8957	.4447	2.014	1.116	2.249	.4964
1.12	64° 10′	.9001	.4357	2.066	1.111	2.295	.4840
1.13	64° 45′	.9044	.4267	2.120	1.106	2.344	.4718
1.14	65° 19′	.9086	.4176	2.176	1.101	2.395	.4596
1.15	65° 53′	0.9128	0.4085	2.234	1.096	2.448	0.4475
1.16	66° 28′	.9168	.3993	2.296	1.091	2.504	.4356
1.17	67° 02′	.9208	.3902	2.360	1.086	2.563	.4237
1.18	67° 37′	.9246	.3809	2.427	1.082	2.625	.4120
1.19	68° 11′	.9284	.3717	2.498	1.077	2.691	.4003
1.20	68° 45′	0.9320	0.3624	2.572	1.073	2.760	0.3888
1.21	69° 20′	.9356	.3530	2.650	1.069	2.833	.3773
1.22	69° 54′	.9391	.3436	2.733	1.065	2.910	.3659
1.23	70° 28′	.9425	.3342	2.820	1.061	2.992	.3546
1.24	71° 03′	.9458	.3248	2.912	1.057	3.079	.3434
1.25	71° 37′	0.9490	0.3153	3.010	1.054	3.171	0.3323
1.26	72° 12′	.9521	.3058	3.113	1.050	3.270	.3212
1.27	72° 46′	.9551	.2963	3.224	1.047	3.375	.3102
1.28	73° 20′	.9580	.2867	3.341	1.044	3.488	.2993
1.29	73° 55′	.9608	.2771	3.467	1.041	3.609	.2884
1.30	74° 29′	0.9636	0.2675	3.602	1.038	3.738	0.2776
1.31	75° 03′	.9662	.2579	3.747	1.035	3.878	.2669
1.32	75° 38′	.9687	.2482	3.903	1.032	4.029	.2562
1.33	76° 12′	.9711	.2385	4.072	1.030	4.193	.2456
1.34	76° 47′	.9735	.2288	4.256	1.027	4.372	.2350
1.35	77° 21′	0.9757	0.2190	4.455	1.025	4.566	0.2245
1.36	77° 55′	.9779	.2092	4.673	1.023	4.779	.2140
1.37	78° 30′	.9799	.1994	4.913	1.021	5.014	.2035
1.38	79° 04′	.9819	.1896	5.177	1.018	5.273	.1931
1.39	79° 38′	.9837	.1798	5.471	1.017	5.561	.1828
1.40	80° 13′	0.9854	0.1700	5.798	1.015	5.883	0.1725
1.41	80° 47′	.9871	.1601	6.165	1.013	6.246	.1622
1.42	81° 22′	.9887	.1502	6.581	1.011	6.657	.1519
1.43	81° 56′	.9901	.1403	7.055	1.010	7.126	.1417
1.44	82° 30′	.9915	.1304	7.602	1.009	7.667	.1315
1.45	83° 05′	0.9927	0.1205	8.238	1.007	8.299	0.1214

TABLE III (*continued*)

θ radians or Real Number x	θ degrees	$\sin \theta$ or $\sin x$	$\cos \theta$ or $\cos x$	$\tan \theta$ or $\tan x$	$\csc \theta$ or $\csc x$	$\sec \theta$ or $\sec x$	$\cot \theta$ or $\cot x$
1.45	83° 05′	0.9927	0.1205	8.238	1.007	8.299	0.1214
1.46	83° 39′	.9939	.1106	8.989	1.006	9.044	.1113
1.47	84° 13′	.9949	.1006	9.887	1.005	9.938	.1001
1.48	84° 48′	.9959	.0907	10.98	1.004	11.03	.0910
1.49	85° 22′	.9967	.0807	12.35	1.003	12.39	.0810
1.50	85° 57′	0.9975	0.0707	14.10	1.003	14.14	0.0709
1.51	86° 31′	.9982	.0608	16.43	1.002	16.46	.0609
1.52	87° 05′	.9987	.0508	19.67	1.001	19.69	.0508
1.53	87° 40′	.9992	.0408	24.50	1.001	24.52	.0408
1.54	88° 14′	.9995	.0308	32.46	1.000	32.48	.0308
1.55	88° 49′	0.9998	0.0208	48.08	1.000	48.09	0.0208
1.56	89° 23′	.9999	.0108	92.62	1.000	92.63	.0108
1.57	89° 57′	1.000	.0008	1256	1.000	1256	.0008

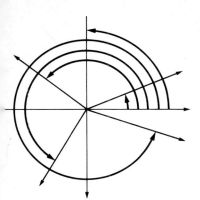

ANSWERS

Exercise Set 1.1

1. $\sin \alpha = 0.61$ $\csc \alpha = 1.64$
$\cos \alpha = 0.79$ $\sec \alpha = 1.26$
$\tan \alpha = 0.77$ $\cot \alpha = 1.30$

3. $\sin A = 0.40$ $\csc A = 2.52$
$\cos A = 0.92$ $\sec A = 1.09$
$\tan A = 0.43$ $\cot A = 2.31$

5. $\sin CBA = 0.33$ $\csc CBA = 3.05$
$\cos CBA = 0.94$ $\sec CBA = 1.06$
$\tan CBA = 0.35$ $\cot CBA = 2.88$

7. $\dfrac{p}{m}$ **9.** $\dfrac{r}{p}$ **11.** $\dfrac{q}{r}$ **13.** $\dfrac{n}{r}$ **15.** $\dfrac{n}{p+q}$ **17.** $\dfrac{p}{m}$ **19.** $\dfrac{q}{r}$ **21.** $\dfrac{q}{s}$ **23.** $\dfrac{p}{q+r}$

25. $\sin DAC = 0.61$ $\csc DAC = 1.65$
$\cos DAC = 0.80$ $\sec DAC = 1.25$
$\tan DAC = 0.76$ $\cot DAC = 1.32$

27. $\sin DBC = 0.78$ $\csc DBC = 1.29$
$\cos DBC = 0.63$ $\sec DBC = 1.59$
$\tan DBC = 1.24$ $\cot DBC = 0.81$

29. $\sin CDB = 0.80$ $\csc CDB = 1.26$
$\cos CDB = 0.59$ $\sec CDB = 1.69$
$\tan CDB = 1.35$ $\tan CDB = 0.74$

31. $\sin DAC = 0.38$ $\csc DAC = 2.62$
$\cos DAC = 0.93$ $\sec DAC = 1.08$
$\tan DAC = 0.41$ $\cot DAC = 2.42$

33. $2, \dfrac{1}{2}, \dfrac{2}{\sqrt{3}}$ **35.** $1, 1, \dfrac{1}{\sqrt{2}}$ **37.** $\dfrac{1}{4}, \dfrac{4}{\sqrt{15}}, \dfrac{1}{\sqrt{15}}$

39. $\sin 45° = \dfrac{1}{\sqrt{2}}$ $\csc 45° = \sqrt{2}$

$\cos 45° = \dfrac{1}{\sqrt{2}}$ $\sec 45° = \sqrt{2}$

$\tan 45° = 1$ $\cot 45° = 1$

Exercise Set 1.2

1. $\dfrac{1}{4}$ **3.** $\dfrac{3\sqrt{3}}{2}$ **5.** $\dfrac{3}{2}$ **7.** $\dfrac{4}{3}$ **9.** $\dfrac{(\sqrt{3}-8)}{4\sqrt{3}}$

11. 4   **13.** 1   **15.** $(\sqrt{3}+1)\sqrt{3}$ **17.** 0.2823   **19.** 2.6325

21. 0.9466   **23.** 0.1086   **25.** 3.9232   **27.** 4.6569   **29.** 2.3486

31.

α	10°	5°	1°	0.1°
$\sin \alpha$	0.17	0.09	0.02	0.00
$\csc \alpha$	5.77	11.47	57.30	572.96

33.

α	10°	5°	1°	0.1°
$\tan \alpha$	0.18	0.09	0.02	0.00
$\cot \alpha$	5.67	11.43	57.29	572.96

As α approaches 0°, $\sin \alpha$ approaches 0 and $\csc \alpha$ becomes very large.

As α approaches 0°, $\tan \alpha$ approaches 0 and $\cot \alpha$ becomes very large.

35.

α	80°	85°	89°	89.9°
$\cos \alpha$	0.17	0.09	0.02	0.00
$\sec \alpha$	5.76	11.47	57.30	572.96

As α approaches 90°, $\cos \alpha$ approaches 0 and $\sec \alpha$ becomes very large.

Exercise Set 1.3

1. $\alpha = \text{Tan}^{-1}\dfrac{1}{\sqrt{3}}$ **3.** $\beta = \text{Csc}^{-1}\, 3.1421$ **5.** $\gamma = \text{Sec}^{-1}\, 2.6312$ **7.** $\sin \alpha = \dfrac{1}{\sqrt{2}}$

9. $\cos \beta = 0.6214$ **11.** $\sec \gamma = 2.6060$ **13.** 30°   **15.** 45°

17. 30°   **19.** 18.4°   **21.** 72.4°   **23.** 37.5°

25. 58.6°   **27.** $\alpha = 21.6°,\ \beta = 68.4°$ **29.** $\alpha = 60.4°,\ \beta = 29.6°$ **31.** $\beta = 14.0°$

33. $(\sin \alpha = 0.4231)$ $\csc \alpha = 2.3635$
$\cos \alpha = 0.9061$ $\sec \alpha = 1.1037$
$\tan \alpha = 0.4670$ $\cot \alpha = 2.1415$

35. $\sin \alpha = 0.8303$ $\csc \alpha = 1.2044$
$\cos \alpha = 0.5574$ $(\sec \alpha = 1.7942)$
$\tan \alpha = 1.4897$ $\cot \alpha = 0.6713$

37. $\sin \gamma = \dfrac{1}{\sqrt{2}}$ $\csc \gamma = \sqrt{2}$

$\left(\cos \gamma = \dfrac{1}{\sqrt{2}}\right)$ $\sec \gamma = \sqrt{2}$

$\tan \gamma = 1$ $\cot \gamma = 1$

39. $\sin \gamma = \dfrac{1}{2}$ $(\csc \gamma = 2)$

$\cos \gamma = \dfrac{\sqrt{3}}{2}$ $\sec \gamma = \dfrac{2}{\sqrt{3}}$

$\tan \gamma = \dfrac{1}{\sqrt{3}}$ $\cot \gamma = \sqrt{3}$

Exercise Set 1.4

1. $\alpha = 48°$ **3.** $a \approx 1.78$

5. $\beta \approx 56.0°$ **7.** $c \approx 7.09$

1

9. $c \approx 7.57$
3. $\alpha \approx 33.7°$, $\beta \approx 56.3°$, $c \approx 3.61$
7. $\alpha \approx 39.0°$, $a \approx 14.8$, $b \approx 18.3$
1. $\beta \approx 39.7°$, $a \approx 18.1$, $c \approx 23.5$
5. 360 meters
9. 27.1 decimeters
3. 61.4 feet
7. 5.88 centimeters

11. $b \approx 2.32$
15. $\alpha \approx 61.9°$, $\beta \approx 28.1°$, $b \approx 8.00$
19. $\alpha \approx 67.8°$, $b \approx 4.49$, $c \approx 11.9$
23. $\alpha \approx 48.4°$, $a \approx 33.8$, $c \approx 45.2$
27. 53.1°
31. 239 meters
35. 12,300 feet
39. $4\sqrt{3}$

Review Exercises

1. $\sin A = 0.9$ $\csc A = 1.1$
   $\cos A = 0.4$ $\sec A = 2.5$
   $\tan A = 2.3$ $\cot A = 0.4$

2. $\sin B = 0.4$ $\csc B = 2.5$
   $\cos B = 0.9$ $\sec B = 1.1$
   $\tan B = 0.4$ $\cot B = 2.3$

3. $\sin \alpha = 0.9$ $\csc \alpha = 1.1$
   $\cos \alpha = 0.4$ $\sec \alpha = 2.6$
   $\tan \alpha = 2.4$ $\cot \alpha = 0.4$

4. $\sin \beta = 0.4$ $\csc \beta = 2.6$
   $\cos \beta = 0.9$ $\sec \beta = 1.1$
   $\tan \beta = 0.4$ $\cot \beta = 2.4$

5. $\sin CAD = 0.4$ $\csc CAD = 2.6$
   $\cos CAD = 0.9$ $\sec CAD = 1.1$
   $\tan CAD = 0.4$ $\cot CAD = 2.4$

6. $\sin CDA = 0.9$ $\csc CDA = 1.1$
   $\cos CDA = 0.4$ $\sec CDA = 2.6$
   $\tan CDA = 2.4$ $\cot CDA = 0.4$

7. $\sin CBD = 0.6$ $\csc CBD = 1.6$
   $\cos CBD = 0.8$ $\sec CBD = 1.3$
   $\tan CBD = 0.8$ $\cot CBD = 1.3$

8. $\sin CDB = 0.8$ $\csc CDB = 1.3$
   $\cos CDB = 0.6$ $\sec CDB = 1.6$
   $\tan CDB = 1.3$ $\cot CDB = 0.8$

9. $\dfrac{1}{6}$
10. $\dfrac{2 + \sqrt{6}}{2\sqrt{2}}$
11. $\dfrac{4\sqrt{3} - 2}{\sqrt{3}}$
12. $\dfrac{8\sqrt{2} + 2}{\sqrt{6}}$
13. 0.9759

14. 0.6899
15. 0.9298
16. 0.7120
17. 1.0255
18. 1.4715

19. $\alpha = \mathrm{Cos}^{-1} \dfrac{1}{2}$; $\alpha = 60°$
20. $\alpha = \mathrm{Tan}^{-1} \sqrt{3}$; $\alpha = 60°$
21. $\alpha = \mathrm{Csc}^{-1} 2$; $\alpha = 30°$

22. $\alpha = \mathrm{Cot}^{-1} 1$; $\alpha = 45°$
23. $\alpha = \mathrm{Sin}^{-1} 0.3124$; $\alpha = 18.2°$
24. $\alpha = \mathrm{Tan}^{-1} 1.6219$; $\alpha = 58.3°$
25. $\alpha = \mathrm{Cot}^{-1} 1.7213$; $\alpha = 30.2°$
26. $\alpha = \mathrm{Sec}^{-1} 2.1462$; $\alpha = 62.2°$
27. 39.8°

28. 25.4°
29. 27.1°
30. 29.4°

31. $\alpha = 39°$, $a \approx 8.8$, $b \approx 10.9$
32. $\alpha \approx 53.7°$, $\beta \approx 36.3°$, $b \approx 4.0$
33. $\alpha = 43.8°$, $a \approx 6.5$, $c \approx 9.4$
34. $\beta = 77.6°$, $a \approx 34.8$, $b = 158.2$
35. $\beta = 57°24'$, $b \approx 20.2$, $c \approx 23.9$
36. $\beta = 18°42'$, $a \approx 28.7$, $c \approx 30.3$
37. 340.0 feet
38. 16.3°
39. 12.2°
40. 56.7 feet

Exercise Set 2.1

1.

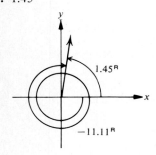

3. 60.0°; Quad. I

5. 840.0°; Quad. II

7. −288.0°; Quad. I

9. −405.0°; Quad. IV

11. 10.3°; Quad. I

13. 295.7°; Quad. IV

15. −15.5°; Quad. IV

17. −231.5°; Quad. II

19. 0.52^R; Quad. I

21. 1.31^R; Quad. I

23. 2.50^R; Quad. II

25. 7.01^R; Quad. I

27. −0.65^R; Quad. IV

29. −1.40^R; Quad. IV

31. −4.19^R; Quad. II

33. −8.90^R; Quad. III

35. 0.44^R

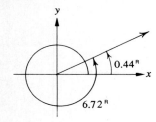

37. 19°

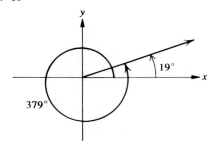

39. 1.45^R

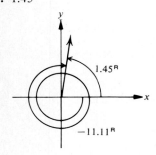

41. 288°

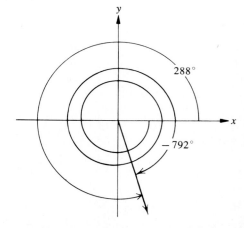

43. $(21 + k \cdot 360)°$, $k \in J$

45. $(242 + k \cdot 360)°$, $k \in J$

47. $(\pi/6 + k \cdot 2\pi)^R$, $k \in J$

9. $(5\pi/3 + k \cdot 2\pi)^R$, $k \in J$ **51.** 6.2 meters   **53.** $5\pi^2/6 \approx 8.2$ decimeters

55. 7.1 feet $(-7.1$ directed length)   **57.** $126.9°$ **59.** $280.7°$

61. $\dfrac{1}{12}$ **63.** -2 **65.** $\dfrac{5}{4}$ **67.** 0.5646

69. 0.2776   **71.** 1.6507   **73.** 0.8^R **75.** 0.9^R

77. Area of the sector is to the area of the circle as the central angle θ is to 2π radians. Thus,

$$\frac{\mathcal{A}\text{(of sector)}}{\pi r^2} = \frac{\theta^R}{2\pi}, \quad \text{from which} \quad \mathcal{A} = \frac{\pi r^2(\theta^R)}{2\pi} = \tfrac{1}{2}r^2\theta^R.$$

79. 4693^R

Exercise Set 2.2

1. $\sin \alpha = \dfrac{2}{\sqrt{29}}$ $\csc \alpha = \dfrac{\sqrt{29}}{2}$ **3.** $\sin \alpha = \dfrac{-3}{\sqrt{13}}$ $\csc \alpha = \dfrac{-\sqrt{13}}{3}$

$\cos \alpha = \dfrac{5}{\sqrt{29}}$ $\sec \alpha = \dfrac{\sqrt{29}}{5}$ $\cos \alpha = \dfrac{-2}{\sqrt{13}}$ $\sec \alpha = \dfrac{-\sqrt{13}}{2}$

$\tan \alpha = \dfrac{2}{5}$ $\cot \alpha = \dfrac{5}{2}$ $\tan \alpha = \dfrac{3}{2}$ $\cot \alpha = \dfrac{2}{3}$

5. $\sin \alpha = 0$ $\csc \alpha$ (undef.)   **7.** $\sin \alpha = 0$ $\csc \alpha$ (undef.)

$\cos \alpha = 1$ $\sec \alpha = 1$ $\cos \alpha = -1$ $\sec \alpha = -1$

$\tan \alpha = 0$ $\cot \alpha$ (undef.) $\tan \alpha = 0$ $\cot \alpha$ (undef.)

9. IV   **11.** I   **13.** III   **15.** II   **17.** $60°$

19. $45°$ **21.** $\dfrac{\pi}{4}^R$ **23.** $\dfrac{\pi}{3}^R$ **25.** 1 **27.** $\dfrac{-1}{2}$

29. 1 **31.** $\dfrac{-1}{\sqrt{2}}$ **33.** $-\sqrt{3}$ **35.** $\dfrac{-2}{\sqrt{3}}$ **37.** -1

39. Undef.   **41.** 0

43. $\sqrt{3}$ **45.** $1/\sqrt{2}$ **47.** $\sqrt{3}$ **49.** 0 **51.** $-1/\sqrt{2}$

53. $\sqrt{3}$ **55.** 0.6428   **57.** 8.6428   **59.** -3.1716 **61.** -1.6390

63. -1.704 **65.** -0.0508 **67.** 6.246   **69.** -0.9780

Exercise Set 2.3

1. $\cos\left(\dfrac{3\pi}{4} + 2\pi\right)$ **3.** $\tan\left(\dfrac{\pi}{4} + \pi\right)$ **5.** $\sin\left(\dfrac{5\pi}{4} + 2\pi\right)$

7. $\tan\left(\dfrac{3\pi}{4} + 3\pi\right)$ **9.** $\sin\left(\dfrac{7\pi}{6} + 2\pi\right)$ **11.** $\sin\left(\dfrac{\pi}{3} + 4\pi\right)$

13. 1.83 centimeters *below* the "at rest" position at $t = 1.2$ seconds; 0.58 centimeters *above* the "at rest" position at $t = 1.8$ seconds.

15. 1.48 centimeters *above* the "at rest" position at $t = \frac{1}{2}$ second; 0.32 centimeter *above* the "at rest" position at $t = \frac{3}{4}$ second.

17. 11.76 centimeters to the *right*.   **19.** 12.00 centimeters to the *left*.   **21.** -106 volts

23. -68.5 volts   **25.** 81.1 volts   **27.** -0.029 ampere

29. 0.000 ampere   **31.** 0.000 ampere   **33.** 0 dyne per square centimete

35. 0 dyne per square centimeter   **37.** 0.242 feet

Review Exercises

1. $72.0°$ **2.** $138.1°$ **3.** 1.31^R **4.** 4.49^R **5.** $218°$

6. 0.95^R **7.** $(131 + k \cdot 360)°, \ k \in J$ **8.** $\left(\frac{7\pi}{4} + k \cdot 2\pi\right)^R, \ k \in J$

9. 2.9 meters   **10.** 0.2^R **11.** $\dfrac{2 - \sqrt{3}}{2\sqrt{3}}$ **12.** $\dfrac{3}{4}$

13. 0.3248   **14.** 1.2383   **15.** 0.9^R **16.** 0.5^R

17. $\sin \alpha = \dfrac{3}{5} \quad \csc \alpha = \dfrac{5}{3}$

$\cos \alpha = \dfrac{-4}{5} \quad \sec \alpha = \dfrac{-5}{4}$

$\tan \alpha = \dfrac{-3}{4} \quad \cot \alpha = \dfrac{-4}{3}$

18. $\sin \alpha = \dfrac{2}{\sqrt{5}} \quad \csc \alpha = \dfrac{\sqrt{5}}{2}$

$\cos \alpha = \dfrac{-1}{\sqrt{5}} \quad \sec \alpha = -\sqrt{5}$

$\tan \alpha = -2 \quad \cot \alpha = -\dfrac{1}{2}$

19. $\sin \alpha = \dfrac{-1}{2} \quad \csc \alpha = -2$

$\cos \alpha = \dfrac{\sqrt{3}}{2} \quad \sec \alpha = \dfrac{2}{\sqrt{3}}$

$\tan \alpha = \dfrac{-1}{\sqrt{3}} \quad \cot \alpha = -\sqrt{3}$

20. $\sin \alpha = \dfrac{-15}{17} \quad \csc \alpha = \dfrac{-17}{15}$

$\cos \alpha = \dfrac{-8}{17} \quad \sec \alpha = \dfrac{-17}{8}$

$\tan \alpha = \dfrac{15}{8} \quad \cot \alpha = \dfrac{8}{15}$

21. III   **22.** III   **23.** $30°$ **24.** $30°$ **25.** $\dfrac{\pi}{4}^R$ **26.** $\dfrac{\pi}{4}^R$

27. $\dfrac{1}{2}$ **28.** $\dfrac{-1}{\sqrt{3}}$ **29.** $\dfrac{-1}{\sqrt{2}}$ **30.** $-\sqrt{2}$ **31.** 0.2924   **32.** -1.8650

33. 0.1291   **34.** -2.4030 **35.** $\cos\left(\dfrac{2\pi}{3} + k \cdot 2\pi\right), k \in J$

36. $\sin\left(\dfrac{\pi}{4} + k \cdot 2\pi\right), k \in J$ **37.** 4 centimeters   **38.** 0 centimeter

39. 0 volt   **40.** 0.0259 amps

Exercise Set 3.1

1.

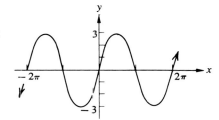

$-2\pi \quad -\dfrac{3\pi}{2} \quad -\pi \quad -\dfrac{\pi}{2} \quad 0 \quad \dfrac{\pi}{2} \quad \pi \quad \dfrac{3\pi}{2} \quad 2\pi$

$-5 \qquad\qquad 0 \qquad\qquad 5$

3.

$0 \quad \dfrac{\pi}{4} \quad \dfrac{\pi}{2} \quad \dfrac{3\pi}{4} \quad \pi \quad \dfrac{5\pi}{4} \quad \dfrac{3\pi}{2} \quad \dfrac{7\pi}{4}$

$0 \qquad\qquad \pi \qquad\qquad 2\pi$

5.

$-2\pi \quad -\dfrac{4\pi}{3} \quad -\dfrac{2\pi}{3}$

$-\dfrac{5\pi}{3} \quad -\pi \quad -\dfrac{\pi}{3} \quad 0 \quad \dfrac{\pi}{3} \quad \dfrac{2\pi}{3} \quad \pi \quad \dfrac{4\pi}{3} \quad \dfrac{5\pi}{3} \quad 2\pi$

$-2\pi \qquad -\pi \qquad 0 \qquad \pi \qquad 2\pi$

7. Zeros: $x = k\pi, \ k \in J$;
amplitude: 3

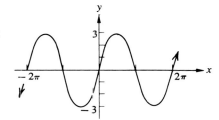

9. Zeros: $x = \pi/2 + k\pi, \ k \in J$;
amplitude: 1/2

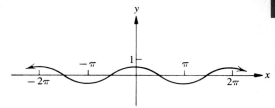

11. Zeros: $x = k\pi, \ k \in J$;
amplitude: 1

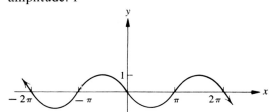

13. Zeros: $x = k\pi/2, \ k \in J$;
amplitude: 1

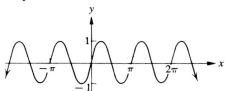

15. Zeros: $x = \pi + k \cdot 2\pi, \ k \in J$;
amplitude: 1

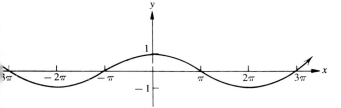

17. Zeros: $x = \pi/2 + k\pi, \ k \in J$;
amplitude: 1

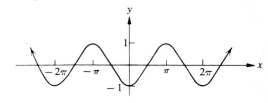

19. Zeros: $\{-3, -1, 1, 3\}$;
   amplitude: 1

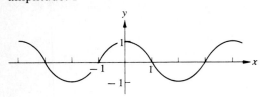

21. Zeros: $\{-2, -1, 0, 1, 2\}$;
   amplitude: 1

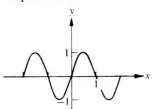

23.

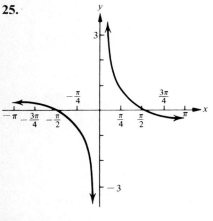

As x approaches zero from the left or right, $\dfrac{\sin x}{x}$ approaches a value of 1. The graph is not periodic because there is no number $a\,(a \neq 0)$ such that $\dfrac{\sin(x - a)}{x + a} = \dfrac{\sin x}{x}$.

25.

As x approaches zero from the left, $\dfrac{\cos x}{x}$ is negative and its absolute value becomes greater. As x approaches zero from right, $\dfrac{\cos x}{x}$ is positive and its absolute value becomes greater. The graph is not periodic because there is no number $a\,(a \neq 0)$ such that $\dfrac{\cos(x + a)}{x + a} = \dfrac{\cos x}{x}$.

27. Period is 1.

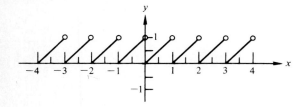

29. Period is 1/2.

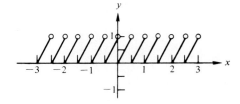

Exercise Set 3.2

1. Zeros: $\{-2\pi, -\pi, 0, \pi, 2\pi\}$; amplitude: 3

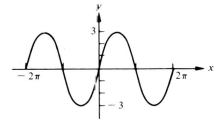

3. Zeros: $\left\{-\dfrac{3\pi}{2}, -\dfrac{\pi}{2}, \dfrac{\pi}{2}, \dfrac{3\pi}{2}\right\}$; amplitude: 2

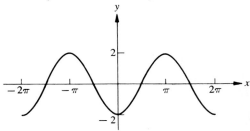

5. Zeros: $\{-2\pi, -\pi, 0, \pi, 2\pi\}$; amplitude: 1/2

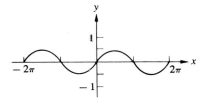

7. Zeros: $\left\{-\dfrac{7\pi}{4}, -\dfrac{5\pi}{4}, -\dfrac{3\pi}{4}, -\dfrac{\pi}{4}, \dfrac{\pi}{4}, \dfrac{3\pi}{4}, \dfrac{5\pi}{4}, \dfrac{7\pi}{4}\right\}$;
amplitude: 1

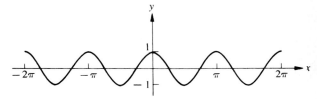

9. Zeros: $\{-2\pi, 0, 2\pi\}$; amplitude: 1

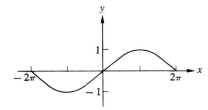

11. Zeros: $\left\{-2\pi, -\dfrac{3\pi}{2}, -\pi, -\dfrac{\pi}{2}, 0, \dfrac{\pi}{2}, \pi, \dfrac{3\pi}{2}, 2\pi\right\}$;
amplitude: 2

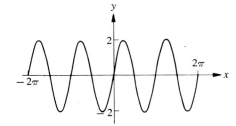

13.

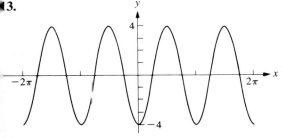

Zeros: $\left\{-\dfrac{7\pi}{4}, -\dfrac{5\pi}{4}, -\dfrac{3\pi}{4}, -\dfrac{\pi}{4}, \dfrac{\pi}{4}, \dfrac{3\pi}{4}, \dfrac{5\pi}{4}, \dfrac{7\pi}{4}\right\}$;
amplitude: 4

15. Zeros: $\left\{\dfrac{1}{2}, \dfrac{3}{2}, \dfrac{5}{2}, \dfrac{7}{2}\right\}$; amplitude: 1

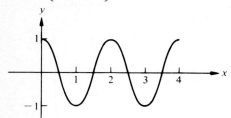

17. Zeros: $\{0, 3, 6, 9, 12\}$; amplitude: 1/2

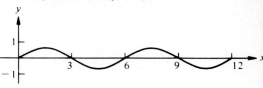

19. 3

21. 1.9

23.

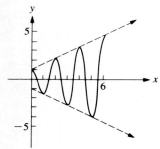

25.

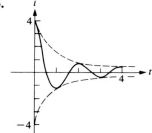

27.

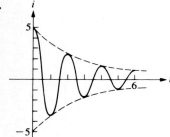

29.

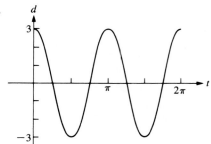

a. At $t = 0$, displacement is 3.0 cm.
b. Maximum displacement is 3.0 cm.
c. Period is 3.1 sec and frequency is
0.3 oscillation per sec.
d. At $t = 1.2$, displacement is -2.2 cm.

1.

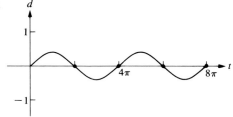

33.

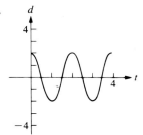

3

a. At $t = 0$, displacement is 0.0 cm.
b. Maximum displacement is 0.5 cm.
c. Period is 12.6 sec and frequency is
   0.1 oscillation per sec.
d. At $t = 2.8$, displacement is 0.5 cm.

5. $d = 2.4 \cos 2\pi(3.2)t$
   $\approx 2.4 \cos 20.1t$

9. $i = e^{-0.4t} \cos 2\pi(60)t$; at $t = 2$, current is
   approximately 2.2 amps.

a. At $t = 0$, displacement is 2.0 cm.
b. Maximum displacement is 2.0 cm.
c. Period is 2.0 sec and frequency is
   0.5 oscillation per sec.
d. At $t = 0.8$, displacement is -1.6 cm.

37. $d = 4.5 \cos 2\pi(1.8)t$ at $t = 3$, $d \approx -3.6$ cm.
   Hence, weight is approximately 3.6 centimeters
   above the resting position.

Exercise Set 3.3

1. Zeros: $\left\{ -\dfrac{3\pi}{2}, -\dfrac{\pi}{2}, \dfrac{\pi}{2}, \dfrac{3\pi}{2} \right\}$;
   amplitude: 1; phase shift: π

3. Zeros: $\left\{ -\dfrac{7\pi}{4}, -\dfrac{3\pi}{4}, \dfrac{\pi}{4}, \dfrac{5\pi}{4} \right\}$;
   amplitude: 2; phase shift: $\pi/4$

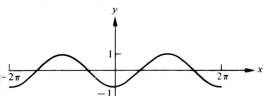

5. Zeros: $\left\{ -\dfrac{4\pi}{3}, -\dfrac{\pi}{3}, \dfrac{2\pi}{3}, \dfrac{5\pi}{3} \right\}$; amplitude: 1/2; phase shift: $\pi/3$

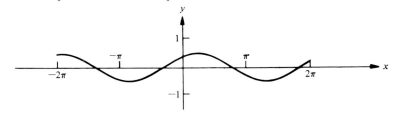

7. Zeros: $\left\{-2\pi, -\dfrac{3\pi}{2}, -\pi, -\dfrac{\pi}{2}, 0, \dfrac{\pi}{2}, \pi, \dfrac{3\pi}{2}, 2\pi\right\}$; amplitude: 2; phase shift: $\pi/4$

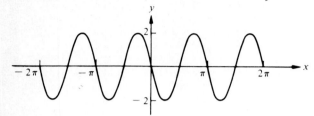

9. Zeros: $\{-\pi, \pi\}$; amplitude: 1/2; phase shift: π

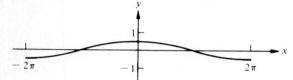

11. Zeros: $\left\{-\dfrac{\pi}{2}\right\}$; amplitude: 3; phase shift: π

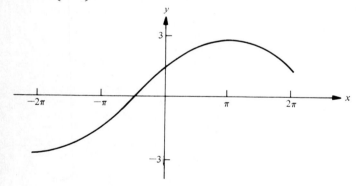

13.

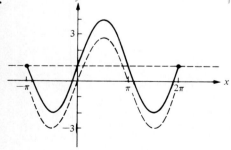

15.

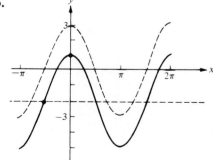

7.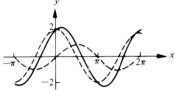

19.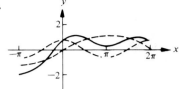

21.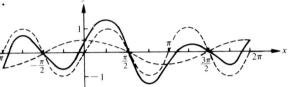

23.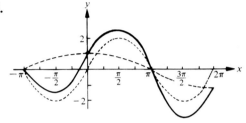

25.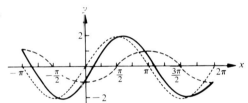

27.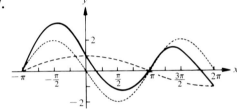

29.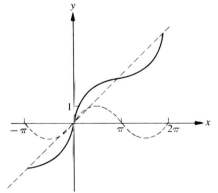

31. a. Amplitude: 0.4 radians
Period: 2 seconds
Frequency: 0.5 oscillations per second

b.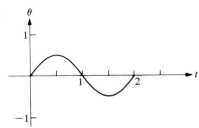

c. At $t = 1$, $\theta = 0$ radians;
at $t = 1.5$, $\theta = -0.4$ radians.

3

33. a. Amplitude: 100 volts
Period: $\pi/2$ seconds
Frequency: 0.64 oscillations per second

c. At $t = 1$, $e \approx 33.5$ volts;
at $t = 3.2$, $e \approx -82.3$ volts

b.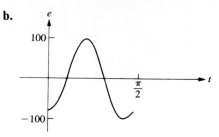

35.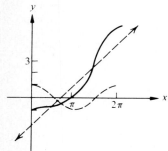

37.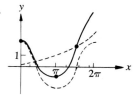

39.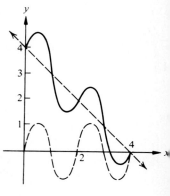

41.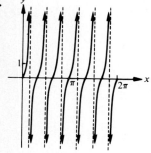

43.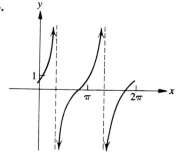

Exercise Set 3.4

1.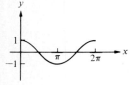

a. $y \in R$
b. $x = \pi/6 + k \cdot \pi/3,\ k \in J$
c. $x = k \cdot \pi/3,\ k \in J$

3.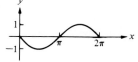

a. $y \in R$
b. $x = \pi/3 + k\pi,\ k \in J$
c. $x = 5\pi/6 + k\pi,\ k \in J$

5.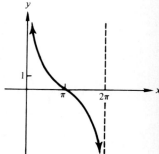

a. $y \in R$
b. $x = k \cdot 2\pi,\ k \in J$
c. $x = \pi + k \cdot 2\pi,\ k \in J$

7.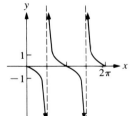

a. $y \in R$
b. $x = \pi/2 + k\pi, \ k \in J$
c. $x = k\pi, \ k \in J$

9.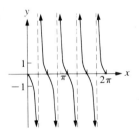

a. $y \in R$
b. $x = \pi/4 + k \cdot \pi/2, \ k \in J$
c. $x = k \cdot \pi/2, \ k \in J$

13. $x = \pi/2 + k \cdot \pi, \ k \in J$

1.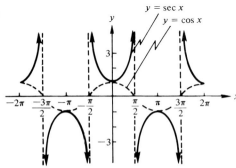

5.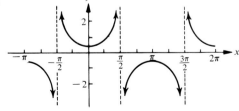

17.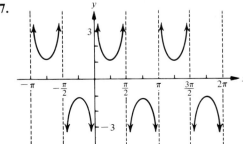

19.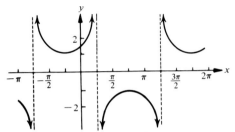

21.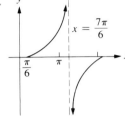

23.

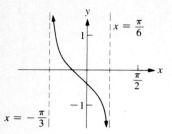

25.

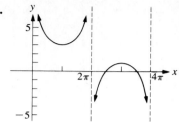

Review Exercises

1.

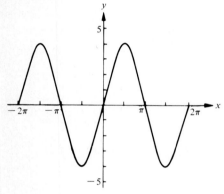

a. amplitude: 4
b. period: 2π

2.

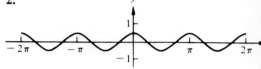

a. amplitude: 1/2
b. period: π

3.

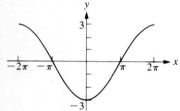

a. amplitude: 3
b. period: 4π

4.

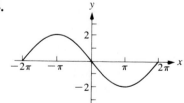

a. amplitude: 2
b. period: 4π

5.

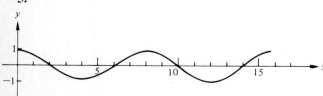

6.

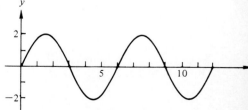

7.

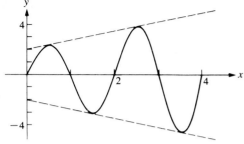

8.

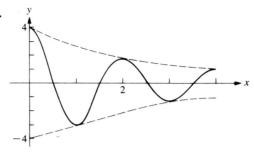

9.

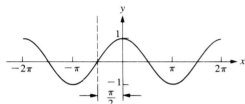

a. amplitude: 1
b. period: 2π
c. phase shift: $\pi/2$ (to the left)

10.

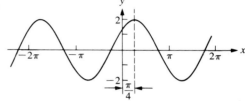

a. amplitude: 2
b. period: 2π
c. phase shift: $\pi/4$ (to the right)

1.

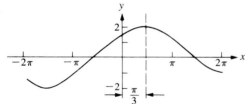

a. amplitude: 2
b. period: 4π
c. phase shift: $\pi/3$ (to the right)

12.

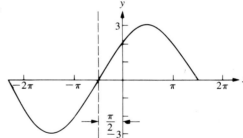

a. amplitude: 3
b. period: 4π
c. phase shift: $\pi/2$ (to the left)

3.

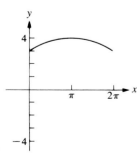

14.

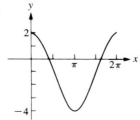

15.

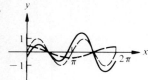

16.

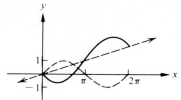

17.

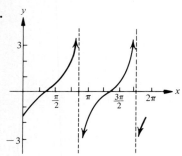

a. range: $y \in R$
b. asymptotes: $x = 5\pi/6 + k \cdot \pi, \ k \in J$
c. zeros: $x = \pi/3 + k \cdot \pi, \ k \in J$

18.

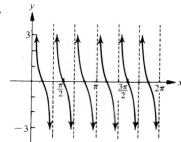

a. range: $y \in R$
b. asymptotes: $x = k \cdot (\pi/3), \ k \in J$
c. zeros: $x = \pi/6 + k \cdot (\pi/3), \ k \in J$

19.

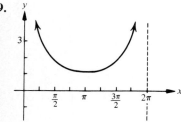

a. range: $y \le -1$ or $y \ge 1$
b. asymptotes: $x = k \cdot 2\pi, \ k \in J$
c. no zeros

20.

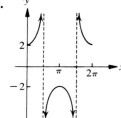

a. range: $y \le -2$ or $y \ge 2$
b. asymptotes: $x = \pi/2 + k \cdot \pi, \ k \in J$
c. no zeros

Exercise Set 4.1

1. $\sec^2 \alpha$

3. $\sin^2 \alpha$

5. $-\sin^2 \alpha$

7. $\tan^3 \alpha$

9. $\dfrac{1}{\sin \alpha}$

11. $\dfrac{\cos \alpha}{\sin^2 \alpha}$

13. $\dfrac{\cos^2 \alpha - \sin^2 \alpha}{\cos^2 \alpha}$

15. $-\dfrac{\sin^2 \alpha}{\cos^2 \alpha}$

17. $\dfrac{1}{\sin \beta \cos \beta}$

19. $\dfrac{\cos \beta}{\sin^3 \beta}$

21. $\dfrac{1}{\sin \gamma}$

23. $\dfrac{\cos^2 \gamma}{\sin \gamma}$

25. $\cos^2 x$

27. $\dfrac{\sin^2 x}{\cos^4 x}$

29. $\dfrac{1}{\sin y \cos y}$

31. $\dfrac{\sin^4 \theta + \cos^4 \theta}{\cos^2 \theta \sin^2 \theta}$ **33.** $\dfrac{1}{\cos \theta \sin^2 \theta}$ **35.** $\dfrac{\cos^2 \alpha}{\sin^3 \alpha}$ **37.** $\dfrac{\sin^2 \alpha - \cos^2 \alpha}{\cos \alpha \sin^3 \alpha}$ **39.** $\dfrac{-1}{\cos \alpha}$

41. $\dfrac{-1}{\cos \beta \sin \beta}$ **43.** $\dfrac{-1}{\cos x}$ **45.** $\dfrac{-1}{\sin y \cos y}$ **47.** $\dfrac{1}{\cos \alpha - \sin \alpha}$ **49.** $\dfrac{\sin \alpha + \cos \alpha}{\sin \alpha}$

Exercise Set 4.2

51. If $x = \pi/4$, then $\sin \dfrac{\pi}{4} + \left(\cos \dfrac{\pi}{4}\right)\left(\tan \dfrac{\pi}{4}\right) = 2;\ \dfrac{1}{\sqrt{2}} + \left(\dfrac{1}{\sqrt{2}}\right) \cdot 1 \neq 2$

53. If $y = \pi/2$, then $2\left(\sin \dfrac{\pi}{2}\right)\left(\cos \dfrac{\pi}{2}\right) + \sin \dfrac{\pi}{2} = 0;\ 2(1)(0) + 1 \neq 0$

Exercise Set 4.3

1. $(\sqrt{6} - \sqrt{2})/4$ **3.** $-(\sqrt{2} + \sqrt{6})/4$ **5.** $(\sqrt{2} + \sqrt{6})/4$ **13.** $\cos 9x$

5. $\cos 2x$ **17.** $\cos x$ **19.** $\cos 2x$ **21.** $16/65$

3. $13/\sqrt{170}$ **25.** $-\cos 60°$ **27.** $-\cos 30°$ **29.** $\cos 45°$

7. If $\alpha = 0°$ and $\beta = 60°$, then

$$\cos(\alpha + \beta) = \cos(0° + 60°) = \cos 60° = \tfrac{1}{2};$$
$$\cos \alpha + \cos \beta = \cos 0° + \cos 60° = 1 + \tfrac{1}{2} = \tfrac{3}{2} \neq \tfrac{1}{2}.$$

9. Using the difference formula and substituting $\pi/2$ for α and α for β, we obtain

$$\cos\left(\dfrac{\pi}{2} - \alpha\right) = \cos \dfrac{\pi}{2} \cos \alpha + \sin \dfrac{\pi}{2} \sin \alpha$$
$$= 0(\cos \alpha) + 1(\sin \alpha) = \sin \alpha.$$

Exercise Set 4.4

1. $(\sqrt{6} + \sqrt{2})/4$ **3.** $(\sqrt{6} - \sqrt{2})/4$ **5.** $(\sqrt{2} + \sqrt{6})/4$ **13.** $\sin 3x$

5. $\sin 4x$ **17.** $\sin 2x$ **19.** $\sin 2x$ **21.** $-44/125$

3. $5/\sqrt{754}$ **25.** $-\sin 60°$ **27.** $\sin 30°$ **29.** $-\sin 30°$

7. If $\alpha = 90°$ and $\beta = 60°$, then

$$\sin(\alpha - \beta) = \sin(90° - 60°) = \sin 30° = \tfrac{1}{2};$$
$$\sin \alpha - \sin \beta = \sin 90° - \sin 60° = 1 - \dfrac{\sqrt{3}}{2} \neq \dfrac{1}{2}.$$

Exercise Set 4.5

1. $(1 + \sqrt{3})/(1 - \sqrt{3})$ **3.** $(\sqrt{3} - 1)/(1 + \sqrt{3})$ **5.** $(\sqrt{3} - 1)/(1 + \sqrt{3})$

3. $\tan 9x$ **15.** $\tan 2x$ **17.** $\tan x$ **19.** $-\tan 60°$ **21.** $\tan 30°$ **23.** $-\tan 45°$

Exercise Set 4.6

1. $\sin 2x$ **3.** $\cos 2x$ **5.** $\sin 4x$ **7.** $\frac{1}{2}\sin x$ **9.** $\cos 10\alpha$ **11.** $2\cos 12\alpha$
13. $\sin \alpha$ **15.** $\cos 2x$ **33.** $24/25$ **35.** $5/\sqrt{34}$ **37.** $-24/7$

Exercise Set 4.7

1. $\frac{1}{2}\sqrt{2-\sqrt{3}}$ **3.** $2-\sqrt{3}$ **5.** $\frac{1}{2}\sqrt{2-\sqrt{3}}$ **7.** $\frac{1}{2}\sqrt{2+\sqrt{2}}$
9. $2+\sqrt{3}$ **11.** $-\frac{1}{2}\sqrt{2-\sqrt{3}}$ **19.** $1/\sqrt{3}$ **21.** $\sqrt{2/3}$
23. $-\frac{1}{2}\sqrt{2+\sqrt{2}}$ **31.** $2\cos 12°\cos 8°$ **33.** $2\sin 4\alpha\cos \alpha$ **35.** $2\cos 3\alpha\sin 2\alpha$

37. $\cos \alpha \cos \beta = \frac{1}{2}[\cos(\alpha + \beta) + \cos(\alpha - \beta)]$
$= \frac{1}{2}[\cos \alpha \cos \beta - \sin \alpha \sin \beta + \cos \alpha \cos \beta + \sin \alpha \sin \beta]$
$= \frac{1}{2}[2\cos \alpha \cos \beta] = \cos \alpha \cos \beta$

39. $\sin \alpha \cos \beta = \frac{1}{2}[\sin(\alpha + \beta) + \sin(\alpha + \beta)]$
$= \frac{1}{2}(\sin \alpha \cos \beta + \cos \alpha \sin \beta + \sin \alpha \cos \beta - \cos \alpha \sin \beta)$
$= \frac{1}{2}(2\sin \alpha \cos \beta) = \sin \alpha \cos \beta$

41. Using Equation (7) we have

$$2\cos\left(\frac{\alpha + \beta}{2}\right)\cos\left(\frac{\alpha - \beta}{2}\right) = 2 \cdot \frac{1}{2}\left[\cos\left(\frac{\alpha + \beta}{2} + \frac{\alpha - \beta}{2}\right) + \cos\left(\frac{\alpha + \beta}{2} - \frac{\alpha - \beta}{2}\right)\right]$$
$$= \cos \alpha + \cos \beta$$

43. Using Equation (9) we have

$$2\sin\left(\frac{\alpha + \beta}{2}\right)\cos\left(\frac{\alpha - \beta}{2}\right) = 2 \cdot \frac{1}{2}\left[\sin\left(\frac{\alpha + \beta}{2} + \frac{\alpha - \beta}{2}\right) + \sin\left(\frac{\alpha + \beta}{2} - \frac{\alpha - \beta}{2}\right)\right]$$
$$= \sin \alpha + \sin \beta$$

Review Exercises

1. $1/\cos \alpha$ **2.** $1/\sin^2 \alpha$ **3.** $\cos^2 \alpha$ **4.** $1/\sin \alpha$ **5.** $-\cos \alpha$ **6.** $1/\cos \alpha$

23. If $\alpha = \frac{\pi}{4}$, then $\sin\frac{\pi}{4}\sec\frac{\pi}{4} = 2\tan\frac{\pi}{4}$; $\frac{1}{\sqrt{2}} \cdot \sqrt{2} \neq 2 \cdot 1$

24. If $\alpha = \frac{\pi}{4}$, then $\frac{1 - \cos \pi/4}{\sin \pi/4} = \frac{1 + \sin \pi/4}{\cos \pi/4}$; $\frac{1 - 1/\sqrt{2}}{1/\sqrt{2}} \neq \frac{1 + 1\sqrt{2}}{1/\sqrt{2}}$; $\sqrt{2} - 1 \neq \sqrt{2} + 1$

25. $-\sqrt{2 - \sqrt{3}}/2$ **26.** $\sqrt{3} - 2$ **29.** $\frac{1}{2}(\cos 6x + \cos 4x)$ **30.** $2\sin 5x \cos x$

Exercise Set 5.1

1. $\cos y = \sqrt{3}/2,\ 0 \leq y \leq \pi;\ \pi/6$ **3.** $\sin y = 1,\ -\pi/2 \leq y \leq \pi/2;\ \pi/2$
5. $\cot y = 1/\sqrt{3},\ -\pi/2 \leq y \leq \pi/2,\ y \neq 0;\ \pi/3$ **7.** $\pi/4$ **9.** $\pi/4$

1. $\pi/3$

9. $-\pi/6$

7. 0.46

5. 1.19

13. 0

21. $-\pi/3$

29. 0.36

37. -0.50

15. $\pi/6$

23. $2\pi/3$

31. 1.24

39. 2.92

17. $\pi/4$

25. 0.52

33. 1.45

41. -1.45

xercise Set 5.2

1. $\sqrt{3}/2$ **3.** 0   **5.** $-1/\sqrt{2}$ **7.** $\pi/4$ **9.** $\pi/6$ **11.** 0

3. π **15.** 1.3983   **17.** 0.9018   **19.** 0.6285   **21.** 1.1728   **23.** 0.3641

5. $x = \text{Cos}^{-1}\dfrac{3}{4}$ **27.** $x = \dfrac{1}{3}\,\text{Sin}^{-1}\dfrac{1}{4}$ **29.** $x = \dfrac{1}{2}\,\text{Tan}^{-1}\dfrac{5}{2}$

1. $x = \dfrac{1}{3}\,\text{Cos}^{-1} 0.62$ **33.** $x = \dfrac{1}{3}\,\text{Tan}^{-1}\dfrac{y}{2}$ **35.** $x = 2\,\text{Sin}^{-1}\dfrac{y}{4}$

7. 4/5   **39.** 2/5   **41.** 3/4   **43.** $\dfrac{\sqrt{9-y^2}}{3}$

5. $\sqrt{1-x^2}$ **47.** 0   **49.** $\dfrac{\sqrt{3}+1}{2\sqrt{2}}$ **51.** 0

3.

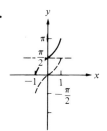

55.

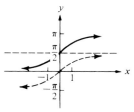

xercise Set 5.3

1a. $30°$ **b.** $\{30°, 330°\}$ **3a.** $150°$ **b.** $\{150°, 330°\}$ **5a.** $30°$ **b.** $\{30°, 210°\}$

7a. $45°$ **b.** $\{45°, 315°\}$ **9a.** $210°$ **b.** $\{210°, 330°\}$ **11a.** $210°$ **b.** $\{210°, 330°\}$

3a. $135°$ **b.** $\{135°, 225°\}$ **15a.** $45°$ **b.** $\{45°, 135°\}$ **17a.** $0°$ **b.** $\{0°, 180°\}$

9a. $150°$ **b.** $\{150°, 210°\}$ **21a.** $0°$ **b.** $\{0°, 180°\}$ **23a.** $0°$ **b.** $\{0°, 180°\}$

5a. $14°$ **b.** $\{14°, 166°\}$ **27a.** $111°$ **b.** $\{111°, 291°\}$ **29a.** $14.3°$ **b.** $\{14.3°, 345.7°\}$

1a. $155.3°$ **b.** $\{155.3°, 204.7°\}$ **33a.** $25.4°$ **b.** $\{25.4°, 205.4°\}$ **35a.** $231.4°$ **b.** $\{231.4°, 308.6°\}$

7a. $213.2°$ **b.** $\{213.2°, 326.8°\}$ **39a.** $66.7°$ **b.** $\{66.7°, 293.3°\}$ **41a.** $45.6°$ **b.** $\{45.6°, 225.6°\}$

3a. 0.70^R **b.** $\{0.70^R, 2.44^R\}$ **45a.** 0.23^R **b.** $\{0.23^R, 3.37^R\}$ **47a.** 0.98^R **b.** $\{0.98^R, 5.30^R\}$

9a. 2.89^R **b.** $\{2.89^R, 3.39^R\}$ **51a.** 3.26^R **b.** $\{3.26^R, 6.16^R\}$ **53a.** 3.63^R **b.** $\{3.63^R, 5.79^R\}$

5. $\{\alpha\,|\,\alpha \approx 16.2° + k\cdot 360°$ or $\alpha \approx 163.8° + k\cdot 360°\}$, $k \in J$

7. $\{\alpha\,|\,\alpha \approx 142.0° + k\cdot 180°\}$, $k \in J$

9. $\{\alpha\,|\,\alpha \approx 231.3° + k\cdot 360°$ or $308.7° + k\cdot 360°\}$, $k \in J$

1. $\{\alpha\,|\,\alpha \approx 16.8° + k\cdot 180°\}$, $k \in J$

5

Exercise Set 5.4

1. $\{\pi/6\}$ **3.** $\{\pi/4\}$ **5.** $\{\pi/2\}$
7. $\{\pi/4, \pi/2\}$ **9.** $\{0, \pi/2\}$ **11.** $\{\pi/4, \pi/2\}$
13. $\{0\}$ **15.** $\{0, 2\pi/3, 4\pi/3\}$ **17.** $\{1.91, 4.37\}$

19. a. $\{270°\}$ **b.** $\{x \mid x = 270° + k \cdot 360°\}, \ k \in J$

21. a. $\{60°, 180°, 300°\}$
b. $x = 60° + k \cdot 360°$ or $x = 180° + k \cdot 360°$ or $x = 300° + k \cdot 360°, \ k \in J$

23. a. $\{30°, 90°, 150°\}$
b. $x = 30° + k \cdot 360°$ or $x = 90° + k \cdot 360°$ or $x = 150° + k \cdot 360°, \ k \in J$

25. a. $\{218.2°, 321.8°\}$
b. $x = 218.2° + k \cdot 360°$ or $x = 321.8° + k \cdot 360°, \ k \in J$

27. a. $\{39.9°, 115.7°, 244.3°, 320.1°\}$
b. $x = 39.9° + k \cdot 360°$ or $x = 115.7° + k \cdot 360°$ or $x = 244.3° + k \cdot 360°$
or $x = 320.1° + k \cdot 360°, \ k \in J$

29. $\left\{\dfrac{\pi}{3}, \pi\right\}$ **31.** $\left\{\dfrac{3\pi}{2}\right\}$ **33.** $x = \dfrac{3\pi}{4} + k \cdot \pi, \ k \in J$

Exercise Set 5.5

1. $\{0, \pi/3\}$ **3.** $\{\pi/6, 5\pi/12\}$ **5.** $\{\pi/15\}$ **7.** $\{\pi/6\}$
9. $\{0\}$ **11.** $\{\pi/8\}$ **13.** $\{0\}$
15. $\{10°, 50°, 130°, 170°, 250°, 290°\}$ **17.** $\{15°, 75°, 135°, 195°, 225°, 315°\}$
19. $\{45°, 135°, 225°, 315°\}$ **21.** $\{15°, 45°, 75°, 105°, 135°, 165°, 195°, 225°, 255°, 285°, 315°, 345°\}$
23. $\{180°\}$ **25.** $\{60°, 180°, 300°\}$
27. $x = \pi/2 + k \cdot 2\pi$ or $x = \pi/6 + k \cdot 2\pi$ or $x = 5\pi/6 + k \cdot 2\pi, \ k \in J$
29. $x = k\pi$ or $x = \pi/6 + k\pi/3, \ k \in J$
31. $x \approx 1.28 + k\pi/2$ or $x = \pi/8 + k\pi/2, \ k \in J$
33. $x = \pi/12 + k(2\pi/3)$ or $x = \pi/4 + k(2\pi/3), \ k \in J$
35. $\{0.5\}$

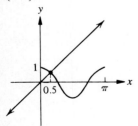

Exercise Set 5.6

1. a. $\{(0, 1), (6.3, 1)\}$

3. a. $\{(2.1, 0.5)\}$

5. a. $\{(2.4, -1), (5.5, -1)\}$

b.

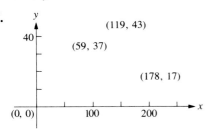

b.

b.

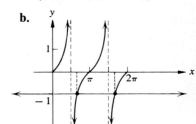

7. a. $\{(0, 0), (1.3, 0.5), (3.1, 0), (5.0, -0.5)(6.3, 0)\}$

9. $2x - 3y = 21$

b.

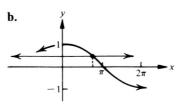

1. $x^2 + y^2 = 4$

13. $x^2 + y = 1, \ y \ge 0$

15. $x^2 + y^2 = 1; \ x \ge 0$

7.

(119, 43)

(59, 37)

40

(178, 17)

(0, 0)   100   200

(components given to the nearest unit)

19.

2

5

21.

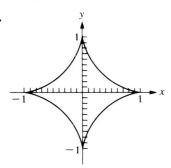

23. $y = \dfrac{8}{x^2 + 4}$

25. $xy = 1; |x| \ge 1, |y| \le 1$

27. $\dfrac{(x - h)^2}{a^2} + \dfrac{(y - k)^2}{b^2} = 1$

Review Exercises

1. $\cos y = 1/\sqrt{2}$; $(\pi/4)^R$

2. $\cot y = -1$; $(-\pi/4)^R$

3. $\csc y = -2$; $(-\pi/6)^R$

4. $\sec y = 2/\sqrt{3}$; $(\pi/6)^R$

5. $(\pi/2)^R$

6. $(-\pi/6)^R$

7. $(3\pi/4)^R$

8. $(\pi/6)^R$

9. 1.45^R

10. 0.95^R

11. 0.47^R

12. -1.02^R

13. $x = \text{Cos}^{-1} \dfrac{3}{5}$, $0 \le x \le \pi$

14. $x = \dfrac{1}{2} \text{Sin}^{-1} \dfrac{1}{5}$, $-\dfrac{\pi}{4} \le x \le \dfrac{\pi}{4}$

15. $x = \dfrac{1}{2} \text{Tan}^{-1} \dfrac{y}{3}$, $-\dfrac{\pi}{4} < x <$

16. $x = 2 \text{Cos}^{-1} \dfrac{y}{2}$, $0 \le x \le 2\pi$

17. $\sqrt{3}$

18. $\dfrac{\sqrt{3}}{2}$

19. $\dfrac{\pi}{2}$

20. $\dfrac{\pi}{4}$

21. a. $\{60°\}$
 b. $\alpha = 60° + k \cdot 360°$ or $\alpha = 120° + k \cdot 360°$, $k \in J$

22. a. $\{135°\}$
 b. $\alpha = 135° + k \cdot 180°$, $k \in J$

23. a. $\{118.0°\}$
 b. $\alpha \approx 118.0° + k \cdot 360°$ or $\alpha \approx 242.0° + k \cdot 360°$, $k \in J$

24. a. $\{36.1°\}$
 b. $\alpha \approx 36.1° + k \cdot 360°$ or $\alpha \approx 143.9° + k \cdot 360°$, $k \in J$

25. a. $\{0.369^R\}$
 b. $\alpha \approx 0.369^R + k \cdot 2\pi^R$ or $\alpha \approx 5.915^R + k \cdot 2\pi^R$, $k \in J$

26. a. $\{1.810^R\}$
 b. $\alpha \approx 1.810^R + k \cdot \pi^R$, $k \in J$

27. a. $\{0.316^R\}$
 b. $\alpha \approx 0.316^R + k \cdot 2\pi^R$ or $\alpha \approx 2.826^R + k \cdot 2\pi^R$, $k \in J$

28. a. $\{0.459\}$
 b. $\alpha \approx 0.459^R + k \cdot 2\pi^R$ or $\alpha \approx 5.824^R + k \cdot 2\pi^R$, $k \in J$

29. $\{\pi/3\}$

30. $\{\pi/6\}$

31. $\{\pi/4\}$

32. $\{\pi/2\}$

33. $\{30°, 150°, 210°, 330°\}$

34. $\{45°, 135°, 225°, 315°\}$

35. $\{15°, 105°, 135°, 225°, 255°, 345°\}$

36. $\{15°, 30°, 105°, 120°, 195°, 210°, 285°, 300°\}$

37. $\{0°, 45°, 135°, 180°, 225°, 315°\}$

38. $\{0°, 180°\}$

39. $\{(1.2, 0.3), (5.0, 0.3)\}$

40. $\{(0.4, 0.8), (2.8, 0.8)\}$

41. $\{(0.3, 0.5), (1.6, 0), (2.9, -0.5), (4.7, 0)\}$

42. $\{(0.6, 0.3), (2.5, 0.3), (3.7, -0.3), (5.7, -0.3)\}$

43. $y = \dfrac{1}{8}(x + 1)^2$

44. $x^2 + \dfrac{y^2}{16} = 1$

Exercise Set 6.1

1. $\gamma = 85.0°$, $b \approx 14.1$, $c \approx 14.6$

3. $\alpha = 42.8°$, $b \approx 52.0$, $c \approx 84.7$

5. $\beta = 52.9°$, $a \approx 0.307$, $c \approx 0.526$

7. $\gamma = 77.5°$, $a \approx 7.05$, $b \approx 13.3$

9. One triangle: $\beta \approx 20.9°$, $\gamma \approx 129.1°$, $c \approx 10.9$
1. Two triangles: $\gamma \approx 50.7°$, $\beta \approx 97.1°$, $b \approx 7.82$ or $\gamma' \approx 129.3°$, $\beta' \approx 18.5°$, $b' \approx 2.50$
3. One right triangle: $\alpha = 60.0°$, $\gamma = 90.0°$, $a \approx 27.7$
5. Two triangles: $\alpha \approx 60.4°$, $\gamma \approx 76.9°$, $c \approx 5.60$ or $\alpha' \approx 119.6°$, $\gamma' \approx 17.7°$, $c' \approx 1.75$
7. 1.51   **19.** 3,049 feet
1. 7.90 centimeters; 14.8 centimeters   **23.** 78.4 meters
5. 57.8 feet   **27.** 64.3 meters

9. From the figure, $\tan \gamma = \dfrac{c}{AC}$, or $AC = \dfrac{c}{\tan \gamma}$.

Using the Law of Sines, $\dfrac{AC}{\sin \beta} = \dfrac{a}{\sin[180° - (\alpha + \beta)]} = \dfrac{a}{\sin(\alpha + \beta)}$.

Hence, $AC = \dfrac{a \sin \beta}{\sin(\alpha + \beta)}$, from which $\dfrac{c}{\tan \gamma} = \dfrac{a \sin \beta}{\sin(\alpha + \beta)}$; $c = \dfrac{a \sin \beta \tan \gamma}{\sin(\alpha + \beta)}$

1. From the Law of Sines, $\dfrac{a}{b} = \dfrac{\sin \alpha}{\sin \beta}$. Adding 1 to both members gives $\dfrac{a}{b} + 1 = \dfrac{\sin \alpha}{\sin \beta} + 1$, from

which $\dfrac{a + b}{b} = \dfrac{\sin \alpha + \sin \beta}{\sin \beta}$.

3. From the results of Exercises 31 and 32 and dividing left members and right members, respectively, gives

$$\frac{\dfrac{a - b}{b}}{\dfrac{a + b}{b}} = \frac{\dfrac{\sin \alpha - \sin \beta}{\sin \beta}}{\dfrac{\sin \alpha + \sin \beta}{\sin \beta}}, \quad \text{from which} \quad \frac{a - b}{a + b} = \frac{\sin \alpha - \sin \beta}{\sin \alpha + \sin \beta}.$$

Exercise Set 6.2

1. $a \approx 4.37$, $\beta \approx 59.9°$, $\gamma \approx 49.1°$ **3.** $c \approx 3.39$, $\alpha \approx 65.9°$, $\beta \approx 38.8°$
5. $b \approx 1.42$, $\alpha \approx 17.9°$, $\gamma \approx 20.6°$ **7.** $a \approx 3.83$, $\beta \approx 24.3°$, $\gamma \approx 55.3°$
9. $\alpha \approx 95.7°$, $\beta \approx 50.7°$, $\gamma \approx 33.6°$ **11.** $\alpha \approx 4.0°$, $\beta \approx 31.6°$, $\gamma \approx 144.4°$
3. $\alpha \approx 26.4°$, $\beta \approx 36.4$, $\gamma \approx 117.2°$ **15.** $\alpha \approx 41.4°$, $\beta \approx 55.8°$, $\gamma \approx 82.8°$
7. 42.7°   **19.** 82.5°   **21.** 10.9   **23.** 5.33 kilometers
5. 11.7 centimeters   **27.** 47.8°   **29.** 1,240 feet   **31.** 875 kilometers

3. Adding 1 to both members of $\cos \alpha = \dfrac{b^2 + c^2 - a^2}{2bc}$ gives

$$1 + \cos \alpha = 1 + \frac{b^2 + c^2 - a^2}{2bc}$$

$$= \frac{2bc + b^2 + c^2 - a^2}{2bc} = \frac{b^2 + 2bc + c^2 - a^2}{2bc}$$

$$= \frac{(b + c)^2 - a^2}{2bc} = \frac{(b + c + a)(b + c - a)}{2bc}$$

6

35. $a + b + c = 2s$ and $b + c - a = 2s - 2a = 2(s - a)$. Substituting $2s$ for $a + b + c$ and $2(s - a)$ fo $b + c - a$ in the results of Exercise 33 gives

$$1 + \cos \alpha = \frac{(2s)(2)(s - a)}{2bc}, \quad \text{from which} \quad \frac{1 + \cos \alpha}{2} = \frac{s(s - a)}{bc}.$$

37. Multiplying left members and right members, respectively, of the result of Exercises 35 and 36 gives

$$\left(\frac{1 + \cos \alpha}{2}\right)\left(\frac{1 - \cos \alpha}{2}\right) = \left[\frac{s(s - a)}{bc}\right]\left[\frac{(s - b)(s - c)}{bc}\right],$$

$$\frac{1 - \cos^2 \alpha}{4} = \frac{s(s - a)(s - b)(s - c)}{b^2 c^2},$$

$$\frac{\sin^2 \alpha}{4} = \frac{s(s - a)(s - b)(s - c)}{b^2 c^2}.$$

Taking the square root of both members gives

$$\frac{1}{2} \sin \alpha = \frac{\sqrt{s(s - a)(s - b)(s - c)}}{bc},$$

$$\frac{1}{2} bc \sin \alpha = \sqrt{s(s - a)(s - b)(s - c)}.$$

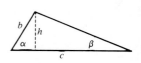

Because $h = b \sin \alpha$ (see the figure), the area $\mathcal{A}$ of a triangle is $\frac{1}{2}(\text{base} \times \text{altitude})$, or $\frac{1}{2}c(b \sin \alpha)$. Thus,

$$\mathcal{A} = \sqrt{s(s - a)(s - b)(s - c)}.$$

39. 3.14

41. Noting in the figure that the radius r is the altitude of the three inner triangles, we have that the total area is $\frac{1}{2}(ra + rb + rc) = \frac{1}{2}(a + b + c)(r) = sr$. From the results of Exercise 37, we have

$$\mathcal{A} = \sqrt{s(s - a)(s - b)(s - c)} = sr,$$

from which $\quad r = \sqrt{\dfrac{s(s - a)(s - b)(s - c)}{s^2}} = \sqrt{\dfrac{(s - a)(s - b)(s - c)}{s}}.$

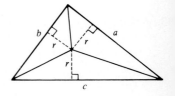

Exercise Set 6.3

1. 2.3; 90°   **3.** 3.5; −60°   **5.** 2.3; −90°

7.

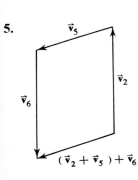

$\vec{v}_1 + \vec{v}_2$ $\vec{v}_2$ $\vec{v}_1$

9.

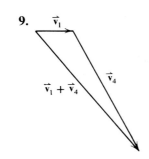

$\vec{v}_1$ $\vec{v}_1 + \vec{v}_4$ $\vec{v}_4$

11.

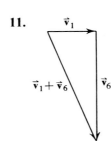

$\vec{v}_1$ $\vec{v}_1 + \vec{v}_6$ $\vec{v}_6$

13.

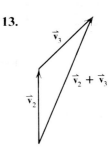

$\vec{v}_3$ $\vec{v}_2 + \vec{v}_3$ $\vec{v}_2$

15.

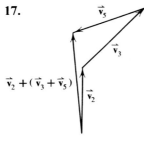

$\vec{v}_5$ $\vec{v}_2$ $\vec{v}_6$ $(\vec{v}_2 + \vec{v}_5) + \vec{v}_6$

17.

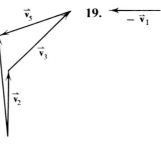

$\vec{v}_5$ $\vec{v}_3$ $\vec{v}_2 + (\vec{v}_3 + \vec{v}_5)$ $\vec{v}_2$

19.

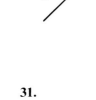

$-\vec{v}_1$

21.

$2\vec{v}_7$

23.

$-2\vec{v}_5$

25.

$\frac{1}{2}\vec{v}_3$

27.

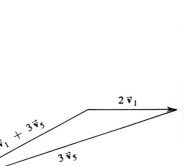

$2\vec{v}_1$ $2\vec{v}_1 + 3\vec{v}_5$ $3\vec{v}_5$

29.

$\frac{1}{2}\vec{v}_4 + \vec{v}_7$ $\frac{1}{2}\vec{v}_4$ $\vec{v}_7$

31.

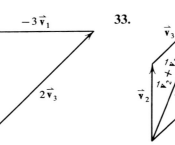

$-3\vec{v}_1$ $2\vec{v}_3 + (-3\vec{v}_1)$ $2\vec{v}_3$

33.

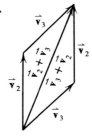

$\vec{v}_3$ $\frac{1}{2}\vec{v}_3$ $\frac{1}{2}\vec{v}_2$ $\frac{1}{2}\vec{v}_2 + \frac{1}{2}\vec{v}_3$ $\vec{v}_2$ $\vec{v}_2$ $\vec{v}_3$

6

35.

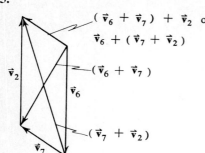

$(\vec{v}_6 + \vec{v}_7) + \vec{v}_2$ or
$\vec{v}_6 + (\vec{v}_7 + \vec{v}_2)$

$(\vec{v}_6 + \vec{v}_7)$

$\vec{v}_2$

$\vec{v}_6$

$(\vec{v}_7 + \vec{v}_2)$

$\vec{v}_7$

37.

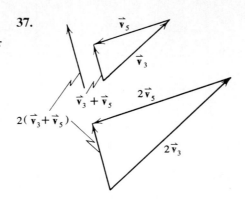

$\vec{v}_5$

$\vec{v}_3$

$\vec{v}_3 + \vec{v}_5$

$2\vec{v}_5$

$2(\vec{v}_3 + \vec{v}_5)$

$2\vec{v}_3$

39.

$\frac{1}{3}\vec{v}_1$ $2\vec{v}_1$

$\frac{7}{3}\vec{v}_1$

Exercise Set 6.4

1. $\|\vec{v}_x\| = 5$, $\|\vec{v}_y\| = 5$ **3.** $\|\vec{v}_x\| = 0$, $\|\vec{v}_y\| = 15$

5. $\|\vec{v}_x\| = 6\sqrt{3}$, $\|\vec{v}_y\| = 6$ **7.** $\|\vec{v}_x\| = 4$, $\|\vec{v}_y\| = 4\sqrt{3}$

9. $\|\vec{v}_3\| = 5$ pounds; 53.1° with 3-pound force, 36.9° with 4-pound force.

11. $\|\vec{v}_3\| = 27.9$ pounds; 43.8° with 15-pound force, 31.2° with 20-pound force.

13. 8.72 newtons, $\alpha \approx 36.6°$, $\beta \approx 83.4°$

15. 40 newtons, 36.9°   **17.** 19.7 pounds, 37.1 pounds

19. 511.3 pounds   **21.** 33.4°

23. $\|\vec{v}_x\| = 1.64$ meters/sec², $\|\vec{v}_y\| = 1.15$ meters/sec²   **25.** 0.967 kilometer per hour

27. Drift angle is 11.3° to the right; ground speed is 612 miles per hour; course is 146.3°.

29. Drift angle is 12° to the right; wind direction is 318° measured from true north; wind speed is 106.5 mile per hour.

31. Heading is 85.2°; air speed is 602 kilometers per hour.

33. 34.6 nautical miles, 265.3°   **35.** 70.7 pounds

Exercise Set 6.5

1. a.

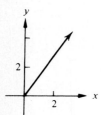

b. 5, 53°6′

3. a.

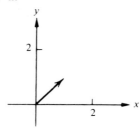

b. $\sqrt{2}$, 45°

5. a.

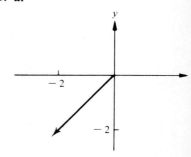

b. $3\sqrt{2}$, 225°

7. a.

b. $4, 0°$

9. $(2, 2)$

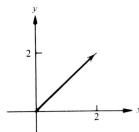

1. $(0, 1)$

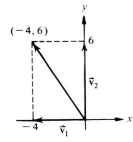

13. $\left(-\frac{5}{2}\sqrt{3}, -\frac{5}{2}\right)$

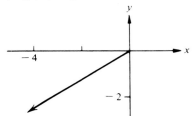

5. $(-4, 6)$

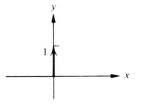

17. $(3, 1)$

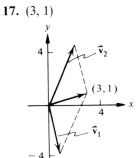

19. $(-4, 1)$

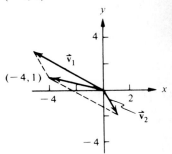

6

21. $(-3, -6)$

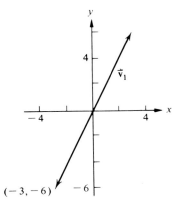

23. $(6, 12)$

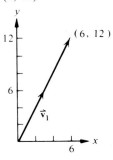

25. $(-13, -16)$

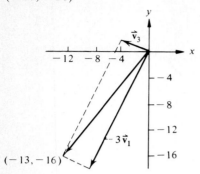

27. $(-9, 15)$

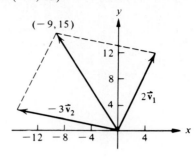

29. $(7, -8)$

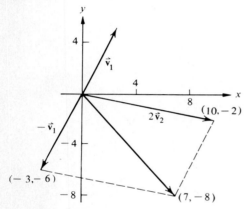

31. $(-7, -1)$

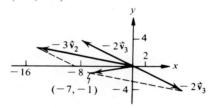

33. $(5/2, -1/2)$

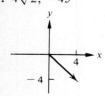

35. $3\mathbf{i} + 5\mathbf{j}$
37. $\mathbf{i} - 4\mathbf{j}$
39. $-3\mathbf{i}$

41. $4\sqrt{2}, -45°$

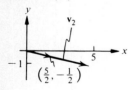

43. $\sqrt{5}, 26°36'$

45. $\sqrt{10}, -108°24'$

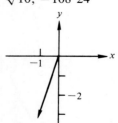

7. 2, 120°

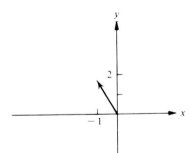

Exercise Set 6.6

1. 48   **3.** −2   **5.** 45   **7.** 13   **9.** $36\sqrt{2}$ **11.** 3/4

3. 156.8°   **15.** 90°   **17.** 24.0°

9. $20/\sqrt{29}$, 5   **21.** $7/\sqrt{41}$, $7/\sqrt{13}$

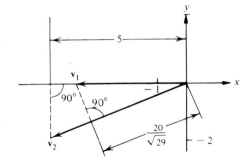

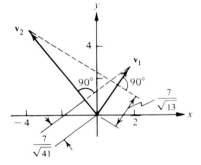

23. $-14/\sqrt{10}$, $-14/\sqrt{34}$ **25.** 0, 0

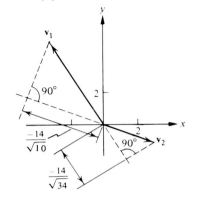

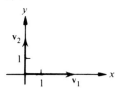

33. 24   **35.** 36   **37.** $1000\sqrt{3}$ meter-newtons

39. 1,000,000 meter-newtons

Review Exercises

1. $\gamma = 77.5°$, $a \approx 7.03$, $b \approx 13.3$
2. $\beta = 10.0°$, $a \approx 1.73$, $b \approx 0.425$
3. $\gamma = 85.1°$, $a \approx 6.99$, $c \approx 7.07$
4. $\beta = 90.5°$, $b \approx 13.91$, $c \approx 6.57$
5. One triangle: $c \approx 8.1$, $\alpha \approx 28.3°$, $\gamma \approx 130.0°$
6. One triangle: $\beta \approx 30.6°$, $\gamma \approx 13.9°$, $c \approx 17.5$
7. $\alpha \approx 51.3$, $\beta \approx 70.5°$, $c \approx 2.07$
8. $\beta \approx 14.1°$, $\gamma \approx 142.8°$, $a = 3.7$
9. $\alpha \approx 35.2°$, $\beta \approx 42.2°$, $\gamma \approx 102.6°$
10. $\alpha \approx 30.8°$, $\beta \approx 56.0°$, $\gamma \approx 93.2°$
11. $a \approx 4.40$, $\beta \approx 144.2$, $\gamma \approx 21.2°$
12. $\beta \approx 38.2$, $\alpha \approx 81.7°$, $\gamma \approx 60.0°$

13. $(\vec{v_1} + \vec{v_2}) + \vec{v_3}$

14.
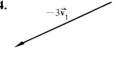

15.
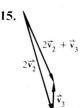

16.
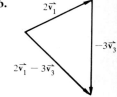

17. $4\sqrt{3}$; 4
18. 5; $36.9°$
19. 120 newtons; $22.6°$
20. 6.71 miles per hour; drift of $63.4°$
21. Wind bearing is $357.7°$; wind speed is 183 kilometers per hour.
22. $36.9°$
23. 5, $143.1°$
24. $5\sqrt{2}$, $-45°$

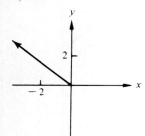

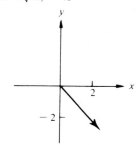

25. $(-5\sqrt{2}, -5\sqrt{2})$
26. $(-\sqrt{3}, 1)$
27. $(12, -4)$
28. $(-10, 10)$
29. $(8, 2)$
30. $(12, 2)$
31. $(-1, 3)$
32. $(5/2\sqrt{13}, 2/\sqrt{13})$
33. $7i - 9j$
34. $-8i - j$
35. $2\sqrt{29}$, $158°12'$
36. $\sqrt{130}$, $-52°6'$
37. -54
38. 75
39. $16\sqrt{3}$
40. $25\sqrt{3}$
41. $136°48'$
42. $56°30'$
43. $3/\sqrt{101}$, $3/5\sqrt{2}$
44. $-82/\sqrt{89}$, $-82/\sqrt{85}$
45. $(-4)(4) + (-5)(-16/5) = -16 + 16 = 0$
46. $(7)(5) + (6)(-35/6) = 35 - 35 = 0$

Exercise Set 7.1

1. $1 + 10i$
3. $-10 - 11i$
5. $-1 + 6i$
7. $0 + 0i$
9. $12 - 12i$
11. $-36 - 12i$
13. $2 + 3i$
15. $-19 + 2i$

7. $62 - 39i$

19. $26 - 19i$

21. $a^2 + b^2$

23. i

5. -1

27. -1

29. $-i$

31. $5i$

3. $2\sqrt{3}\,i$

35. $1 + 3i$

37. $3 - 2\sqrt{5}\,i$

39. $8 + 2i$

1. $-8 + 8i$

43. $-10\sqrt{5} + 2\sqrt{10}\,i$

45. $23 + 2i$

47. $9 + \sqrt{3}\,i$

9. If $x = 2i$, then $(2i)^2 + 4 = 4i^2 + 4$
$$= -4 + 4 = 0;$$
if $x = -2i$, then $(-2i)^2 + 4 = -4i^2 + 4$
$$= -4 + 4 = 0.$$

1. a. $\left(-\dfrac{1}{2} + i\dfrac{\sqrt{3}}{2}\right)^2 = \dfrac{1}{4} - i\dfrac{\sqrt{3}}{2} + i^2\left(\dfrac{3}{4}\right) = -\dfrac{1}{2} - i\dfrac{\sqrt{3}}{2};$

$\left(-\dfrac{1}{2} + i\dfrac{\sqrt{3}}{2}\right)^3 = \left(-\dfrac{1}{2} + i\dfrac{\sqrt{3}}{2}\right)^2\left(-\dfrac{1}{2} + i\dfrac{\sqrt{3}}{2}\right)$

$\qquad = \left(-\dfrac{1}{2} - i\dfrac{\sqrt{3}}{2}\right)\left(-\dfrac{1}{2} + i\dfrac{\sqrt{3}}{2}\right) = \dfrac{1}{4} - i^2\left(\dfrac{3}{4}\right) = 1$

b. $\left(-\dfrac{1}{2} - i\dfrac{\sqrt{3}}{2}\right)^2 = \dfrac{1}{4} + i\dfrac{\sqrt{3}}{2} + i^2\left(\dfrac{3}{4}\right) = -\dfrac{1}{2} + i\dfrac{\sqrt{3}}{2};$

$\left(-\dfrac{1}{2} - i\dfrac{\sqrt{3}}{2}\right)^3 = \left(-\dfrac{1}{2} - i\dfrac{\sqrt{3}}{2}\right)^2\left(-\dfrac{1}{2} - i\dfrac{\sqrt{3}}{2}\right)$

$\qquad = \left(-\dfrac{1}{2} + i\dfrac{\sqrt{3}}{2}\right)\left(-\dfrac{1}{2} - i\dfrac{\sqrt{3}}{2}\right) = \dfrac{1}{4} - i^2\left(\dfrac{3}{4}\right) = 1$

3. $a = 0, b = 1$ or $a = 0, b = -1$

55. $e^{\pi i} = \cos \pi + i \sin \pi$
$$= -1 + i(0) = -1$$

57. $\dfrac{e^{ix} - e^{-ix}}{2i} = \dfrac{\cos x + i \sin x - [\cos(-x) + i \sin(-x)]}{2i}$

$\qquad = \dfrac{\cos x + i \sin x - \cos x + i \sin x}{2i} = \dfrac{2i \sin x}{2i} = \sin x$

59. Using the results of Exercises 57 and 58,

$$\sin^2 x + \cos^2 x = \left(\dfrac{e^{ix} - e^{-ix}}{2i}\right)^2 + \left(\dfrac{e^{ix} + e^{-ix}}{2}\right)^2$$

$$= \dfrac{e^{2ix} - 2e^0 + e^{-2ix}}{4i^2} + \dfrac{e^{2ix} + 2e^0 + e^{-2ix}}{4}$$

$$= \dfrac{-e^{2ix} + 2 - e^{-2ix}}{4} + \dfrac{e^{2ix} + 2 + e^{-2ix}}{4} = \dfrac{4}{4} = 1.$$

Exercise Set 7.2

1. $13 - 4i$

3. $2 + 2i$

5. $5 - i$

7. $6 - 3i$

9. $-3 + 3i$

11. $-3i$

13. $\dfrac{3}{2} - \dfrac{9}{4}i$

15. $\dfrac{8}{85} + \dfrac{36}{85}i$

7

17. $\frac{9}{17} + \frac{2}{17}i$ **19.** $-\frac{6}{5} + \frac{11}{10}i$ **21.** $\dfrac{-i}{\sqrt{5}}$ **23.** $-\dfrac{2i}{\sqrt{2}}$

25. $\dfrac{4}{7} - \dfrac{\sqrt{5}}{7}i$ **27.** $\dfrac{18}{11} - \dfrac{6\sqrt{2}}{11}i$ **29.** $\dfrac{6\sqrt{2}}{7} - \dfrac{6\sqrt{5}}{7}i$ **31.** $\dfrac{2\sqrt{5}}{13} - \dfrac{4\sqrt{2}}{13}i$

Exercise Set 7.3

1. $(-1, 6)$ **3.** $(8, 1)$ **5.** $(0, 6)$ **7.** $-8 + 7i$ **9.** $-4 + 0i$ **11.** $7 - \sqrt{2}\,i$

13.

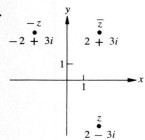

15.

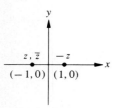

17.

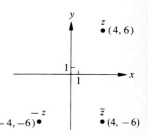

19.

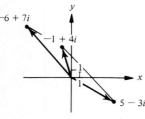

21. $-1 + 6i$

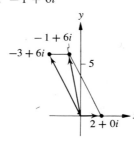

23. $-4 - 5i$

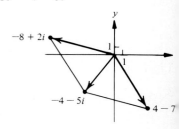

25. $-1 + 4i$

27. $\sqrt{29}$; $21.8°$ **29.** $2\sqrt{2}$; $225°$

31. 7; $0°$ **33.** 5; $270°$ **35.** $3\sqrt{10}$; $288.4°$
37. 4; $0°$ **39. a.** $a \neq 0, b = 0$ **b.** $a = 0, b \neq 0$ **c.** $b > 0$ **d.** $b < 0$
41. 10.8 ohms; **43.** 12.6 ohms; **45.** 7.2 ohms;
$\quad$ $21.8°$ $\quad$ $-71.6°$ $\quad$ $33.7°$

47.

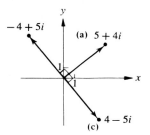

(b) $-4 + 5i$

(a) $5 + 4i$

(c) $4 - 5i$

49.

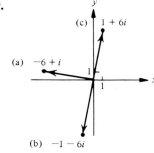

(c) $1 + 6i$

(a) $-6 + i$

(b) $-1 - 6i$

51.

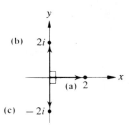

(b) $2i$

(a) 2

(c) $-2i$

Exercise Set 7.4

1. $4(\cos 120° + i \sin 120°)$

3. $10(\cos 210° + i \sin 210°)$

5. $2\sqrt{2}(\cos 315° + i \sin 315°)$

7. $8(\cos 180° + i \sin 180°)$

9. $-1 - \sqrt{3}\,i$

11. $-\sqrt{2} + \sqrt{2}\,i$

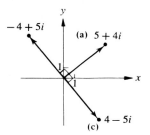

13. $\dfrac{1}{\sqrt{2}} - \dfrac{1}{\sqrt{2}i}$

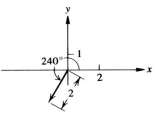

15. $-\dfrac{\sqrt{3}}{4} + \dfrac{1}{4}\,i$

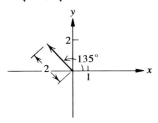

17. $-5\sqrt{2} + 5\sqrt{2}\,i$

19. $\dfrac{3}{2}\sqrt{3} - \dfrac{3}{2}\,i$

21. $-\sqrt{2} + \sqrt{2}\,i$

23. $-4 + 0i$

25. $-\dfrac{1}{5} - \dfrac{\sqrt{3}}{5}\,i$

27. $\dfrac{3}{2} - \dfrac{3\sqrt{3}}{2}\,i$

29. $4 + 4\sqrt{3}\,i$

31. $(a + bi)^2 = \rho(\cos \theta + i \sin \theta) \cdot \rho(\cos \theta + i \sin \theta)$; using Equation (1) in this section gives

$$(a + bi)^2 = \rho \cdot \rho[\cos(\theta + \theta) + i \sin(\theta + \theta)]$$
$$= \rho^2[\cos 2\theta + i \sin 2\theta].$$

33. $(a + bi)^4 = (a + bi)^3(a + bi)$; using the results of Exercise 32 and Equation (1) in this section, gives

$$\rho^3(\cos 3\theta + i \sin 3\theta) \cdot \rho(\cos \theta + i \sin \theta) = \rho^3 \cdot \rho[\cos(3\theta + \theta) + i \sin(3\theta + \theta)]$$
$$= \rho^4(\cos 4\theta + i \sin 4\theta).$$

35. $\sin(-\alpha) = -\sin \alpha$ and $\cos(-\alpha) = \cos \alpha$. Thus,

$$\cos(-\theta) + i \sin(-\theta) = \cos \theta - i \sin \theta,$$

the conjugate of $\cos \theta + i \sin \theta$.

Exercise Set 7.5

1. $\dfrac{27}{2} + \dfrac{27\sqrt{3}}{2} i$

3. $0 + 8i$

5. $-\dfrac{1}{8} + 0i$

7. $0 + \dfrac{1}{81} i$

9. $-\dfrac{1}{32} + \dfrac{1}{32} i$

11. $-\dfrac{1}{64} - \dfrac{\sqrt{3}}{64} i$

13. a. $\cos 40° + i \sin 40°, \cos 220° + i \sin 220°$
　　b. $\cos 30° + i \sin 30°, \cos 210° + i \sin 210°$

15. a. $2(\cos 40° + i \sin 40°), 2(\cos 160° + i \sin 160°), 2(\cos 280° + i \sin 280°)$
　　b. $\cos 20° + i \sin 20°, \cos 140° + i \sin 140°, \cos 260° + i \sin 260°$

17. a. $\cos 60° + i \sin 60°, \cos 150° + i \sin 150°, \cos 240° + i \sin 240, \cos 330° + i \sin 330°$
　　b. $\cos 75° + i \sin 75°, \cos 165° + i \sin 165°, \cos 255° + i \sin 255°, \cos 345° + i \sin 345°$

19. $\left\{\dfrac{1}{2} + \dfrac{\sqrt{3}}{2} i, -1, \dfrac{1}{2} - \dfrac{\sqrt{3}}{2} i\right\}$

21. $\left\{1, -\dfrac{1}{2} + \dfrac{\sqrt{3}}{2} i, -\dfrac{1}{2} - \dfrac{\sqrt{3}}{2} i\right\}$

23. $\left\{\dfrac{1}{2^{1/3}} (\cos 110° + i \sin 110°), \dfrac{1}{2^{1/3}} (\cos 230° + i \sin 230°), \dfrac{1}{2^{1/3}} (\cos 350° + i \sin 350°)\right\}.$

25. In all cases, the graphs are spaced equally on the unit circle.

a.

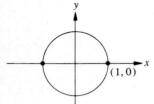

b.

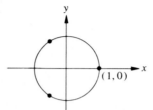

c.

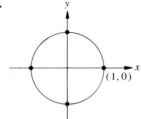

d.

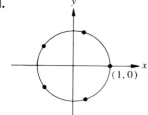

e.

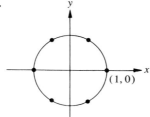

f.

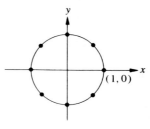

7. $\cos 3\theta = 4\cos^3 \theta - 3\cos \theta$

xercise Set 7.6

1. $(3, 20°), (-3, 200°), (-3, -160°), (3, -340°)$ 3. $(5, -90°), (-5, -270°), (-5, 90°), (5, 270°)$

5. $(-4, 60°), (4, 240°), (4, -120°), (-4, -300°)$

7. $(2\sqrt{3}, 2)$ 9. $(-3/\sqrt{2}, -3/\sqrt{2})$ 11. $(3\sqrt{3}/2, -3/2)$ 13. $(4, 0^R)$

5. $\left(2\sqrt{2}, \dfrac{\pi}{4}^R\right)$ 17. $\left(1, \dfrac{5\pi}{6}^R\right)$ 19. $\rho \cos \theta = 4$ 21. $\rho = \dfrac{8 \sin \theta}{\cos^2 \theta}$

3. $\rho = 4$ 25. $\rho = 3 \sin \theta$ 27. $x^2 + y^2 = 36$ 29. $x^2 + y^2 = 4y$

1. $y = 2$ 33. $x^2 - 8y - 16 = 0$

xercise Set 7.7

1.

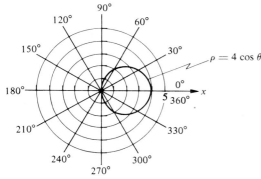

3.

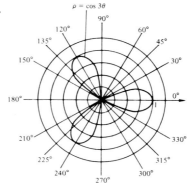

7

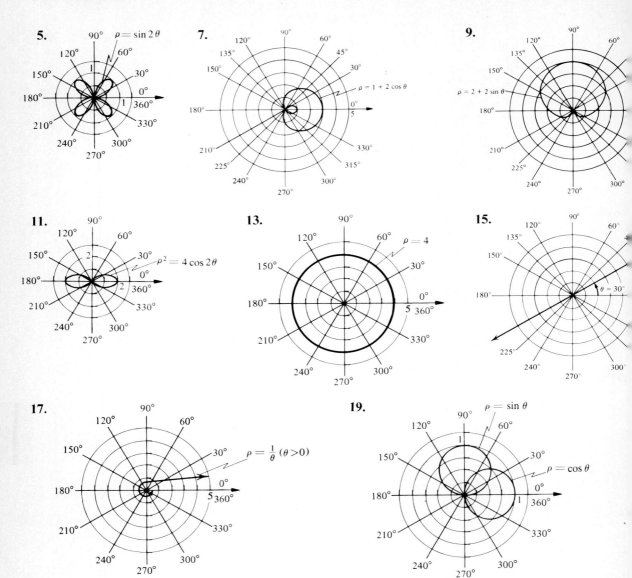

5. $\rho = \sin 2\theta$

7. $\rho = 1 + 2\cos\theta$

9. $\rho = 2 + 2\sin\theta$

11. $\rho^2 = 4\cos 2\theta$

13. $\rho = 4$

15. $\theta = 30°$

17. $\rho = \dfrac{1}{\theta}\ (\theta > 0)$

19. $\rho = \sin\theta$ $\rho = \cos\theta$

$(0.7, 45°)$ and $(0, \theta)$ where θ has any measure

1.

90° $\rho = 1 + \sin \theta$

120°   60°

150°   30°

$\rho = 1 + \cos \theta$

180°   0°
                        360°

210°   330°

240°   270°   300°

$(1.7, 45°), (0.3, 225°)$ and $(0, \theta)$ where θ has any measure

3. $\{(1/\sqrt{2}, 45°)\}$; since ρ is a distance measured from the origin, it may be zero for more than one value of θ. In this case, $\rho = 0$ when $\theta = 0°$ in $\rho = \sin \theta$, and $\rho = 0$ $\theta = 90°$ in $\rho = \cos \theta$.

5. $\left\{ \left(\dfrac{\sqrt{2} + 1}{\sqrt{2}}, 45° \right), \left(\dfrac{\sqrt{2} - 1}{\sqrt{2}}, 225° \right) \right\}$

7. ρ_1 and ρ_2 are the lengths of sides with included angle $\theta_1 - \theta_2$. Using the Law of Cosines, $d^2 = \rho_1^2 + \rho_2^2 - 2\rho_1\rho_2 \cos(\theta_1 - \theta_2)$,

$$d = \sqrt{\rho_1^2 + \rho_2^2 - 2\rho_1\rho_2 \cos(\theta_1 - \theta_2)}.$$

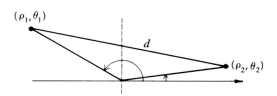

7

Review Exercises

1. $8 - i$	**2.** $4 - 2i$	**3.** $-8 + 8i$	**4.** $0 - i$	**5.** $-46 + 28i$
6. $22 - 14i$	**7.** i	**8.** $-i$	**9.** $\sqrt{10} + 3\sqrt{5}\,i$	**10.** 14

1. If $x = 3i$, then $(3i)^2 + 9 = 9i^2 + 9 = -9 + 9 = 0$;
if $x = -3i$, then $(-3i)^2 + 9 = 9i^2 + 9 = -9 + 9 = 0$.

2. If $x = 1 + 3i$, then $(1 + 3i)^2 - 2(1 + 3i) + 10 = 1 + 6i + 9i^2 - 2 - 6i + 10 = 0$;
if $x = 1 - 3i$, then $(1 - 3i)^2 - 2(1 - 3i) + 10 = 1 - 6i + 9i^2 - 2 + 6i + 10 = 0$.

3. $-15 - i$	**14.** $3 - 5i$	**15.** $\dfrac{7}{5} - \dfrac{13}{10}i$	**16.** $-\dfrac{13}{34} - \dfrac{8}{17}i$	**17.** $\dfrac{2i}{\sqrt{6}}$
8. $\dfrac{8}{9} - \dfrac{4\sqrt{5}}{9}i$	**19.** $(-4, 1)$	**20.** $(2\sqrt{3}, -2)$	**21.** $9 - 6i$	**22.** $3 + 4i$

23.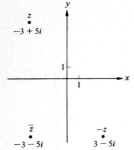

24.

25.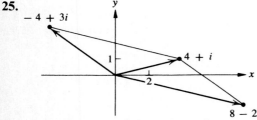

26.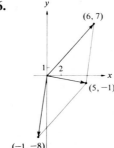

27. $2\sqrt{5}$, $26° 30'$

28. $\sqrt{29}$, $248° 12'$

29. $\sqrt{2}(\cos 45° + i \sin 45°)$

30. $6(\cos 120° + i \sin 120°)$

31. $-\dfrac{7}{2} + \dfrac{7\sqrt{3}}{2} i$

32. $\sqrt{2} + \sqrt{2} i$

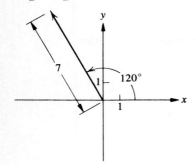

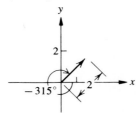

33. $-3\sqrt{2} - 3\sqrt{2} i$

34. $3\sqrt{3} + 3i$

35. $-\dfrac{8}{9} + 0i$

36. $-2\sqrt{2} + 2\sqrt{2} i$

37. $\dfrac{81}{2} + \dfrac{81\sqrt{3}}{2} i$

38. $0 - 8i$

39. $-\dfrac{1}{32} - \dfrac{\sqrt{3}}{32} i$

40. $-\dfrac{1}{256} + \dfrac{1}{256} i$

41. $\cos 15° + i \sin 15°$, $\cos 195° + i \sin 195°$

42. $5(\cos 110° + i \sin 110°)$, $5(\cos 230° + i \sin 230°)$, $5(\cos 350° + i \sin 350°)$

43. $\{\cos 45° + i \sin 45°, \cos 135° + i \sin 135°, \cos 225° + i \sin 225°, \cos 315° + i \sin 315°\}$

44. $\{10^{1/5}(\cos 48° + i \sin 48°), 10^{1/5}(\cos 120° + i \sin 120°), 10^{1/5}(\cos 192° + i \sin 192°),$
$10^{1/5}(\cos 264° + i \sin 264°), 10^{1/5}(\cos 336° + i \sin 336°)\}$

45. $(2, 250°), (-2, 70°), (2, -110°), (-2, -290°)$ **46.** $(-3, -330°), (-3, 30°), (3, 210°), (3, -150°)$

47. $\left(\dfrac{3}{2}, \dfrac{3\sqrt{3}}{2}\right)$ **48.** $(\sqrt{2}, -\sqrt{2})$ **49.** $\left(2\sqrt{2}, \dfrac{7\pi^{R}}{4}\right)$ **50.** $\left(2, \dfrac{7\pi^{R}}{6}\right)$

51. $\rho \cos \theta = 3$ **52.** $\rho^2(\cos^2 \theta + 3 \sin^2 \theta) = 5$

53. $x^2 + y^2 = 5y$ **54.** $y^2 - 4x - 4 = 0$

55.

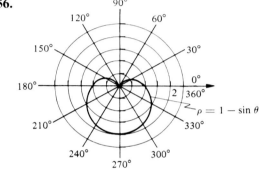

56.

57.

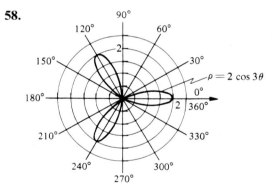

58.

7

INDEX

$$F_1 = \|F_1\|(\cos 135 \, i + \sin 45 \, j)$$

$$F_2 = \|F_2\|(\cos 45 \, i + \sin 45 \, j)$$

$$F_3 = 100 \, i - 100 \, j$$

$$F_1 + F_2 + F_3 = 0$$

$$-\frac{\sqrt{2}}{2}\|F_1\| \, i + \frac{\sqrt{2}}{2}\|F_1\| \, j + \frac{\sqrt{2}}{2}\|F_2\| \, i + \frac{\sqrt{2}}{2}\|F_2\| \, j - 100 \, j = 0$$

$$-\frac{\sqrt{2}}{2}\|F_1\| \, i + \left(4 \, \frac{\sqrt{2}}{2}\|F_2\| \, i\right) = 0$$

$$\frac{\sqrt{2}}{2}\|F_1\| + \frac{\sqrt{2}}{2}\|F_2\| - 100 = 0$$

$$\|F_1\| = \frac{100 - \frac{\sqrt{2}}{2}\|F_2\|}{\frac{\sqrt{2}}{2}}$$

$$-\frac{\sqrt{2}}{2}\left(\frac{100 - \frac{\sqrt{2}}{2}\|F_2\|}{\frac{\sqrt{2}}{2}}\right) - \frac{\sqrt{2}}{2}\|F_2\| = 0$$

$$-100 + \frac{\sqrt{2}}{2}\|F_2\| - \frac{\sqrt{2}}{2}\|F_2\| = 0$$

SUMMARY

1.1 Trigonometric Ratios in a Right Triangle

$$\sin \alpha = \frac{\text{length of side opposite } \alpha}{\text{length of hypotenuse}}$$

$$\cos \alpha = \frac{\text{length of side adjacent to } \alpha}{\text{length of hypotenuse}}$$

$$\tan \alpha = \frac{\text{length of side opposite } \alpha}{\text{length of side adjacent to } \alpha}$$

$$\csc \alpha = \frac{\text{length of hypotenuse}}{\text{length of side opposite } \alpha}$$

$$\sec \alpha = \frac{\text{length of hypotenuse}}{\text{length of side adjacent to } \alpha}$$

$$\cot \alpha = \frac{\text{length of side adjacent to } \alpha}{\text{length of side opposite } \alpha}$$

1.2 Basic Identities

$$\csc \alpha = \frac{1}{\sin \alpha}; \quad \sec \alpha = \frac{1}{\cos \alpha}; \quad \cot \alpha = \frac{1}{\tan \alpha}$$

1.2, 2.1–2.2 Functions for Special Angles

$$\frac{1}{2} = 0.500$$

$$\sqrt{2} \approx 1.414$$

$$\frac{1}{\sqrt{2}} \approx 0.707$$

$$\sqrt{3} \approx 1.732$$

$$\frac{\sqrt{3}}{2} \approx 0.866$$

$$\frac{1}{\sqrt{3}} \approx 0.577$$

$$\frac{2}{\sqrt{3}} \approx 1.155$$

2.1 Conversion Formulas

$$\frac{\alpha^{\circ}}{360} = \frac{\alpha^{R}}{2\pi}; \quad \alpha^{\circ} = \frac{180}{\pi} \alpha^{R}; \quad \alpha^{R} = \frac{\pi}{180} \alpha^{\circ}$$

Circular Motion (arc length s)

$$s = r \cdot \alpha^{R}$$

2.2 Trigonometric Ratios ($r = \sqrt{x^2 + y^2}$, $r \neq 0$)

$$\sin \alpha = \frac{y}{r} \qquad \csc \alpha = \frac{r}{y} \quad (y \neq 0)$$

$$\cos \alpha = \frac{x}{r} \qquad \sec \alpha = \frac{r}{x} \quad (x \neq 0)$$

$$\tan \alpha = \frac{y}{x} \quad (x \neq 0) \qquad \cot \alpha = \frac{x}{y} \quad (y \neq 0)$$

2.2 Signs of Trigonometric Ratios

Quadrant:	I	II	III	IV
	$x > 0,$	$x < 0,$	$x < 0,$	$x > 0,$
	$y > 0$	$y > 0$	$y < 0$	$y < 0$
$\sin \alpha$ or $\csc \alpha$	+	+	−	−
$\cos \alpha$ or $\sec \alpha$	+	−	−	+
$\tan \alpha$ or $\cot \alpha$	+	−	+	−

3.1–3.4 Graphs of Trignometric Functions

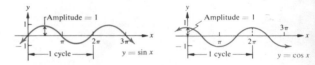

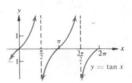

4.1–4.7 Additional Identities

$$\cos^2 x + \sin^2 x = 1$$

$$\tan^2 x + 1 = \sec^2 x; \quad \cot^2 x + 1 = \csc^2 x$$

$$\sin(-x) = -\sin x; \quad \cos(-x) = \cos x; \quad \tan(-x) = -\tan x$$

$$\cos(\alpha + \beta) = \cos \alpha \cos \beta - \sin \alpha \sin \beta$$

$$\cos(\alpha - \beta) = \cos \alpha \cos \beta + \sin \alpha \sin \beta$$

$$\cos\left(\frac{\pi}{2} - x\right) = \sin x; \quad \sin\left(\frac{\pi}{2} - x\right) = \cos x$$

$$\sin(\alpha + \beta) = \sin \alpha \cos \beta + \cos \alpha \sin \beta$$

$$\sin(\alpha - \beta) = \sin \alpha \cos \beta - \cos \alpha \sin \beta$$

$$\tan(\alpha + \beta) = \frac{\tan \alpha + \tan \beta}{1 - \tan \alpha \tan \beta}$$

$$\tan(\alpha - \beta) = \frac{\tan \alpha - \tan \beta}{1 + \tan \alpha \tan \beta}$$

$$\cos 2x = \cos^2 x - \sin^2 x = 1 - 2\sin^2 x$$
$$= 2\cos^2 x - 1$$

$$\cos^2 x = \frac{1 + \cos 2x}{2}; \quad \cos^2 \frac{x}{2} = \frac{1 + \cos x}{2}$$

$$\sin 2x = 2 \sin x \cos x$$

$$\sin^2 x = \frac{1 - \cos 2x}{2}; \quad \sin^2 \frac{x}{2} = \frac{1 - \cos x}{2}$$

$$\tan 2x = \frac{2 \tan x}{1 - \tan^2 x}; \quad \tan \frac{x}{2} = \frac{1 - \cos x}{\sin x}$$